Privacy, Big Data, and the Public Good

Frameworks for Engagement

Massive amounts of data on human beings can now be analyzed. Pragmatic purposes abound, including selling goods and services, winning political campaigns, and identifying possible terrorists. Yet "big data" can also be harnessed to serve the public good: scientists can use big data to do research that improves the lives of human beings, improves government services, and reduces taxpayer costs. In order to achieve this goal, researchers must have access to this data – raising important privacy questions. What are the ethical and legal requirements? What are the rules of engagement? What are the best ways to provide access while also protecting confidentiality? Are there reasonable mechanisms to compensate citizens for privacy loss?

The goal of this book is to answer some of these questions. The book's authors paint an intellectual landscape that includes legal, economic, and statistical frameworks. The authors also identify new practical approaches that simultaneously maximize the utility of data access while minimizing information risk.

Julia Lane is Senior Managing Economist for the American Institutes for Research in Washington, DC. She holds honorary positions as professor of economics at the BETA University of Strasbourg CNRS, chercheur associée at Observatoire des Sciences et des Techniques, Paris, and professor at the University of Melbourne's Institute of Applied Economics and Social Research.

Victoria Stodden is Assistant Professor of Statistics at Columbia University and is affiliated with the Columbia University Institute for Data Sciences and Engineering.

Stefan Bender is head of the Research Data Center at the German Federal Employment Agency in the Institute for Employment Research.

Helen Nissenbaum is Professor of Media, Culture and Communication, and Computer Science at New York University, where she is also director of the Information Law Institute.

Privacy, Big Data, and the Public Good

Frameworks for Engagement

Edited by

Julia Lane
American Institutes for Research, Washington DC

Victoria Stodden
Columbia University

Stefan Bender
Institute for Employment Research of the German Federal Employment Agency

Helen Nissenbaum
New York University

CAMBRIDGE
UNIVERSITY PRESS

CAMBRIDGE
UNIVERSITY PRESS

32 Avenue of the Americas, New York, NY 10013-2473, USA

Cambridge University Press is part of the University of Cambridge.

It furthers the University's mission by disseminating knowledge in the pursuit of education, learning, and research at the highest international levels of excellence.

www.cambridge.org
Information on this title: www.cambridge.org/9781107637689

First published 2014

Printed in the United States of America

A catalog record for this publication is available from the British Library.

Library of Congress Cataloging in Publication Data
Privacy, big data, and the public good : frameworks for engagement / [edited by] Julia Lane, American Institutes for Research, Washington DC, Victoria Stodden, Columbia University, Stefan Bender, Institute for Employment Research of the German Federal Employment Agency, Helen Nissenbaum, New York University.
pages cm
Includes bibliographical references and index.
ISBN 978-1-107-06735-6 (hardback : alk. paper) –
ISBN 978-1-107-63768-9 (pbk. : alk. paper)
1. Privacy, Right of. 2. Research – Moral and ethical aspects. 3. Big data – Social aspects. 4. Common good. I. Lane, Julia I.
JC596.P747 2015
323.44′8–dc23 2014009737

ISBN 978-1-107-06735-6 Hardback
ISBN 978-1-107-63768-9 Paperback

For
Mike Holland
and
Diana Gillooly

CONTENTS

CONTRIBUTORS

Alessandro Acquisti Professor of Information Technology and Public Policy at the Heinz College, Carnegie Mellon University; Co-director of CMU Center for Behavioral Decision Research; acquisti@andrew.cmu.edu

Solon Barocas Doctoral student in the Department of Media, Culture, and Communication, New York University; solon@nyu.edu

Cynthia Dwork Distinguished Scientist, Microsoft Research Silicon Valley; dwork@microsoft.com

Peter Elias Professor at the Warwick Institute for Employment Research, University of Warwick; peter.elias@warwick.ac.uk

Robert M. Goerge Senior Research Fellow at Chapin Hall, University of Chicago; Director of the Joint Program on Public Policy and Computing and Lecturer, Harris School for Public Policy Studies, University of Chicago; rgoerge@chapinhall.org

Daniel Greenwood Lecturer and Research Scientist in the Human Dynamics Group, Media Lab, Massachusetts Institute of Technology; dazza@civics.com

Thomas Hardjono Technical Lead and Executive Director of the MIT Kerberos & Internet Trust Consortium; hardjono@mit.edu

Michael J. Holland Chief of Staff at the Center for Urban Science & Progress, New York University; mike.holland@nyu.edu

Alan F. Karr Director of the National Institute of Statistical Sciences; Professor of Statistics and Biostatistics, University of North Carolina at Chapel Hill; karr@niss.org

Steven E. Koonin Director of the Center for Urban Science & Progress, New York University; steven.koonin@nyu.edu

Frauke Kreuter Associate Professor in the Joint Program in Survey Methodology, University of Maryland; Professor of Statistics at Ludwig-Maximilians-Universität; Head of the Statistical Methods Research Department, Institute for Employment Research, Germany; fkreuter@survey.umd.edu

Carl Landwehr Lead Research Scientist at the Cyber Security Policy and Research Institute, George Washington University; carl.landwehr@gmail.com

Helen Nissenbaum Professor of Media, Culture, and Communication, Professor of Computer Science, and Director of the Information Law Institute, New York University; helen.nissenbaum@nyu.edu

Paul Ohm Associate Professor at the University of Colorado Law School; paul.ohm@colorado.edu

Roger D. Peng Associate Professor in the Department of Biostatistics, Johns Hopkins Bloomberg School of Public Health; rpeng@jhsph.edu

Alex Pentland Toshiba Professor of Media, Arts, and Sciences; Director of the Human Dynamics Laboratory; and Director of the Media Lab Entrepreneurship Program, Massachusetts Institute of Technology; sandy@media.mit.edu

Jerome P. Reiter Mrs. Alexander Hehmeyer Professor of Statistical Science at Duke University; Principal Investigator, Triangle Census Research Network; Associate Director of the Information Initiative, Duke University; jerry@stat.duke.edu

Victoria Stodden Assistant Professor of Statistics and Affiliated Member of the Institute for Data Science, Columbia University; victoria@stodden.net

Arkadiusz Stopczynski Doctoral Student at the Technical University of Denmark; Visiting Student in the Human Dynamics Group, Media Lab, Massachusetts Institute of Technology; arks@media.mit.edu

Katherine J. Strandburg Alfred B. Engelberg Professor of Law at New York University School of Law; katherine.strandburg@nyu.edu

Brian Sweatt Research Staff Member in the Human Dynamics Group, Media Lab, Massachusetts Institute of Technology; brian717@media.mit.edu

John Wilbanks Chief Commons Officer at Sage Bionetworks; Senior Fellow at the Ewing Marion Kauffman Foundation, Washington DC

EDITORS' INTRODUCTION

Julia Lane, Victoria Stodden, Stefan Bender, and Helen Nissenbaum

Massive amounts of data on human beings can now be accessed and analyzed. And the new 'big data'[1] are much more likely to be harvested from a wide variety of different sources. Much has been made of the many uses of such data for pragmatic purposes, including selling goods and services, winning political campaigns, and identifying possible terrorists. Yet big data can also be harnessed to serve the public good in other ways: scientists can use new forms of data to do research that improves the lives of human beings; federal, state, and local governments can use data to improve services and reduce taxpayer costs; and public organizations can use information to advocate for public causes, for example.

Much has also been made of the privacy and confidentiality issues associated with access. Statisticians are not alone in thinking that consumers should worry about privacy issues, and that an ethical framework should be in place to guide data scientists;[2] the European Commission and the U.S. government have begun to address the problem. Yet there are many unanswered questions. What are the ethical and legal requirements for scientists and government officials seeking to use big data to serve the public good without harming individual citizens? What are the rules of engagement with these new data sources? What are the best ways to provide access while protecting confidentiality? Are there reasonable mechanisms to compensate citizens for privacy loss?

The goal of this book is to answer some of these questions. The book's authors paint an intellectual landscape that includes the legal, economic, and statistical context necessary to frame the many privacy issues, including the value to the public of data access, clarifying personal data ownership questions, and raising issues of agency in personal data. The authors also identify core practical approaches that use new technologies to simultaneously maximize the utility of data access while minimizing information risk. As is appropriate for such a new and evolving field, each chapter also identifies important questions that require future research.

The work in this book is also intended to be accessible to an audience broader than those in the research and policy spheres. In addition to informing the public, we hope that the book will be useful to people trying to provide data access within confidentiality constraints in their roles as data custodians for federal, state, and local agencies, or decision makers on Institutional Review Boards.

Historical and Future Use of Data for the Public Good

Good data are critically important for good public decisions. For example, national and international government policies depend on GDP estimates – indeed, international crises have been exacerbated when statistical agencies have cooked the data books.[3] Good data are also important for good science – as Daniel Kahneman famously pointed out, the first big breakthrough in our understanding of the mechanism of association was an improvement in a method of measurement.[4]

Historically the leading producers of high-quality data were statistical government agencies engaged in collecting data through large-scale statistically representative surveys. There are several reasons for this. One was the sheer scale of the necessary activity: generating representative samples required an expensive, constantly updated population frame and extensive investments in survey methodology and data storage, cleaning, and dissemination. The second was that the public trusted the government to protect confidentiality, and statistical agencies invested heavily in the appropriate statistical disclosure limitation methodologies. The third was that the statistical agencies were seen to be objective, and not trying to sell a product. The U.S. Census Bureau's mission statement reflects all three of these reasons:

> The Census Bureau's mission is to serve as the **leading source of quality** data about the nation's people and economy. We honor **privacy, protect confidentiality**, share our expertise globally, and conduct our work openly. We are guided on this mission by **scientific objectivity**, our strong and capable workforce, our devotion to research-based innovation, and our abiding commitment to our customers.[5] (*Emphases added*)

The public good has clearly been served by the creation and careful dissemination of data by both the government and the research community. Of course, the nature of the data, as well as the dissemination modality, has evolved over time. The development and release of large-scale public use datasets like the Current Population Survey and, later, the National Longitudinal Surveys of Youth and the Panel Study of Income Dynamics and the German Socio-Economic Panel have transformed our understanding of labor markets while protecting respondent confidentiality. The development of large-scale administrative datasets and their access through secure data enclaves have lowered costs, increased sample size, and reduced respondent burden,[6] as well as created completely new classes of information.[7]

Big data, by which we mean the data necessary to support new types of data-intensive research,[8] hold the promise of even more profound contributions to the public good. Knowledge derived from big data is likely to become one of the foundational elements in the functioning of society by, for example, generating real-time information about economic and social activity or generating new insights into human behavior.[9] Yet the pathway to developing this foundation is not clear, because experience provides little guidance. The data currently used to inform decisions – survey and administrative data – have benefited from decades of statistical research, as well as clear rules defining ownership and responsibility. Statistical agencies, the primary custodians, have developed clear ways to both protect and access the data. By contrast, the value of big data in informing evidence-based policy is still being established, and the ownership of big data, typically distributed across multiple entities, is not well defined. Big data have many elements of a natural resource, and sensible rules must be developed in order to avoid a tragedy of the commons, and to create a commonly pooled resource for improving scientific understanding for the public good.[10]

Privacy, Big Data, and the Public Good: The Contributions of This Book

The vast changes in the data world have brought with them changes in the nature of data collectors and producers. Data on human beings, though now more likely to be collected, are much *less* likely to be purposively collected by researchers and government agencies, and are thus less often held by organizations with traditional knowledge about how to protect privacy. There are serious consequences: the lack of dissemination experience of non-governmental collectors can lead to massive privacy breaches (the 2006 AOL data release is but one famous example).[11] Even worse, if no dissemination is allowed, the quality of privately held data is largely unknown[12] absent detailed researcher inspection and validation. Similarly, Institutional Review Boards with few reference guidelines are likely to slow or prevent research on human subjects with complex data.[13]

Because of the importance of the topic, there is a rich and vibrant literature; the contributors to this book have provided, for the first time in one place, an accessible summary of existing research on many of the important aspects of balancing access to data with protection of privacy. They have also identified practical suggestions – to help guide practitioners and Institutional Review Boards – and identified important areas for future research.

Opening Part I, on the conceptual framework, *Strandburg* argues that the acquisition, transfer, and aggregation of data on a massive scale for data mining and predictive analysis raises questions that simply are not answered by the paradigms that have dominated privacy law to date. She develops a taxonomy of current U.S. privacy law and uses that taxonomy to elucidate the mismatch between current law and big data privacy concerns. *Barocas and Nissenbaum* argue that big data involves practices that have radically disrupted entrenched information flows. From modes of acquisition to aggregation, analysis, and application, these disruptions affect actors, information types, and transmission principles. Privacy and big data are simply incompatible, and the time has come to reconfigure choices that we made decades ago to enforce certain constraints. They argue that it is time for the background of rights, obligations, and legitimate expectations to be explored and enriched so that notice and consent can do the work for which it is best suited. *Acquisti* discusses how the economics and behavioral economics of privacy can be applied to investigate the implications of consumer data mining and business analytics. An important insight is that personal information, when shared, can become a public good whose analysis can reduce inefficiencies and increase economic welfare; when abused, it can lead to transfer of economic wealth from data subjects to data holders. The interesting economic question then becomes who will bear the costs if privacy-enhancing technologies become more popular in the age of big data: data subjects (whose benefits from business analytics and big data may shrink as they restrict the amount of information they share), data holders (who may face increasing costs associated with collecting and handling consumer data), or both?

There are practical implications. *Ohm* provides an overview of how information privacy laws regulate the use of big data techniques, if at all. He discusses whether these laws strike an appropriate balance between allowing the benefits of big data and protecting individual privacy and, if not, how the laws might be extended and amended to better strike this balance. He notes that most information privacy law focuses on collection or disclosure and not use. Once data has been legitimately obtained, few laws dictate what may be done with the information. The chapter offers five general proposals for change. *Stodden* sets out the scientific rationale for access to data and computational methods, to enable the verification and validation of published research findings. She describes the legal landscape in the context of big data research and suggests two guiding principles to facilitate reproducibility and reuse of research data and code within and beyond the scientific context.

Koonin and Holland open Part II, on the practical framework, by addressing the motivations of the new urban science and the value for cities in big data – particularly with respect to analysis of the infrastructure, the environment, and the people. They discuss the key technical issues in building a data infrastructure for curation, analytics, visualization, machine learning, and data mining, as well as modeling and simulation to keep up with the volume and speed of data. *Goerge* describes the creation of a data warehouse that links data on multiple services provided by the public sector to individuals and families as a way to highlight both the opportunities and the challenges in a city's use of data. He identifies the key issues that need to be addressed – what data to develop and access from counties, states, the federal government, and private sources; how to develop the capacity to use data; how to present data and be transparent; and how best to keep data secure so that individuals and organizations are protected – as well as the key barriers. *Elias* provides a broader perspective than simply the United States by noting that many of the legal and ethical issues associated with big data have wider relevance. Much can be learned from examining the progress made in Europe toward developing a harmonized approach to legislation designed to provide individuals and organizations with what has become known as the 'right to privacy'. The legislative developments have had and are continuing to have substantial impact on cross-border access to microdata for research purposes; that impact is also examined.

Greenwood, Stopczynski, Sweatt, Hardjono and Pentland explore the emergence of the Big Data society, arguing that the 'personal data sector' of the economy needs productive collaboration between the government, the private sector, and the citizen to create new markets – just as the automobile and oil industries did in prior centuries. They sketch a model of data access that is governed by 'living informed consent', whereby the user is entitled to know what data are being collected about her by which entities, empowered to understand the implications of data sharing, and finally is put in charge of any data-sharing authorizations. They envision the establishment of a New Deal on Data, grounded in principles such as the opt-in nature of data provision, the framing of data usage boundaries, and the credentialing of parties authorized to access data.

Landwehr takes a very pragmatic approach. He notes that, regardless of what data policies have been agreed, access must be allowed through controls engineered into the data infrastructure. Without sound technical enforcement, incidents of abuse, misuse, theft of data, and even invalid scientific conclusions based on undetectably altered data can be expected. He discusses what features access controls might have – delineating the characteristics of subjects, objects, and access modes – and notes that advances in practical cryptographic solutions to computing on encrypted data could change the picture in the future by reducing the need to trust hardware and system software. Advances in methods for building systems in which information flow, rather than access control, is the basis for policy enforcement could also open the door to better enforcement of comprehensible policies.

Wilbanks is similarly practical. He provides an overview of frameworks that are available to permit data reuse and discusses how legal and technical systems can be structured to allow people to donate their data to science. He argues that traditional frameworks to permit data reuse have been left behind by the mix of advanced techniques for re-identification and cheap technologies for the creation of data about individuals. He surveys the approaches developed in technological and organizational systems to 'create' privacy where it has been eroded while allowing data reuse, but also discusses a new approach of 'radical honesty' toward data contribution and the development of 'portable' approaches to informed consent that could potentially support a broad range of research without the unintended fragmentation of data created by traditional consent systems

Kreuter and Peng open Part III, on the statistical framework, with a discussion of the new statistical challenges associated with inference in the context of big data. They begin by noting that reliable statistical inference requires an understanding of the data-generating process. That process is not well understood in the case of big data, so it is important that researchers be given access to the source data so that coverage and quality issues can be identified and addressed. Standard statistical disclosure limitations are unlikely to work, because an important feature of big data is the ability to examine different, targeted populations, which often have unique and easily re-identifiable characteristics. *Karr and Reiter* explore the interactions between data dissemination, big data, and statistical inference. They identify a set of lessons that stewards of big data can learn from statistical agencies' experiences with the measurement of disclosure risk and data utility. Their conclusion is that the sheer scale and potential uses of big data will require that analysis be taken to the data rather than the data to the analyst or the analyst to the data. They argue that a viable way forward for big data access is an integrated system including (i) unrestricted access to highly redacted data, most likely some version of synthetic data, followed with (ii) means for approved researchers to access the confidential data via remote access solutions, glued together by (iii) verification servers that allow users to assess the quality of their inferences with the redacted data so as to be more efficient with their use (if necessary) of the remote access to the confidential data. *Dwork* concludes with a vision for the future. She shows how differential privacy provides a mathematically rigorous theory of privacy, a theory amenable to measuring (and minimizing) cumulative privacy loss, as data are analyzed and re-analyzed, shared, and linked. There are trade-offs – differential privacy requires a new way of interacting with data, in which the analyst accesses data only through a privacy mechanism, and in which accuracy and privacy are improved by minimizing the viewing of intermediate results. But the approach provides a measure that captures cumulative privacy loss over multiple releases; it offers the possibility that data usage and release could be accompanied by publication of privacy loss.

Thanks As with any book, we have benefited enormously from the support and help of many people. Our editor, Diana Gillooly, has worked tirelessly and efficiently at all phases – going well beyond the call of duty. Our referees took time out of their busy schedules to give thoughtful, constructive guidance to the authors. They include Micah Altman, Mike Batty, Jason Bobe, Aleksandra Bujnowska, Fred Conrad, Josep Domingo-Ferrer, Stephanie Eckman, Mark Elliot, Martin Feldkircher, Simson Garfinkel, Bob Goerge, Eric Grosse, Patricia Hammar, David J. Hand, Dan Harnesk, Kumar Jayasuriya, Gary King, Frauke Kreuter, Tom Kvan, Bethany Letalien, William Lowrance, Lars Lyberg, Tim Mulcahy, Kobbi Nissim, Onora O'Neill, Kathleen Perez Lopez, Carlo Reggiani, Jerome H. Reichman, Guy Rothblum, Subu R. Sangameswar, Fred Schneider, Aleksandra Slavkovic, Tom Snijders, Omer Tene, Vincenc Torra, Paul Uhlir, Richard Valliant, and Felix Wu.

The book was sponsored by New York University, through the Center for Urban Science and Progress. Konstantin Baetz, Veronika Zakrocki, Felicitas Mittereder, Reinhard Sauckel, Dominik Braun, Shaylee Nielson, and Christian Rafidi provided superb administrative support, Mark Righter in NYU General Counsel's office provided legal assistance, and Ardis Kadiu of Spark451 gave assistance with our project website. The book was also supported by the Privacy and Confidentiality Subcommittee of the American Statistical Association and the Institute for Employment Research of the German Federal Employment Agency.

We are also very grateful to Steve Pierson, American Statistical Association; Kim Alfred, CUSP; and Kelly Berschauer, Microsoft Research, for their help with outreach to key stakeholders.

NOTES

1. 'Big data' is given many definitions. One fairly representative version says that the term "refers to large, diverse, complex, longitudinal, and/or distributed data sets generated from instruments, sensors, Internet transactions, email, video, click streams, and/or all other digital sources available today and in the future" (NSF BIGDATA solicitation http://www.nsf.gov/funding/pgm_summ.jsp?pims_id=504767&org=CISE). Another uses the velocity, volume, and variety rubric: "data is now available faster, has greater coverage and scope, and includes new types of observations and measurements that previously were not available" (L. Einav and J. D. Levin, "The Data Revolution and Economic Analysis," National Bureau of Economic Research Working Paper, No. 19035, Cambridge, MA, 2013, retrieved from http://www.nber.org/papers/w19035).
2. For a survey of statisticians' opinions on the privacy and ethics related to data collections, see http://blog.revolutionanalytics.com/2013/09/statistician-survey-results.html.
3. For a good review of the role of statistics in the Greek financial crisis, see http://www.ft.com/cms/s/0/82b15932-18fe-11e1-92d8-00144feabdc0.html#axzz2g7W3pWOJ.
4. D. Kahneman, *Thinking Fast and Slow* (New York: Farrar, Straus and Giroux, 2011).
5. See http://www.census.gov/aboutus/#.
6. J. Groen, "Sources of Error in Survey and Administrative Data: The Importance of Reporting Procedures," *Journal of Official Statistics* 28, no. 2 (2012): 173–198.
7. J. M. Abowd and L. Vilhuber, "National Estimates of Gross Employment and Job Flows from the Quarterly Workforce Indicators with Demographic and Industry Detail," *Journal of Econometrics* 161, no. 1 (2011): 82–99, retrieved from http://ideas.repec.org/a/eee/econom/v161y2011i1p82-99.html.
8. T. Hey, S. Tansley, and K. Tolle, *The Fourth Paradigm: Data Intensive Scientific Discovery* (Redmond, WA: Microsoft Research, 2009).
9. Einav and Levin, "The Data Revolution and Economic Analysis."
10. E. Ostrom, Governing the Commons: The Evolution of Institutions for Collective Action (Cambridge, UK: Cambridge University Press, 1990).
11. See http://en.wikipedia.org/wiki/AOL_search_data_leak.

12. See; e.g., http://www.theguardian.com/commentisfree/2013/may/04/adp-forecasting-monthly-bls-jobs-reports.
13. Many of the issues associated with the role of Institutional Review Boards are highlighted in a recent National Academies' workshop http://sites.nationalacademies.org/DBASSE/BBCSS/CurrentProjects/DBASSE_080452.

Part I Conceptual Framework

This part begins by considering the existing legal constraints on the collection and use of big data in the privacy and confidentiality context. It then identifies gaps in the current legal landscape and issues in designing a coherent set of policies that both protect privacy and yet permit the potential benefits that come with big data. Three themes emerge: that the concepts used in the larger discussion of privacy and big data require updating; that how we understand and assess harms from privacy violations needs updating; and that we must rethink established approaches to managing privacy in the big data context.

The notion of 'big data' is interpreted as a change in paradigm, rather than solely a change in technology. This illustrates the first central theme of this part of the book. Barocas and Nissenbaum define big data as a "paradigm, rather than a particular technology," while Strandburg differentiates between collections of data, and collections of data that have been "datafied," that is, "aggregated in a computationally manipulable format." She claims that such datafication is a key step in heightening privacy concerns and creating a greater need for a coherent regulatory structure for data acquisition. Traditional regulatory tools for managing privacy – notice and consent – have failed to provide a viable market mechanism allowing a form of self-regulation governing industry data collection. Strandburg elucidates the current legal restrictions and guidance on data collection in the industrial setting, including the Fair Information Practice Principles (FIPPs) dating from 1973 and underlying the Fair Credit Reporting Act (FCRA) from 1970 and the Privacy Act from 1974. Strandburg advocates a more nuanced assessment of trade-offs in the big data context, moving away from individualized assessments of the costs of privacy violations. The privacy law governing the collection of private data for monitoring purposes should be strengthened, in particular, a substantive distinction should be made between datafication and the repurposing of data that was collected as a byproduct of providing services. Additionally, she suggests taking a substantive approach to the ideas of notice and consent in particular to clarify their meaning for large entities.

The inadequacy of assessment of harm from privacy breaches is another major theme of this part, extending it from the level of the individual to that of groups or classes, and even society as a whole. Acquisti notes that this requires a greater understanding of the breadth of types of privacy breaches and of the nature of harm as diffused over time, and an improved valuation of privacy in the big data context. Consumers may value their own privacy in variously flawed ways. They may have incomplete information, for example, or be given an overabundance of information that renders processing impossible, or use heuristics that systematize deviations from rational decision making. Acquisti notes that privacy protection can both potentially increase and decrease economic efficiency in the marketplace, and that deriving benefits from big data may not conflict with benefits from assuring privacy protection.

In order to address these issues, several authors ask us to rethink traditional approaches to privacy. This is the third overarching theme of this part of the book. Barocas and Nissenbaum argue that the concepts of anonymity and informed consent do not create solutions to the privacy issue. As datasets become increasingly linked, anonymity is largely impossible to guarantee in the future. This also implies that it is impossible to give truly informed consent, since we cannot, by definition, know what the risks are from revealing personal data either for individuals or for society as a whole.

The use of privately collected data is largely unregulated. Ohm describes the few regulations that do apply – such as the Health Insurance Portability and Accountability Act (HIPAA), the Privacy Act, and the Fair Credit Reporting Act (FCRA) – and explains that the United States employs a 'sectoral' approach to privacy regulation, in that different economic areas have separate privacy laws. Ohm also calls into question the traditional notion of notice in the case of big data. To whom are you to give notice, and for what? The results of big data analysis can be unpredictable and sometimes unexplainable, another reason it is difficult to assess privacy risks accurately in the big data context.

Ohm advocates a new conceptualization of legal policy regarding privacy in the big data context, guided by five principles for reform: (1) rules must take into account the varying levels of inherent risk to individuals across different datasets, (2) traditional definitions of personally identifiable information need to be rethought, (3) regulation has a role in creating and policing walls between datasets, (4) those analyzing big data must be reminded, with the frequency in proportion to the sensitivity of the data, that they are dealing with people, and (5) the ethics of big data research must be an open topic for continual reassessment.

In the final chapter, Stodden focuses on the theme of research integrity in the big data context. She notes a conflict between research requirements regarding the replication of computational results, which can require data access, and traditional methods of privacy protection via sequestration. She advocates establishing 'middle ground' solutions whenever possible that maximize verification of computational findings, while taking into account any legal and ethical barriers. Permitting authorized researchers access to confidential data within a 'walled garden' can increase the ability of others to independently replicate big data findings, for example. Two principles are presented to help guide thinking regarding reproducibility and verification in big data research: the Principle of Scientific Licensing and the Principle of Scientific Data and Code Sharing. That is, in the scientific context, legal encumbrances to data sharing for purposes of independent verification should be minimized wherever possible, and access to the data and methods associated with published findings should be maximized subject to legal and ethical restrictions.

1 Monitoring, Datafication, and Consent: Legal Approaches to Privacy in the Big Data Context

Katherine J. Strandburg

Introduction

Knowledge is power. 'Big data' has great potential to benefit society. At the same time, its availability creates significant potential for mistaken, misguided, or malevolent uses of personal information. The conundrum for law is to provide space for big data to fulfill its potential for societal benefit, while protecting citizens adequately from related individual and social harms. Current privacy law evolved to address different concerns and must be adapted to confront big data's challenges. This chapter addresses only one aspect of privacy law: the regulation of private sector acquisition, aggregation, and transfer of personal information.[1] It provides an overview and taxonomy of current law, highlighting the mismatch between current law and the big data context, with the goal of informing the debate about how to bring big data practice and privacy regulation into optimal harmony.

Part I briefly describes how privacy regulation in the United States has evolved in response to a changing technological and social milieu. Part II introduces a taxonomy of privacy laws relating to data acquisition, based on the following features: (1) whether the law provides a rule- or a fact-based standard; (2) whether the law is substantive or procedural, in a sense defined below; and (3) which mode(s) of data acquisition are covered by the law. It also argues that the recording, aggregation, and organization of information into a form that can be used for data mining, here dubbed 'datafication', has distinct privacy implications that often go unrecognized by current law.[121] Part III provides a selective overview of relevant privacy laws in light of that taxonomy. Section A discusses the most standards-like legal regimes, such as the privacy torts, for which determining liability generally involves a fact-specific analysis of the behavior of both data subjects and those who acquire or transfer the data ('data handlers'). Section B discusses the Federal Trade Commission's (FTC's) 'unfair and deceptive trade practices' standard,[2] which depends on a fact-specific inquiry into the behavior of data handlers, but makes general assumptions about data subjects. Section C discusses rule-like regimes, such as the Privacy Rule[3] of the Health Insurance Portability and Accountability Act[4] (HIPAA Rule). Part IV points out some particular features of the mismatch between current law's conceptualization of the issues and the big data context, using the taxonomy developed in Part II as an aid to the analysis. It then makes several suggestions about how to devise a better fit.

I. The Evolution of U.S. Privacy Law

Outside of the law enforcement context, privacy law was erected on the foundation of Warren and Brandeis's famous 1890 article, *The Right to Privacy*.[5] The privacy torts built on that foundation were concerned primarily with individualized harms of emotional distress, embarrassment, and humiliation arising out of 'intrusion upon seclusion' or 'public disclosure of private facts'. Privacy law also aimed to protect confidentiality in certain kinds of relationships, often involving professional expertise, in which information asymmetry and power imbalances create a potential for exploitation. These torts provide compensation for individualized injuries caused by egregious deviations from social norms. In principle, and often in fact, the tort paradigm employs a highly contextualized analysis of the actions of plaintiff and defendant and the relationship between them.

In the 1970s, the development of digital computers and the increasing complexity of the administrative state led to an expansion in 'computer-based record-keeping operations' by governments and certain other large institutions, such as banks. This expansion raised fears of misuse, unfairness, lack of transparency in decision making, and chilling of autonomous behavior distinct from the concerns about emotional distress and reputation at the heart of the privacy torts. Fair Information Practice Principles (FIPPs), which have become the mainstay of data privacy law, were developed during this period as an approach to those issues. The Fair Credit Reporting Act (FCRA),[6] adopted in 1970, and the Privacy Act of 1974,[7] regulating data use by government agencies, were based on FIPPs.

A set of five FIPPs were proposed in 1973 in a report commissioned by the Department of Health, Education, and Welfare (HEW Report):

1. There must be no personal data record-keeping systems whose very existence is secret.
2. There must be a way for a person to find out what information about the person is in a record and how it is used.

3. There must be a way for a person to prevent information about the person that was obtained for one purpose from being used or made available for other purposes without the person's consent.
4. There must be a way for a person to correct or amend a record of identifiable information about the person.
5. Any organization creating, maintaining, using, or disseminating records of identifiable personal data must assure the reliability of the data for their intended use and must take precautions to prevent misuses of the data.[8]

These principles, along with three sets of related 'safeguard requirements',[9] attempted to cope with the scale of data collection by substituting transparency and consent for the individualized fact-specific approach of the privacy torts. The HEW Report recognized the difficulty of legislating substantive rules in light of the "enormous number and variety of institutions dealing with personal data," arguing that institutions should be "deterred from inappropriate practices, rather than being forced by regulation to adopt specific practices."[10]

Another important version of FIPPs was formulated by the OECD in 1980 (OECD FIPPs), articulating eight principles, which expanded on the HEW FIPPs and include a "Collection Limitation Principle" that there "should be limits to the collection of personal data," a "Data Quality Principle" that data collected should be "relevant to the purposes for which they are to be used" and "accurate, complete and kept up-to-date," and a "Purpose Specification Principle" that purposes should be specified in advance and "subsequent use limited to the fulfillment of those purposes or such others as are not incompatible with those purposes."[11]

At the turn of the 21st century, the FTC, which has taken the primary role in commercial privacy regulation, recommended a rather slimmed down set of FIPPs for online privacy:

(1) Notice – Web sites would be required to provide consumers clear and conspicuous notice of their information practices. . . .
(2) Choice – Web sites would be required to offer consumers choices as to how their personal identifying information is used beyond the use for which the information was provided (e.g., to consummate a transaction). . . .
(3) Access – Web sites would be required to offer consumers reasonable access to the information a Web site has collected about them, including a reasonable opportunity to review information and to correct inaccuracies or delete information.

(4) Security – Web sites would be required to take reasonable steps to protect the security of the information they collect from consumers.[12]

In practice, outside of a few sectors, such as health care, the FIPPs approach in the United States has been whittled down to a focus on procedures for ensuring notice and consent. What explains this progression? Cheap and ubiquitous information technology makes large-scale data aggregation and use possible for a wide variety of private sector entities in addition to the government agencies and large institutions, such as banks, that were the subject of privacy concerns in the 1970s and 1980s. The resulting expansion of private sector data collection and use has competing implications for privacy regulation. On the one hand, the acquisition and use of an increasing quantity and scope of personal information by an increasingly large and various set of private entities heightens privacy concerns relating to data security and breach, accountability and transparency, and unpredictable data uses. On the other hand, substantive regulation of a large and diverse array of private sector entities is politically controversial, regulations that effectively span the field of data handlers are hard to devise, and monitoring the data practices of such a large number of players is difficult and expensive.

As a result, the trend was to assume (or at least hope) that notice and consent would provide a market mechanism for encouraging industry self-regulation of data privacy. In light of recent acceleration in data collection and the development of big data approaches to mining aggregated data, it now is widely recognized that the notice and consent paradigm is inadequate to confront the privacy issues posed by the big data explosion.[13] The notice and consent paradigm assumes that citizens are able to assess the potential benefits and costs of data acquisition sufficiently accurately to make informed choices. This assumption was something of a legal fiction when applied to data collected by government agencies and regulated industries in the 1970s. It is most certainly a legal fantasy today, for a variety of reasons including the increasing use of complex and opaque predictive data-mining techniques, the interrelatedness of personal data, and the unpredictability of potential harms from its nearly ubiquitous collection.[14]

II. A Taxonomy of Privacy Laws Relevant to Data Acquisition

As mentioned in the introduction, it is useful to organize this selective overview of U.S. privacy law relating to data acquisition and transfer according to a taxonomy focusing on three characteristics: (A) whether the law takes a rule-like or fact-specific standards-like approach; (B) whether the law regulates substance or procedure, in a sense defined below; and (C) what modes of data acquisition are covered by the law. While these distinctions are not bright lines, an approximate categorization is useful in analyzing the uncomfortable fit between current privacy law and big data projects.

A. Rules or Standards

Privacy law regimes vary according to the extent to which they impose flexible fact-specific standards or generally applicable rules, but can be divided roughly into three groups. Laws in the first group, such as the torts of intrusion upon seclusion and public disclosure of private facts, assess liability using standards that depend on detailed fact-intensive inquiry into the activities of both subjects and acquirers of personal information. Laws in the second group, such as Section 5 of the FTC Act, employ standards-based assessment of the activities of data holders, while relying on presumptions about data subjects. Laws in the third group, such as the HIPAA Rule, mandate compliance with rules.

Trade-offs between rules and standards are endemic to law.[15] Ex ante, rules provide clearer direction for behavior, while standards provide leeway so behavior can be tailored more optimally to specific contexts. Ex post, rules are cheaper to enforce and leave less room for bias, while standards leave more discretion for crafting context-sensitive and fair outcomes. These tensions are dynamic. Because of these trade-offs, courts and legislatures often attempt to draw at least some bright lines (such as a line between private and public) to guide the application of standards, while rule-like regimes often become complex (or even byzantine) as lawmakers try to anticipate all relevant contingencies.

B. Substance or Procedure

Privacy law also grapples with trade-offs between substantive and procedural regulation. Compliance with substantive regulation, as I use the term here, is determined by asking: Was it legally acceptable for Data Handler A to acquire or transfer this information about Data Subject B in this situation and in this way? Compliance with procedural regulation, on the other hand, is determined by asking: Did Data Handler A follow the appropriate procedures in acquiring or transferring information about Data Subject B?

Though substantive regulation is preferable when goals are well defined and outcomes are observable by enforcement agencies, procedural regulation is advantageous in some situations. Regulated entities may be better situated than lawmakers, by virtue of expertise or superior information, to make substantive determinations, especially when circumstances are expected to evolve, perhaps by technological advance, but data holders may have incentives to use their discretion in socially undesirable ways. Procedural regulation may be used to limit their ability to do so. Substantive outcomes may be difficult to specify or to observe. Procedural regulations may structure behavior so as to make desirable outcomes more likely and may make compliance easier to audit. Procedural regulation also may help to prevent negligence and mistake. Procedural and substantive approaches often are combined in privacy regulation. For example, some laws require different procedures based on substantive distinctions between data-handling entities, types of data, or purposes of data acquisition.

C. Modes of Data Acquisition

There are three basic avenues for acquiring big data: monitoring, acquisition as a byproduct of another activity, and transfer of pre-existing information. Monitoring, as I use the term here, applies broadly to the recording of information in plain view and information acquisition by means such as wiretapping and spyware. Acquisition as a byproduct of another activity is common for service providers such as telecommunications providers, utilities, online websites and apps, search engines, and governments.

However it is acquired, if information is to be used in a big data project it must be recorded, quantified, formatted, and stored digitally to make it usable for computational knowledge discovery. Note that what I will call 'datafication' is distinct from digitization. Cellphone photos are digital, but they are not datafied unless they are aggregated in a computationally manipulable format. Datafication has independent privacy implications because recording and organizing information changes the uses to which it can be put, both for good and for ill.[16] Importantly, because computation methods are continually developed and refined, datafication is likely to open the door to uses that were not feasible (and hence obviously not anticipated) at the time the data was acquired.

To illustrate the role of datafication in monitoring, consider video surveillance. Without cameras, the only record of what happens on city streets is in the minds of the human beings who pass by. Those memories are scattered, degrade quickly, may be inaccurate, and are very costly to aggregate and to search in retrospect, making it difficult for police to capture burglars, muggers, and murderers. Installing surveillance cameras watched by security guards provides more focused attention on what is occurring in view of the camera, making it more likely that a given event will be noticed and recalled. Security cameras also may deter crime (at least in their vicinity). Adding an analog video recorder makes a more accurate reconstruction of events taking place in view of the camera possible and decreases the cost of surveillance. Still, the difficulty of manually reviewing videotape limits the circumstances under which tapes are likely to be reviewed and makes aggregation of information from different cameras difficult. The storage costs of archiving videotape also constrain how far back one can look. Substituting a digital recorder increases the efficiency of manual review, while storage and transmission become easier and cheaper, meaning that recordings are likely to be archived for longer periods.

As each of these changes is made, monitoring becomes more efficient and effective for law enforcement purposes. The potential for privacy invasion also increases. A security guard might happen to see someone she knows heading into a psychiatrist's office. A building employee might review videotapes to see who recently visited the psychiatrist or search through digital archives to see whether a particular person visited the psychiatrist five years ago. But digital recording also facilitates datafication – long-term storage in a format that is searchable, computationally manipulable, and may be aggregated with information from other security cameras. Datafication opens up the potential for uses that may have been unanticipated or even technologically infeasible at the time of collection and are qualitatively different from the original purposes of the surveillance. For example, the development of facial recognition algorithms might make it possible to aggregate the data from surveillance cameras to track particular individuals' movements, to investigate correlations between the places people go, or to combine the video surveillance data with other data to create profiles of the types of people who visit psychiatrists. Some of these uses may be socially beneficial and some socially harmful on balance. The data can be used to track a political activist or ex-lover as well as a robber fleeing the scene of a mugging. Once datafied, the information can be used by law enforcement officials, researchers, or hackers if they have access. The point thus is that datafication heightens privacy concerns and changes the trade-offs involved in monitoring and its regulation.

When service providers acquire information as a byproduct of providing service, there are several conceptually distinct possibilities. Data may be (1) acquired for service provision but not datafied; (2) acquired, datafied, and used only for service provision; (3) acquired for service provision but datafied for other purposes; or (4) acquired and datafied for service provision but repurposed for other uses, perhaps by transferring it to third parties. Service providers also can leverage their access to their users' computers, mobile phones, or other property to monitor information that would not be acquired as a byproduct of providing service.

For example, consider a provider of a subscription streaming video service. The provider will obtain subscription and payment information as a byproduct of providing its services and might datafy that information to streamline its services. It might also repurpose that information by using it to advertise its own other services or transferring it to third parties for advertising purposes. The provider also will necessarily receive and process information about which videos are requested by which customers, when they are streamed, and so forth. It may have no need to datafy that information to provide the service, but might choose to do so in order to use or transfer the data for some other purpose, such as creating advertising profiles. If the service provider uses cookies to record what webpages its customers visits and what ads they click on, it is no longer obtaining information as a byproduct of service provision. It is engaging in monitoring.

As a normative matter we might want the law to distinguish carefully between these different modes of data acquisition. As we shall see in Part III, current law has been slow to recognize these distinctions, especially in the service provider context.

III. Selective Overview of Privacy Laws

This section considers a sampling of important U.S. privacy laws in light of the taxonomy developed in Part II. Section A deals with legal regimes that rely primarily on standards based on fact-specific analysis of the behavior of both data subjects and data handlers. Section B deals with standards based primarily on data handler behavior, relying on general assumptions about data subjects. Section C deals with legal regimes that employ a mostly rule-like approach. Of course, these categorizations are rough approximations, since the mixture of rules and standards can vary in many different ways.

A. Two-Sided Standards

Privacy laws that apply two-sided fact-specific standards include the intrusion upon seclusion tort and California's state constitutional right to privacy. Two-sided fact-specific standards are most effectively applied in situations involving specific and traceable individualized harms. Like torts more generally, privacy laws employing two-sided standards have difficulty handling small widespread harms, probabilistic harms, and harms that are difficult to trace to a particular relationship, transaction, or incident.

1. The Intrusion upon Seclusion Tort

The intrusion upon seclusion tort is a substantive legal regime that has been employed primarily to regulate monitoring. Privacy torts are usually creatures of state common law (though sometimes they are codified by state legislatures). They were strongly influenced by William Prosser, author of a famous torts treatise. During the first half of the 20th century,[17] Prosser studied and attempted to systematize the common law of invasion of privacy that had developed in the wake of Warren and Brandeis's 1890 article. Prosser's definitions of the privacy torts were incorporated into the 1967 draft of the Second Restatement of Torts, for which he was the chief reporter.[18] Restatements attempt to guide and structure common law discretion, thus importing rule-like aspects into the analysis. Though states' adoption and interpretation of the privacy torts vary, the Second Restatement's definitions have been adopted in most states that recognize the torts.

This discussion focuses on the tort of intrusion upon seclusion, which is most relevant to data acquisition (though the tort of public disclosure of private facts might also apply to data acquisition in some circumstances). The Second Restatement defines intrusion upon seclusion as follows:

> One who intentionally intrudes, physically or otherwise, upon the solitude or seclusion of another or his private affairs or concerns, is subject to liability to the other for invasion of his privacy, if the intrusion would be highly offensive to a reasonable person.[19]

An intrusion claim thus requires (1) an actionable intrusion that is (2) highly offensive.[20]

a. Actionable Intrusion As might be expected from its definition, actionable intrusion traditionally involves monitoring. The Restatement's definition of actionable intrusion seeks to draw a rule-like line between protectable "solitude or seclusion" and "private affairs and concerns" and unprotectable public activity. The Restatement's examples of actionable intrusion mostly involve traditionally "private" zones, such as the home, mail or telephone communications, and financial matters:

The invasion may be by physical intrusion into a place in which the plaintiff has secluded himself, as when the defendant forces his way into the plaintiff's room in a hotel or insists over the plaintiff's objection in entering his home. It may also be by the use of the defendant's senses, with or without mechanical aids, to oversee or overhear the plaintiff's private affairs, as by looking into his upstairs windows with binoculars or tapping his telephone wires. It may be by some other form of investigation or examination into his private concerns, as by opening his private and personal mail, searching his safe or his wallet, examining his private bank account, or compelling him by a forged court order to permit an inspection of his personal documents.[21]

The Restatement also describes what is *not* actionable because it can be observed in public:

[T]here is no liability for the examination of a public record concerning the plaintiff. . . . Nor is there liability for observing him or even taking his photograph while he is walking on the public highway. . . . Even in a public place, however, there may be some matters about the plaintiff, such as his underwear or lack of it, that are not exhibited to the public gaze.[22]

Despite this attempt to draw a line, the Restatement's commentary recognizes the importance of specific facts about the social context, relationships between the parties, and means of intrusion in determining whether an actionable intrusion has occurred. Thus, the Restatement distinguishes "calling [an individual] to the telephone on one occasion or even two or three, to demand payment of a debt"[23] from repeated phone calls "at meal times, late at night and at other inconvenient times."[24] If "the telephone calls are repeated with such persistence and frequency as to amount to a course of hounding the plaintiff," the caller may be liable.[25] By referring to the use of binoculars, telephone taps, and forged court orders, the Restatement also recognizes that the *means* of intrusion must be taken account in determining whether intrusion has occurred.

The case law further illustrates the doubly fact-specific nature of the intrusion definition. There is no bright line rule, for example, as to whether plaintiffs can complain of intrusion upon seclusion in the workplace. The assessment of actionable intrusion may depend on factors such as the degree to which the location was open to others, the use of deception, hidden cameras, or microphones, and the purpose of the intrusion. For example, spying on an employee from a space above the restroom[26] and secretly videotaping conversations between telephone psychics for television broadcast in a shared workplace have been deemed actionable.[27] On the other hand, intrusive investigations are sometimes justified if they are tailored to an appropriate purpose. Thus in *Hernandez v. Hillsides, Inc.*, the employer invaded privacy expectations by "secretly install[ing] a hidden video camera that was both operable and operating (electricity-wise), and that could be made to monitor and record activities inside plaintiffs' office, at will, by anyone who plugged in the receptors, and who had access to the remote location in which both the receptors and recording equipment were located" in order to investigate use of its computers for viewing pornography. Nonetheless, the invasion was not actionable because the employer had aimed the camera at a particular computer, controlled access to the surveillance equipment, and operated the equipment only during limited times.[28] Similarly, while drug testing of employees generally has not been deemed an intrusion, courts have found particular methods of collecting urine specimens intrusive.[29]

Many intrusion upon seclusion cases involve journalists or researchers. In such cases, First Amendment values often outweigh claims to intrusion, depending upon the nature of the intrusion and the egregiousness of its means. For example, while mere deception in the course of journalistic investigation is not actionable, intrusion into a home,[30] the use of hidden cameras and recording devices,[31] or taking advantage of intimate or confidential relationships[32] sometimes leads to liability. In *Shulman v. Group W Productions*, for example, a reporter on a 'ride-along' with an emergency medical helicopter team videotaped an accident scene and recorded conversations between a victim and one of the team's nurses, who was wearing a microphone. While the reporter's general presence at and video recording of the accident scene was not an intrusion, the court refused to dismiss claims based on the recording of the victim's utterances picked up by the nurse's microphone and on the reporter's presence in the helicopter while the accident victim was transported to the hospital. As the court explained: "[The reporter], perhaps, did not intrude into that zone of privacy merely by being present at a place where he could hear such conversations with unaided ears. But by placing a microphone on Carnahan's person, amplifying and recording what she said and heard, defendants may have listened in on conversations the parties could reasonably have expected to be private."[33] In allowing this claim to proceed, the court emphasized the traditional expectation of confidentiality in conversations with providers of medical care.[34]

Courts also rely on the factual context to determine whether defendants intrude upon seclusion by obtaining records from others. In *Hall v. Harleysville Insurance Co.*,[35] for example, the court sustained an intrusion claim alleging that workers' compensation investigators had improperly obtained a claimant's credit reports. In doing so, the court distinguished *Chicarella v. Passant*,[36] which had dismissed a similar claim alleging that an automobile insurance company had obtained the plaintiff's hospital records by pretext while investigating his general credibility. *Hall* opined that credit reports are "a thorough and complete analysis of [an individual's] financial position,"[37] while the hospital records in *Chicarella* contained only "brief descriptions of claimant's medical treatment."[38]

Courts sometimes have allowed intrusion claims based on excessive and prolonged surveillance in public, despite the tort's purported inapplicability to public behavior. A 1913 case found liability for "rough or open shadowing," opining that "[a]ctual pursuit and public surveillance of person and home are suggestive of criminality fatal to public esteem and productive of public contempt or ridicule."[39] In *Galella v. Onassis*, the court held that prolonged and intrusive monitoring of the public activities of Jacqueline Onassis and her family by a paparazzo amounted to a "torrent of almost unrelieved abuse into everyday activity" and constituted actionable intrusion.[40]

Nader v. General Motors[41] considered allegations that General Motors had intruded upon Nader's seclusion via a pattern of tactics including following the well-known consumer advocate in public, interviewing his friends and associates in ways intended to harm his reputation, seeking to entrap him into illicit sexual relationships, harassing him with phone calls, and tapping his phones. The court focused on an allegation that a GM agent had "followed [Nader] into a bank, getting sufficiently close to him to see the denomination of the bills he was withdrawing from his account." As the court explained:

> A person does not automatically make public everything he does merely by being in a public place, and the mere fact that Nader was in a bank did not give anyone the right to try to discover the amount of money he was withdrawing. On the other hand, if the plaintiff acted in such a way as to reveal that fact to any casual observer, then, it may not be said that the appellant intruded into his private sphere.

While noting that "it is manifest that the mere observation of the plaintiff in a public place does not amount to an invasion of his privacy," the court opined that public surveillance nonetheless could be "so overzealous as to render it actionable." A concurring judge would have gone farther, arguing that

> it does not strain credulity or imagination to conceive of the systematic "public" surveillance of another as being the implementation of a plan to intrude on the

privacy of another. Although acts performed in "public," especially if taken singly or in small numbers, may not be confidential, at least arguably a right to privacy may nevertheless be invaded through extensive or exhaustive monitoring and cataloguing of acts normally disconnected and anonymous.[42]

Recently, in the Fourth Amendment context, courts have begun to recognize that prolonged monitoring of public activities, especially when accompanied by datafication, can intrude upon a "reasonable expectation of privacy," as required for a Fourth Amendment violation.[43] In *United States v. Jones*,[44] five concurring justices agreed that there could be a reasonable expectation of privacy under the Fourth Amendment in aggregated location data.[45] While the "reasonable expectation of privacy" test is not identical to the test for actionable intrusion, a similar approach could be imported into intrusion analysis. Justice Sotomayor's concurrence noted that GPS tracking "generates a precise, comprehensive record of a person's public movements that reflects a wealth of detail about her familial, political, professional, religious, and sexual associations." Her concurrence also implicitly recognized the privacy implications of datafication. Thus, the GPS location record invaded reasonable expectation of privacy because it was "precise" and "comprehensive" and because the "Government can store such records and efficiently mine them for information years into the future."[46] Justice Sotomayor also emphasized the fact that the GPS monitoring at issue is "cheap in comparison to conventional surveillance techniques."[47] It is too soon to tell what influence, if any, these developments will have on tort law.

The intrusion upon seclusion tort rarely has been extended to data acquisition as a byproduct of providing service. *Dwyer v. American Express* rejected such an attempt, emphasizing that "[b]y using the American Express card, a cardholder is voluntarily, and necessarily, giving information to defendants that, if analyzed, will reveal a cardholder's spending habits and shopping preferences," and concluding that "[w]e cannot hold that a defendant has committed an unauthorized intrusion by compiling the information voluntarily given to it and then renting its compilation."[48]

In *Pulla v. Amoco*,[49] however, the court upheld a jury finding of intrusion upon seclusion when a credit card company employee improperly accessed and reviewed the company's records to investigate another employee's use of sick leave. The court emphasized the repurposing of the data: "whatever Amoco's interest may have been, its methods were unduly intrusive when balanced against that interest. Amoco used confidential credit card records for a purpose for which they were never intended, and did so surreptitiously, without notice to, or consent or knowledge of, the credit card holder." The fact that the records had been obtained and compiled by Amoco as a byproduct of providing credit card services was not decisive.

Because of increasing concerns about the privacy implications of ubiquitous data collection and aggregation, it is possible that more courts will be inclined to take a nuanced approach to intrusion upon seclusion claims involving the datafication, repurposing, or transfer of data acquired as a byproduct of service provision.[50]

b. The Requirement that the Intrusion Be "Highly Offensive" In addition to requiring an actionable intrusion, the Restatement's version of the intrusion tort requires that the intrusion be "highly offensive," suggesting that the tort is aimed at harms involving mental or emotional distress. The extent to which tort law should recognize emotional and mental injuries was a subject of controversy in the late 19th and early 20th centuries, in part because of concerns that courts would be flooded with trivial claims. To address those concerns, the tort of intentional infliction of emotional distress was limited to extreme and outrageous behavior resulting in severe emotional distress. Prosser, who also wrote extensively about intentional infliction of emotional distress,[51] viewed the intrusion upon seclusion tort as a gap-filler aimed at addressing similar harms:

> It appears obvious that the interest protected by this branch of the tort is primarily a mental one. It has been useful chiefly to fill in the gaps left by trespass, nuisance, the intentional infliction of mental distress, and whatever remedies there may be for the invasion of constitutional rights.[52]

The requirement tends to stand in the way of intrusion claims in modern data contexts, where different privacy concerns are salient. For example, the court in *Hall*, while allowing an intrusion claim based on acquisition of credit reports to go forward, left it to the jury to determine "whether disclosure of a person's credit report would cause shame or embarrassment to a person of ordinary sensibilities."[53] The court in *Chicarella* determined that "the information disclosed by the hospital records is not of the sort which would cause mental suffering, shame or humiliation to a person of ordinary sensibilities."[54] In *Busse v. Motorola*, the court dismissed an intrusion claim alleging that mobile phone carriers had transferred customer information to outside researcher consultants for a study of the effects of cellphone use on mortality. The court determined that "none of the 'personal' information furnished by the customers, standing alone – names, telephone numbers, addresses or social security numbers – have been held to be private facts," because those pieces of information were not "facially revealing, compromising or embarrassing."[55]

The privacy concerns raised by big data stem from the risks associated with data collection and aggregation and do not revolve solely around fear of shame or embarrassment. If the intrusion upon seclusion tort is to play any significant role in regulating data acquisition in the big data context, the conception of a highly offensive intrusion will have to evolve.

2. California's Constitutional Right to Information Privacy

Several state constitutions incorporate explicit rights to privacy, but most, like the federal Fourth Amendment, apply only to government action. California, however, enforces its constitutional right to privacy against both public and private actors. The elements of a claim are (1) a legally protected privacy interest, (2) a reasonable expectation of privacy under the circumstances, and (3) conduct amounting to a serious invasion of protected privacy interests.[56] Though there is a requirement of seriousness, offensiveness and mental distress are not elements of the constitutional claim, perhaps reflecting the fact that informational privacy is "the core value furthered by" California's constitutional right to privacy.[57] Courts have just begun to flesh out the scope of the right's application to modern data acquisition contexts.

In 2012, *Goodman v. HTC America, Inc.* allowed a claim based on allegations that a mobile phone weather app collected location data that was far more accurate than necessary to provide weather information.[58] While phone users "may have expected their phones to transmit fine GPS data occasionally for certain purposes, they did not expect their phones to continually track them for reasons not related to consumer needs." The opinion compared the app's collection of fine location data to GPS monitoring, quoting Justice Sotomayor's observation that "GPS monitoring generates a precise, comprehensive record of a person's public movements that reflects a wealth of detail about her familial, political, professional, religious, and sexual associations."[59]

Goodman may be contrasted with *Fogelstrom v. Lamps Plus*[60] and *In re iPhone Application Litigation.*[61] In *Fogelstrom*, the court dismissed a California constitutional privacy claim based on a retailer's asking customers for their ZIP codes and using them to find customers' home addresses for advertising purposes. The court opined that the retailer's behavior was "not an egregious breach of social norms, but routine commercial behavior." In *iPhone*, the court relied on *Fogelstrom* in ruling that the alleged disclosure to third parties of "the unique device identifier number, personal data, and geolocation information from Plaintiffs' iDevices," even if surreptitious, was insufficiently egregious to state a claim.

The law in this area is in flux and it remains to be seen how it will evolve.

3. Statutory Prohibitions of Eavesdropping and Recording Conversations

Statutes on both the federal and state levels prohibit eavesdropping and unauthorized interception of communications. Many of these statutes make no inquiry into the data subject's behavior and hence are rule like. Some, however, apply only under specified conditions of privacy, employing fact-specific standards to determine whether monitoring is permitted.

The federal Electronic Communications Privacy Act (ECPA) imposes a rule against non-consensual monitoring of wire and electronic communications.[62] ECPA's ban on monitoring "oral communications," however, applies only when there is "an expectation that such communication is not subject to interception under circumstances justifying such expectation."[63] Many decisions essentially turn that standard into a rule that monitoring is permitted if a conversation "could be overheard, unaided by any mechanical device, from an uncontrived position."[64] Some decisions are more nuanced, however. For example, one court determined that an employee may have had an expectation that his conversations near his workstation were free from electronic interception by his supervisors even if other employees standing nearby might have overheard them.[65]

A California statute prohibits the recording of "confidential communications."[66] The California Supreme Court has held that a conversation is confidential if a party has "an objectively reasonable expectation that the conversation is not being overheard or recorded."[67] A later court applied this test in a nuanced analysis of a news show's secret recording of actors' conversations during casting workshops. The court considered factors such as whether the conversing actors had their backs turned to the undercover reporter, whether they were "chatting amongst themselves in a corner," and whether the nature of the gathering was such that attendees would expect their conversations to be recorded or disseminated.[68] Courts also have considered whether the speakers took steps to ensure that their conversations would be confidential.[69]

In Wisconsin, a prohibition on recording a conversation depends on "(1) the volume of the statements; (2) the proximity of other individuals to the speaker, or the potential for others to overhear the speaker; (3) the potential for the communications to be reported; (4) the actions taken by the speaker to ensure his or her privacy; (5) the need to employ technological enhancements for one to hear the speaker's statements; and (6) the place or location where the statements are made."[70] In New York, where the eavesdropping statute prohibits "mechanical overhearing of a conversation," courts have debated whether the prohibition applies when the speaker has no expectation of privacy.[71]

4. *Contracts*

Though contract law is not privacy law per se, contractual agreements often provide substantive terms for the acquisition or transfer of information. For example, a data holder might transfer data to a researcher, subject to an agreement that the researcher keep the data confidential or disclose it only in de-identified form. Or an individual might agree to participate in a survey based on an agreement to keep specific personal information confidential.

With rare exceptions for terms deemed unconscionable or contrary to public policy, contract law has little to say about the appropriateness of substantive contractual terms. Instead, contract law primarily regulates contract formation, sometimes inquiring into the circumstances surrounding a contract's formation to assess its validity.[72] Though standardized 'contracts of adhesion' are nothing new,[73] contract law is challenged by modern data acquisition contexts, where double-sided fact-specific inquiry into contract formation is not practical.[74] In such contexts, contract validity tends to be determined by adherence to procedural formalities, such as requiring users to scroll through pages of text before clicking 'I agree'.

B. One-Sided Standards-based Laws

Many privacy laws, especially in the service provider context, focus on the activities of the data handler, while making general assumptions about data subjects' behavior. This section discusses Section 5 of the FTC Act, which prohibits "unfair or deceptive acts or practices in or affecting commerce."[75] The FTC enforces Section 5. It also holds hearings and issues reports. The FTC's reports, along with decisions from previous enforcement actions, provide guidance and structure to the interpretation of Section 5's very general standard. As discussed in Part I, the FIPPs of notice and consent have been the linchpins of the FTC's approach to enforcing Section 5.

In its 1998 Report, the FTC explained the notice principle in seemingly substantive terms, suggesting that compliance would depend on whether a data subject had an actual understanding of a data handler's information practices before consenting to collection:

Without notice, a consumer cannot make an informed decision as to whether and to what extent to disclose personal information. Moreover, three of the other principles discussed below – choice/consent, access/participation, and enforcement/redress – are only meaningful when a consumer has notice of an entity's policies, and his or her rights with respect thereto.[76]

The Report immediately shifted its focus away from a substantive effectiveness, however, providing a list of types of information that should be included in a notice and asserting (rather glibly in retrospect):

In the Internet context, notice can be accomplished easily by the posting of an information practice disclosure describing an entity's information practices on a company's site on the Web. To be effective, such a disclosure should be clear and conspicuous, posted in a prominent location, and readily accessible from both the site's home page and any Web page where information is collected from the consumer. It should also be unavoidable and understandable so that it gives consumers meaningful and effective notice of what will happen to the personal information they are asked to divulge.[77]

The Report's discussion of choice/consent also quickly turned procedural, asserting (again glibly in retrospect) that "[i]n the online environment, choice easily can be exercised by simply clicking a box on the computer screen that indicates a user's decision with respect to the use and/or dissemination of the information being collected.[78]

In practice, FTC enforcement has been mostly procedural, focusing on ensuring that online companies have privacy policies, that the policies are not hidden in obscure places on company websites, and so forth. Though the FTC does concern itself with whether privacy policies are accurate, it does not make substantive inquiries into whether companies are meeting the goals of meaningful consumer awareness of and consent to their information practices. The defects of this procedural notice and consent approach have been noted by scholars, consumer advocates, and the FTC itself. As the FTC observed in a 2010 report, "Protecting Consumer Privacy in an Era of Rapid Change" (2010 Framework):[79]

[T]he notice-and-choice model, as implemented, has led to long, incomprehensible privacy policies that consumers typically do not read, let alone understand.... In addition, [the model has] struggled to keep pace with the rapid growth of technologies and business models that enable companies to collect and use consumers' information in ways that often are invisible to consumers.[80]

The FTC's attempts to move beyond notice and consent have, for the most part, stayed within a procedural rubric. Thus, the 2010 Framework endorses a 'privacy by design' approach, under which companies should incorporate privacy considerations into all stages of developing and implementing their products, services, and interactions with consumers.[81] The FTC has mandated 'privacy by design' practices in the consent decrees settling some of its enforcement actions. Most notably, the consent decree in an action against Google mandated that Google implement a "comprehensive privacy program" that would, among other things "address privacy risks related to the development and management of new and existing products and services for consumers" by conducting "privacy risk assessments" and implement "reasonable privacy controls and procedures" to address the identified risks.[82]

In a few recent cases involving spyware or smartphone apps that gain access to personal information stored on the phone, the FTC has taken steps toward putting substantive teeth into its interpretation of "unfair and deceptive trade practices" in the privacy context. While most such actions have been based primarily on allegations of specific misrepresentations in companies' privacy policies, a few have relied on a policy's sin of omission in failing to explain clearly that the company was engaging in monitoring that would not be expected in the context of a particular service.[83]

C. Rule-based Laws

In several sectors, privacy laws are far more rule-like than the laws described in the previous two sections. Rule-like approaches tend to have been adopted most often in sectors dealing with traditionally sensitive information, where the stakes seem especially high. Three of the most important of these regimes at the federal level cover electronic communications, financial information, and health information. A rule-like approach is desirable only if interactions between data subjects and data handlers are sufficiently standardized to mitigate the costs of lost flexibility, explaining why rule-like laws tend to be sector specific. Attempts to retain some contextual sensitivity often lead to highly complex sets of rules accounting for various contingencies. Perhaps for this reason, rule-like regimes often apply only to specific types of data handlers, leaving closely related entities unregulated.

1. Electronic Communications Privacy Act

Except with regard to oral communications, ECPA generally imposes a substantive rule against private monitoring.[84] ECPA also generally prohibits entities that provide telecommunication services to the public from disclosing the contents of those communications to private parties.[85] The most important exception to these prohibitions is consent. Unfortunately, however, ECPA suffers from a problem that commonly affects rule-like regimes: as times have changed, the terms that define the rule – such as 'intercept', 'electronic storage', and 'remote computing service' – have become increasingly out of step with reality and hence more and more difficult to interpret.[86] The result is a legal regime that has the benefits of neither of clear rules nor flexible standards. As a result, both privacy advocates and telecommunications companies have advocated ECPA reforms.[87]

2. Computer Fraud and Abuse Act

The CFAA prohibits "intentionally access[ing] a computer without authorization or exceed[ing] authorized access" to obtain "information."[88] Since enacting it in 1984 to criminalize hacking into government computers, Congress has expanded the CFAA's scope of application to virtually all computers. Like ECPA, the CFAA is rule like, but suffers from controversy over the interpretation of key terms, such as what it means to "exceed authorized access."

Comparing the CFAA to the tort of intrusion upon seclusion illustrates the tension between rules and standards. Tort liability depends on contextual factors, such as the type of information stored on the computer, the type of unauthorized access, and the relationship between the person accessing the computer and the computer's owner. While more likely to get the nuances right, such a standards-based approach provide less clear ex ante guidance and thus might be an unsuitable basis for criminal prosecution. The CFAA attempts to impose a bright line prohibition instead. A bright line rule is likely to be either overly broad or overly narrow, leading to controversy over the terms defining the boundary. A relatively narrow interpretation, under which unauthorized access means technical hacking – the use of technical measures to gain access to computer files – misses some blameworthy intrusions. A broad interpretation, under which one can exceed authorized access by using a fake name in violation of a social network service's terms of service, sweeps in a huge swath of ordinary social behavior, thereby leaving room for abuse by prosecutors or litigious private parties.

3. Laws Governing Acquisition and Transfer of Financial Information

Financial institutions, such as banks, obtain sensitive financial information as a byproduct of providing services to individuals. The Gramm-Leach-Bliley Act (GLBA)[89] and related rules[90] regulate covered financial institutions' transfers of "personally identifiable financial information" to outside entities. The GLBA permits virtually unrestricted transfer to certain "affiliated" companies and essentially bans transfer to non-affiliated companies without consent. It mandates very specific notice and consent procedures, including annual privacy notices and specific mechanisms for offering consumers the opportunity to "opt out" of transfer to non-affiliated companies. Essentially, the GLBA approach is a rule-like version, tailored to the financial sector of the FTC's notice and consent approach.

The Fair Credit Reporting Act (FCRA)[91] places substantive restrictions on the circumstances under which "consumer reporting agencies" are permitted to provide "credit reports" to third parties. Unlike banks, credit reporting agencies do not obtain consumer data as a byproduct of providing services to them. Instead, they essentially monitor financial information for the benefit of third parties. FCRA is a rule-like substantive regulation of credit reporting practices. It specifies the types of information that can be included in credit reports under various circumstances and the purposes, which include the extension of credit, employment, insurance, government benefits, licensing, and other "legitimate business needs," for which credit reports may be obtained.

Like many complex rule-like regimes, FCRA and the GLBA apply only to specifically defined entities and circumstances. Newer entities, such as data brokers and companies like Google, which now collect large amounts of financial data about consumers, are largely unregulated unless their activities come within FCRA's specific definitions.

4. Laws Regulating the Acquisition and Transfer of Health Information

Health care institutions traditionally acquire health information in two basic ways: as a byproduct of treatment or through research. Under the taxonomy presented here, acquiring information for research purposes is a type of monitoring. The regulations governing the acquisition and transfer of personal information in the traditional health care context are extremely complex and comprehensive and are generally rule like, though they include some standards-based elements. The health privacy regime combines detailed procedural elements with important substantive distinctions between different types of transfer and use.

In acquiring and transferring personal information obtained as a result of either treatment or research, medical professionals are bound by professional ethical regulations and obligations, by rules governing medical research, and by HIPAA's Privacy Rule, which was updated in 2012. Insurers are "covered entities" under HIPAA, while certain "business associates" providing services such as claims processing and data administration also are regulated.[92] HIPAA does not apply to the increasing variety of online entities, such as medical information websites, online medical forums and chat rooms, online genetic testing sites and medical 'apps', that now handle large quantities of health information. Many are governed only by consumer privacy law, primarily through the FTC's Section 5 enforcement discussed in Section B.

The HIPAA Rule applies only to "protected health information," which generally means "individually identifiable health information" (unless such information is regulated by certain other statutory regimes). Information is considered de-identified, and hence unprotected, if it meets one of two criteria.[93] The first, which is rule like, considers information to be de-identified if 18 particular types of data are removed. The second, which is standards like, requires that an expert deploy generally accepted and documented procedures to ensure that there is a very small risk of re-identification. These two alternatives illustrate the trade-off between the flexibility and adaptability of standards and the clarity and lower administrative costs of rules. The rule-based definition is easy to deploy and provides regulated entities with certainty that they have met legal standards, yet has been subject to criticism in light of studies demonstrating the ease with which many records now can be re-identified, especially when aggregated with other datasets. The standards-based criterion is flexible and can, in principle, be adapted to particular circumstances and to changes in re-identification technology, but is expensive and risky for health care providers to use.

Health information privacy law depends substantively on the purpose for which information is to be acquired, used, or transferred.

a. Acquisition, Use, and Transfer for Treatment Medical professionals are afforded substantial discretion in determining appropriate treatments and sharing information with other professionals for treatment purposes. They are, however, bound by professional ethics and norms in these determinations. Both legally and ethically, medical professionals are required to obtain 'informed consent' to treatment.[94] In principle, the informed consent requirement is a substantive and doubly fact-specific standard: "Informed consent is more than simply getting a patient to sign a written consent form. It is a process of communication between a patient and physician that results in the patient's authorization or agreement to undergo a specific medical intervention."[95] State law and institutional requirements govern procedures for obtaining informed consent and generally require an in-person explanation of treatment benefits and risks, with an opportunity for patients to ask questions, and sometimes require signed consent forms.[96] Once a patient has consented to treatment, the HIPAA Privacy Rule imposes few constraints on the acquisition and transfer of protected health information for purposes of treatment and obtaining payment.[97]

b. Acquisition, Use, and Transfer for Research Medical research projects usually are subject to approval by an Institutional Review Board (IRB). For example, the HHS regulations governing federally funded research (which are applied by many IRBs generally) apply whenever an investigator obtains "identifiable private information,"[98] which includes "information about behavior that occurs in a context in which an individual can reasonably expect that no observation or recording is taking place, and information which has been provided for specific purposes by an individual and which the individual can reasonably expect will not be made public (for example, a medical record)." IRBs evaluate proposed research according to criteria[99] that include minimization of risk, reasonable relationship of risk to benefits, and equitable selection of subjects.

IRBs must consider whether "there are adequate provisions to protect the privacy of subjects and to maintain the confidentiality of data." Informed consent to the research is required and the explanation provided to research subjects must include a "statement describing the extent, if any, to which confidentiality of records identifying the subject will be maintained." These requirements usually apply both to information obtained during a research project and to existing data that was acquired as a byproduct of treatment. Research based on "existing data" is exempt only if the data "sources are publicly available or if the information is recorded by the investigator in such a manner that subjects cannot be identified, directly or through identifiers linked to the subjects."[100]

The HIPAA Rule generally requires written authorization from a data subject for use or transfer of protected health information for purposes other than treatment and payment, including research. The authorization must provide specifics, including the identity of the recipient and the purpose of the transfer or use. The authorization requirement may be waived only for research activities that an IRB or a 'privacy board'[101] determines, among other things, poses "no more than a minimal risk to the privacy of individuals." The rule specifies procedural mechanisms for that evaluation.

c. Use and Transfer for Marketing and Other Purposes HIPAA also regulates the use and transfer of protected health information for purposes other than treatment, payment, and research.[102] Use and transfer of genetic information for underwriting purposes is prohibited entirely.[103] Otherwise, protected health information generally may be used or transferred only with the data subject's signed authorization,[104] which must include specific elements, including a "specific and meaningful" description of the information to be disclosed, the "specific identification" of the persons to whom the information will be disclosed, a description of each purpose of the use or disclosure[105] and an explanation that the authorization may be revoked in most circumstances.[106] If the covered entity will receive financial remuneration from a third party as result of a sale of the information or a use or transfer for marketing purposes, the authorization must state that fact.[107]

IV. Analysis and Discussion

This part begins with a few observations based on the taxonomy and overview in Parts II and III. It then presents thoughts about how to move toward resolving the challenges big data poses to privacy law.

A. Observations about Current Privacy Law

Applying the taxonomy developed here to the privacy laws briefly described in Part III suggests the following observations:

1. Substantive Regulation and Monitoring

For the most part, privacy law tends to engage in substantive regulation of monitoring and procedural regulation (if any) of data acquisition as a byproduct of service provision. The intrusion upon seclusion tort is a standards-based example of this tendency, since intrusion claims based on use of information obtained as a byproduct of service provision have not been successful for the most part. ECPA, the CFAA, and other surveillance statutes provide rule-like substantive regulation of monitoring. FCRA, which regulates entities that provide credit reports to parties that wish to monitor financial behavior, is significantly substantive, while the GLBA, which concerns the transfer of information about individuals obtained as a byproduct of providing financial services to them, is predominantly procedural. In the health care arena, there is significant substantive regulation of research, which is a form of monitoring.

Though privacy law has not always disentangled monitoring by service providers from data acquisition as a byproduct of service provision, there appears to be a trend toward making this distinction. Thus, *Goodman*, which recognized a claim against a service provider under California's substantive constitutional right to privacy, emphasized that the fine location data at issue was acquired for reasons unrelated to service provision. The FTC also has begun to bring substantive Section 5 claims when service providers have sought out information that was unrelated to service provision. The distinction is well recognized in the medical context, where regulations not only distinguish research from treatment, but specify in detail how to deal with the fact that research and treatment often overlap.

2. Substantive Regulation and Datafication and Repurposing

U.S. law for the most part has not regulated the datafication or repurposing of information that is acquired as a byproduct of providing service. Cases including *Dwyer*, *Busse*, *Fogelstrom* and *In re iPhone* have denied intrusion upon seclusion and California right to privacy claims based on datafication and repurposing of customer records; indeed *Fogelstrom* and *In re iPhone* characterized the behavior at issue as "routine commercial behavior." The GLBA distinguishes data transfer to affiliated companies from transfer to non-affiliated companies. That distinction bears some relation to a distinction between use in providing service and repurposing, though it permits affiliated companies to repurpose data for advertising purposes.

Though the idea of purpose limitation as a FIPP goes back to the 1967 HEW Report and "purpose specification" is included in the OECD FIPPs, the FTC's version does not include a purpose limitation principle. The FTC's 2010 Framework recognizes the privacy concerns raised by unanticipated uses of previously collected data and emphasizes that "[u]nder well-settled FTC case law and policy, companies must provide prominent disclosures and obtain opt-in consent before using consumer data in a materially different manner than claimed when the data was collected, posted, or otherwise obtained."[108] The difficulty with this notice-and-consent-based approach to repurposing is that companies easily can get around it by phrasing their privacy policies vaguely as to the uses they make of the data. Indeed, privacy policies rarely, if ever, distinguish between data acquired as a byproduct of service provision and data acquired or datafied for other purposes. For example, Google, which has one of the most detailed privacy policies, distinguishes "information you give us" (for example, by creating a Google account) and "information we get from your use of our services." Nowhere, however, does the policy inform consumers about which of this information is collected to provide the service and which is collected only for other purposes, such as profiling. Similarly, though Google generally describes "how we use information we collect" to offer services, provide tailored advertising, and so forth, its description makes no distinction between data collected for the purpose of providing services and data collected for other purposes.[109] Users are not given the option of opting out of data collection that is unnecessary for service provision.

There are a few exceptions. Most notably, the HIPAA Rule imposes substantive restrictions as to whether information acquired in the course of treatment may be used for other purposes. Indeed, the research use of existing information is treated essentially equivalently to the acquisition of data specifically for research. The court in *Pulla* made a similar distinction in ruling that intrusion upon seclusion applied when an employee's credit card records were pulled from the company's database and used to investigate his use of sick leave.

3. Consent in the Big Data Context

Nearly all privacy law provides a consent exception. When those acquiring personal information deal with data subjects on an individual basis, the question of consent can be assessed in a double-sided factual inquiry. In the big data context, however, such highly contextual inquiry usually is unmanageable due to the sheer number of individuals involved.

Indeed, when big data is acquired by monitoring, perhaps by video surveillance of city streets, it often would be impossible to obtain the consent of data subjects, who have no direct interaction with the data handlers. This is a very different situation from that assumed in medical research regulation, for example, where informed consent plays a central role. In such circumstances, consent exceptions become virtually irrelevant and everything depends on the substantive law.

When data is acquired as a byproduct of service provision, consent remains problematic in the big data context. Notice of what information is being acquired for what purposes must be provided in a standardized format. The more detailed and granular the information provided, the more time-consuming it becomes for data subjects to wade through it, to understand it, and to figure out how it relates to their concerns. The big data context generally offers little opportunity for data subjects to ask individualized questions to clear up any ambiguity or confusion. The validity of the consent process must then be assessed in a one-sided fashion, with relevant facts about data subjects reduced to trivialities, such as whether or not they clicked 'I agree'. The question generally becomes "Did the data handler follow the right procedures to obtain consent?" rather than "Did the data subject consent?"

4. Bad Fit with Big Data Harms

Current privacy laws are not well adapted to the privacy concerns that arise in the big data context, in which risks tend to be probabilistic and data is aggregated. The connection between particular harms and particular data handler behavior becomes attenuated and hard to trace. Harms often are societal and spread over large numbers of individuals. The privacy torts' emphasis on outrageous behavior is clearly inapropos to the privacy concerns associated with large aggregations of data. Moreover, the fact that big data analysis relies heavily on inference means that privacy law's categorical treatment of certain pieces of information as private or public is unlikely to correlate with big data privacy harms. Recent judicial recognition of this problem in the context of location tracking is only the tip of the iceberg of the data aggregation issue. The ubiquity of data collection also undermines the effectiveness of regulatory regimes that are tailored to in particular sectors.

B. Where Do We Go from Here?

This section suggests five approaches that might help to improve the fit between big data and privacy law, most of which build on trends that are already in motion. First, we should recognize that the market cannot make the necessary normative assessments of the trade-offs between big data's benefits and costs and look seriously for alternatives. Second, we should attempt to disentangle the issue of government surveillance from questions of big data use by the private sector by reforming surveillance law so as to raise and individualize the showing required for government access to private sector data. Third, we should strengthen substantive privacy laws with respect to monitoring in the big data context. Fourth, we should adopt a purpose-based framework for regulation in the service provider context. Fifth, we should consider a substantive approach to notice and consent.

1. Assessing Trade-offs in the Big Data Context

The most important mismatch between current privacy law and big data is current law's virtually exclusive focus on individualized assessments of privacy costs. This focus, which is common to both consent-based and fact-specific regimes, is out of sync with the privacy concerns raised by big data for several reasons. First, as I have discussed in more detail elsewhere,[110] the interrelatedness of data renders the expected costs of particular decisions to consent to data collection impossible assess in practice, and perhaps even in principle. (To put this in more familiar economic language, the acquisition and use of large amounts of data is fraught with externalities.) This fact, combined with various other factors, including the probabilistic nature and unpredictability of harms, the opacity and complexity of big data practices, and various standard biases familiar from behavioral economics, means that there is no functioning market for assessing citizen's preferences in this regard. (See Chapter 3 in this volume, by Acquisti.)

As a result, we need to devise other ways to assess the trade-offs that must be made in the big data context. To do this well, we need more information about both the privacy impacts of data acquisition and use and the potential benefits of big data analyses. Substantive assessment of the weight to accord privacy concerns in a particular data acquisition context may demand a combination of technical expertise, metrics and statistics for assessing the likelihood and expected extent of privacy impact, and transparency about data collection practices. Research on topics including the frequency and severity of individualized harms such as identity theft, reputational injury, stalking and fraud, the psychological and sociological effects of ubiquitous data collection and monitoring, and the ways in which personal data are collected, datafied, used, and exchanged in practice undoubtedly is needed.

While better information about the big data ecosystem is important, it may be slow in coming. Moreover, the costs and benefits of big data practices are complex, unpredictable, and difficult to compare. In the end, normative assessments of the trade-offs will be required.[111] Because both the benefits and the privacy implications of big data's inferences and predictions are interdependent, those assessments should be made in some collective fashion.

In thinking about how to handle the big data issue, we should look to other arenas in which law has grappled with similar problems of unpredictability, externalities, probabilistic harms, and valuation difficulties. As others have suggested, environmental regulation faces somewhat similar issues of balancing ephemeral broad-based values against the shorter term benefits of private economic activity.[112] In the mass tort context, the law has grappled with the difficulty of assigning responsibility for increases in statistical risk and harms that are difficult or impossible to trace back to particular tortfeasors. The law also has many mechanisms beyond direct regulation and tort law for encouraging the internalization of externalities, including targeted taxes, subsidies, audit requirements, and safe harbors (see Chapter 9 in this volume, by Greenwood et al.). Better metrics for assessing privacy impact or substantive standards might be used in conjunction with procedural regulation.[113]

For some big data projects, we also should consider institutional mechanisms for case-by-case collective normative assessment. One such mechanism would be review by a privacy board involving deliberation and input from experts and members of the public who are likely be affected, both positively and negatively, by the projects. Such a mechanism might be modeled on IRBs. Another (possibly complementary) mechanism for generating a community assessment of potential privacy impact would be a privacy impact notice, prepared by experts but shared with the public.[114]

2. Disentangle Government Surveillance from Private Sector Data Acquisition and Use

Current U.S. surveillance law sets low or nonexistent barriers to government acquisition of data from private entities.[115] Thus concern about government access to data held by service providers and other private entities colors many discussions of big data and privacy. If we believe that it is particularly important to ensure that government officials do not intrude into the private lives of citizens without good reason, we need to reform surveillance law. Instituting higher thresholds and requiring more individualized reasons for government demands for data from private entities would permit the trade-offs involved in private sector data acquisition to be considered on their own merits.

3. Strengthen Substantive Privacy Law with Respect to Monitoring in the Big Data Context

There is relatively strong consensus in current law that monitoring of private behavior without meaningful consent should be substantively regulated via tort law, surveillance statutes, and regulations governing human subjects research. Part III discussed cases in which this body of law is being extended to cover prolonged surveillance in public and monitoring outside of the scope of service provision. This trend is likely to continue (and in my view should). Recognizing the intrusiveness of aggregating otherwise 'non-private' information and distinguishing between data acquisition by monitoring and data acquisition as a byproduct of providing service is normatively attractive and would clarify current debates about big data.

This need not mean that all monitoring should require consent, but it does mean that monitoring is not necessarily acceptable simply because activity takes place nominally in public. The substantive law might provide mechanisms for authorizing monitoring for particular purposes on a case-by-case basis. For example, there might be an exception from privacy liability for monitoring as a part of research that has been approved by an IRB or privacy board and meets specific criteria for waiver of the consent requirement.[116]

In its 2010 Report, the FTC proposed that data acquisition by companies be limited to "information needed to fulfill a specific, legitimate business need."[117] That proposal is somewhat similar to my suggestion here, but the distinction between monitoring and service provision is more intuitive, less amorphous, and better in line both with current privacy law and with the goals of privacy law. To instantiate the distinction between monitoring and service provision, data handlers should be required to specify in their privacy policies what services they provide, whether they acquire information by monitoring (i.e. not as a byproduct of providing service), what information they acquire as a result of monitoring, whether they datafy that information, and what they do with it. Service providers should bear a heavy burden in demonstrating consent to monitoring, which should be both opt in and explicit. In essence, service providers should not be able to use their access to users' computers, phones, and so forth as means to avoid substantive restrictions on monitoring any more than home repairpersons would be permitted to install listening devices or photocopy documents.

4. Make Substantive Distinctions Based on Datafication and Repurposing for Data Acquired as a Byproduct of Providing Service

The FTC's 2010 proposed Framework suggests eliminating consent requirements for the collection and use of data for "commonly accepted practices,"[118] including "product and service fulfillment," "internal operations," "fraud prevention," "legal compliance and public purpose," and "first-party marketing."[119] The reasoning behind this proposal is that some of these uses are obvious to users, some are necessary for policy reasons, and others are "sufficiently accepted" such that requiring consent would be more burdensome than beneficial. Like "legitimate business need," however, "commonly accepted practices" is a vague and elastic term. Moreover, it has an unfortunate circularity, especially in view of the market failures endemic to notice and choice. Practices may become "commonly accepted" simply because companies adopt them and data subjects are unaware of or do not understand them.

The Framework also suggests increasing transparency about data practices by providing "clearer, shorter, and more standardized" privacy notices. It is not at all clear how this is to be accomplished, however. For data subjects to understand what data handlers are doing with their data, they need more detail than privacy policies currently provide, not less. Given the dearth of substantive privacy law that might provide a common foundation on which standardized notices could be based, it is challenging to provide the necessary detail in a way that is "clearer, shorter, and more standardized" than current policies.

Distinguishing between service provision and other uses and between mere acquisition of information and datafication would be a better way to make a distinction that would simplify and standardize privacy notices. The concept of service provision is intuitive and does not have the vagueness or circularity of "commonly accepted practices." Companies could be required to specify the services they provide, which might include recommendation services that are integral to the company's services, such as Amazon's product recommendations, but would not generally include advertising, whether first or third party. Service provision could presumptively include 'internal operations', 'fraud prevention', and 'legal compliance and public purpose', which would not have to be spelled out in the description of services provided.

Modern data privacy concerns are triggered primarily when information is datafied. While the concept of 'datafication' would have to be fleshed out to be employed in a legal regime, the difference between information that is ephemeral or is kept in a file cabinet and data that is formatted and aggregated for computational manipulation is also intuitive and significant to data subjects. Because datafication increases the risk of certain privacy harms even if the information is used only to provide services, users should be notified of datafication even if the information is used only to provide services.

These distinctions should be embodied in a substantive provision that would (1) permit information acquisition for service provision without notice or consent, (2) permit datafication for purposes of service provision with notice, and (3) prohibit use, datafication, or transfer for other purposes (including first-party advertising) unless there is notice and explicit and specific opt-in consent.[120] Where notice is required, it should specify what information is involved and what it is used for, including the identities or specific categories of entities to which any information is transferred.

Such a legal regime would not solve all of the problems of notice and consent intrinsic to data aggregation, and might not even result in shorter privacy policies. Nonetheless, it would improve over the current regime by relying on intuitive distinctions that correspond to the privacy concerns that users have regarding datafication and repurposing. Because the privacy notice will be short and no consent will be required if a company is using personal information only for providing services, the form of the notice will immediately give users a rough idea of whether there is something for them to worry about. Moreover, since opt-in consent and specificity both would be required, companies would have incentives to explain what they are doing both accurately and clearly enough to overcome user privacy concerns.

5. Adopt a Substantive Approach to Notice and Consent, at Least for Large Entities

Another way to improve the functioning of notice and consent in the big data context would be to impose substantive requirements of clear and understandable notice, at least on larger entities. The FTC might be able to take this step immediately on the basis of its Section 5 authority by deeming a notice and consent process "unfair and deceptive" if it cannot be demonstrated that it is understandable by a reasonable consumer. Techniques such as consumer surveys and experiments, similar to those employed in trademark law, could be used to demonstrate understandability. While a substantive evaluation of this sort might be unreasonably expensive for smaller companies, the necessary testing would be well within the means of the larger companies that collect the bulk of personal information.

Acknowledgement The generous support of the Filomen D'Agostino and Max E. Greenberg Research Fund is gratefully acknowledged.

NOTES

1. See Chapter 4 in this volume, by Ohm, for a discussion of laws regulating data use.
2. 15 U.S.C. § 45 (commonly known as Section 5 of the FTC Act).
3. 45 CFR §§160, 164(A) and 164(E), available at http://www.hhs.gov/ocr/privacy/hipaa/administrative/combined/hipaa-simplification-201303.pdf.
4. 110 Stat. 1936 (1996). The latest version of the Rule accounts for various later statutory additions and revisions. See 78 Fed. Reg. 5566 (January 25, 2013) for an accounting of these authorities.
5. Samuel D. Warren and Louis D. Brandeis, *The Right to Privacy*, 4 HARV. L. REV. 193 (1890).
6. 15 U.S.C. § 1681 et seq.
7. 5 U.S.C. § 552a.
8. Records, Computers and the Rights of Citizens: Report of the Secretary's Advisory Committee on Automated Personal Data Systems (HEW Report) at 4, available at http://www.rand.org/content/dam/rand/pubs/papers/2008/P5077.pdf.
9. HEW Report at 6–8.
10. HEW Report at 5–6.
11. OECD Guidelines on the Protection of Privacy and Transborder Flows of Personal Data (September 23, 1980), available at http://www.oecd.org/internet/ieconomy/oecdguidelinesontheprotectionofprivacyandtransborderflowsofpersonaldata.htm#guidelines. The OECD adopted a revised set of guidelines in 2013, available at http://www.oecd.org/sti/ieconomy/privacy.htm#newguidelines, which are not discussed here.
12. Privacy Online: Fair Information Practices in the Electronic Marketplace: A Report to Congress (May 2000; 2000 FTC Report), available at http://www.ftc.gov/sites/default/files/documents/reports/privacy-online-fair-information-practices-electronic-marketplace-federal-trade-commission-report/privacy2000.pdf. See also Privacy Online: A Report to Congress (June 1998; 1998 FTC Report), available at http://www.ftc.gov/sites/default/files/documents/public_events/exploring-privacy-roundtable-series/priv-23a.pdf.
13. Chapter 2 in this volume, by Barocas and Nissenbaum, discusses this issue. See also e.g. Daniel J. Solove, *Privacy Self-Management and the Consent Dilemma*, 126 HARV. L. REV. 1880 (2013); James P. Nehf, OPEN BOOK: THE FAILED PROMISE OF INFORMATION PRIVACY IN AMERICA 191 (2012) (discussing reasons for failure

of the self-policing model for privacy); Richard Warner, *Undermined Norms: The Corrosive Effect of Information Processing Technology on Informational Privacy*, 55 St Louis L.J. 1047, 1084–86 (2011) (raising questions about the viability of using consent to limit mass surveillance). See also Fred H. Cate, *Protecting Privacy in Health Research: The Limits of Individual Choice*, 98 Calif. L. Rev. 1765 (2010) (arguing that the requirement of individual choice stands in the way of health research and arguing for alternative approaches to protecting privacy).

14. See Katherine J. Strandburg, *The Online Market's Consumer Preference Disconnect*, 2013 Chi. Legal Forum 95 for a more detailed discussion of these concerns.
15. See e.g. Paul M. Schwartz and Daniel J. Solove, *The PII Problem: Privacy and a New Concept of Personally Identifiable Information*, 86 N.Y.U. L. Rev. 1814 (2011) (applying a rules versus standards analysis); Derek E. Bambauer, *Rules, Standards, and Geeks*, 5 Brook. J. Corp. Fin. & Com. L. 49 (2010) (discussing the "age-old debate between rules and standards" in the context of information security). See also, generally, Louis Kaplow, *Rules versus Standards: An Economic Analysis*, 42 Duke L.J. 557 (1992); Cass R. Sunstein, *Problems with Rules*, 83 Cal. L. Rev. 953 (1995).
16. For a few of the many discussions of these issues by legal scholars, see Roger A. Clarke, *Information Technology and Dataveillance*, 31 Comm. ACM 498 (1988); Ryan M. Calo, *The Boundaries of Privacy Harm*, 86 Ind. L.J. 1131 (2011); David Gray and Danielle Citron, *The Right to Quantitative Privacy*, 98 Minn. L. Rev. 62 (2013); Lior Strahilevitz, *Toward a Positive Theory of Privacy Law*, 126 Harv. L. Rev. 2010 (2013); Daniel J. Solove, The Digital Person (2004); Adam Thierer, *A Framework for Benefit-Cost Analysis in Digital Privacy Debates*, 20 Geo. Mason L. Rev. 1055 (2013); Omer Tene and Jules Polonetsky, *Privacy in the Age of Big Data: A Time for Big Decisions*, 64 Stan. L. Rev. Online 63 (2012); Jane Yakowitz, *The Tragedy of the Data Commons*, 25 Harv. J. Law & Tech. 1 (2011); Omer Tene and Jules Polonetsky, *Big Data for All: Privacy and User Control in the Age of Analytics*, 11 Nw. J. Tech. & Intell. Prop. 239 (2013); Julie M. Cohen, *What Privacy Is For*, 126 Harv. L. Rev. 1904 (2013); Tal Z. Zarsky, *Transparent Predictions*, 2013 U. Ill. L. Rev. 1503; Paul Ohm, *The Underwhelming Benefits of Big Data*, 161 U. Pa. L. Rev. Online 339 (2013); Ira S. Rubinstein, *Big Data: The End of Privacy or a New Beginning?* 3 Int'l Data Privacy L. 74 (2013); Katherine J. Strandburg, *Free Fall: The Online Market's Consumer Preference Disconnect*, 2013 Chi. Legal Forum 95.
17. See e.g. William Prosser, *Privacy*, 48 Cal. L. Rev. 383 (1960).
18. Restatement of the Law, Second, Torts § 652 (finalized in 1977). See also Neil M. Richards and Daniel J. Solove, *Prosser's Privacy Law: A Mixed Legacy*, 98 Cal. L. Rev. 1887, 1891–99 (2010) (discussing the development of Prosser's thinking about privacy torts and tracing his definition of the intrusion tort back to his 1941 treatise).
19. Restatement of the Law, Second, Torts, § 652B.
20. Note that courts parse these factors variously, with some including factors relating to the seriousness and justification of the intrusion under the second factor. Here, I focus the discussion of offensiveness on the requirement of a mental or emotional

injury, which often stands in the way of intrusion claims based on modern data acquisition, and include all other factors in the discussion of actionable intrusion.

21. Id., comment b.
22. Id., comment c.
23. Id., comment d.
24. Id., Illustration 5.
25. Id.
26. Speer v. Dept. of Rehabilitation and Correction, 646 N.E.2d 273 (Ct. Cl. Ohio 1994).
27. Sanders v. ABC, 978 P.2d 67 (Ca. 1999). But see Marrs v. Marriott Corp., 830 F. Supp. 274 (D. Md. 1992) (no intrusion by videotaping common area open to other employees).
28. Hernandez v. Hillsides, Inc., 211 P.3d 1063 (Ca. 2009) (extensive discussion of the scope of intrusion upon seclusion in the workplace). But see Johnson v. K Mart Corp., 723 N.E.2d 1192 (1st Dist. Ill. 2000) (possible intrusion where employers' agents investigating theft, vandalism and drug use reported about employees' family problems, health problems, sex lives, future work plans, and attitudes about the employer).
29. See e.g. Borse v. Piece Goods Shop, Inc. 611, 622–28 (3rd Cir. 19) (reviewing cases in various jurisdictions and discussing situations under which drug testing might constitute intrusion); Mares v. Conagra Poultry Co., 971 F.2d 492 (10th Cir. 1992) (no intrusion where employer requested information about prescription drug use in order to improve accuracy of drug test results).
30. Dietmann v. Time, Inc., 449 F.2d 245 (9th Cir. 1971).
31. See e.g. Webb v. CBS Broad. Inc., 37 Media L. Rep. 1792 (2009).
32. See e.g. Taus v. Loftus, 151 P.3d 1185, 1213–24 (Ca. 2007) (analyzing and discussing cases).
33. 955 P.2d 469 (Cal. 1998).
34. Id. at 491–92.
35. 896 F. Supp. 478 (E.D. Pa. 1995).
36. 494 A.2d 1109 (Pa. 1985).
37. 896 F. Supp. at 484.
38. Id., citing 494 A.2d at 1114.
39. Schultz v. Frankfurt, 151 Wis. 537, 545 (1913).
40. Galella v. Onassis, 353 F. Supp. 196 (SDNY 1972). The Second Circuit affirmed the substantive ruling, though it limited the remedy on First Amendment grounds, 487 F.2d 986 (2d Cir. 1973).
41. Nader v. General Motors, 25 N.Y.2d 560, 570 (1970).
42. Id. at 572 (Breitel, J., concurring).
43. See e.g. United States v. Maynard, 615 F.3d 544 (D.C. Cir. 2010). But see In re Application of the United States for Historical Cell Site Data, 724 F.3d 600 (5th Cir. 2013) (no reasonable expectation privacy in cellphone service provider location records).
44. 132 S. Ct. 945 (2012).
45. Id. at 954–57 (Sotomayor, J., concurring); Id. at 963–64 (Alito, J., concurring). The majority opinion rested on other grounds. Id. at 954.

46. Id. at 955–56 (Sotomayor, J., concurring).
47. Id.
48. Dwyer v. Am. Ex., 273 Ill. App. 3d 742 (1st Dist. Ill. 1995).
49. 882 F. Supp. 836 (S.D. Iowa 1994).
50. See Jane Yakowitz Bambauer, *The New Intrusion*, 88 Notre Dame L. Rev. 205 (2012). But see Diane L. Zimmerman, *The New Privacy and the Old: Is Applying the Tort Law of Privacy Like Putting High-Button Shoes on the Internet?* 17 Comm. L. & Pol'y 107 (2012) for a skeptical view of the extensibility of the privacy torts to modern data privacy issues.
51. See e.g. William L. Prosser, *Intentional Infliction of Mental Suffering: A New Tort*, 37 Mich. L. Rev. 874, 884 (1939). See also Richards and Solove, 98 Cal. L. Rev. at 1895–1900 (discussing how Prosser's views about the tort of intentional infliction of emotional distress affected his views of the privacy torts).
52. Prosser, 48 Cal. L. Rev. at 392.
53. 869 F. Supp. at 484.
54. 494 A.2d at 1114.
55. Busse v. Motorola, 351 Ill. App. 3d 67 (1st Dist. Ill. 2004). See also Fogelstrom v. Lamps Plus, 882 F. Supp. 836 (S.D. Iowa 1994) (dismissing claim that a retailer intruded upon plaintiff's seclusion by asking for his ZIP code in the context of a credit card transaction so that it could obtain his home address for marketing purposes where the intrusion was not "highly offensive" because there was no allegation that the information was used for an offensive or improper purpose).
56. Hill v. NCAA, 865 P.2d 633, 654–57 (1994).
57. Id. at 654.
58. 2012 U.S. Dist. LEXIS 88496 (W.D. Wash. June 26, 2012).
59. 132 S. Ct. at 955 (2012).
60. 195 Cal. App. 4th 986 (2d Dist. 2011).
61. 844 F. Supp. 2d 1040 (N.D. Cal. 2013). See also In re Google Android Consumer Privacy Litig., 2013 U.S. Dist. LEXIS 42724 (N.D. Cal. Mar. 26, 2013).
62. See Section III.C.1.
63. 18 U.S.C. § 2510.
64. United States v. Carroll, 337 F. Supp. 1260. (D.D.C. 1971).
65. Walker v. Darby, 911 F.2d 1573 (11th Cir. 1990).
66. Cal. Penal Code § 632.
67. Flanagan v. Flanagan, 27 Cal. 4th 766, 768 (2002).
68. Turnbull v. ABC, 32 Media L. Rep. 2442 (C.D. Cal. 2004).
69. Vera v. O'Keefe, 40 Media L. Rep. 2564 (S.D. Cal. 2012).
70. W.S.A. § 968.27(12); State v. Duchow, 749 N.W.2d 913 (Wis. 2008).
71. N.Y. Penal Law § 250.00; People v. Fata, 159 A.D.2d 180 (1990).
72. Restatement of the Law, Second, Contracts, §5, comments a and b.
73. See e.g. Restatement of the Law, Second, Contracts, § 211.
74. See e.g. Florencia Marotta-Wurgler, *Online Markets vs. Traditional Markets: Some Realities of Online Contracting,* 19 S. Ct. Econ. Rev. 11 (2011).
75. 15 U.S.C. § 45.

76. 1998 FTC Report at 7.
77. Id. at 7–8.
78. Id. at 9.
79. Protecting Consumer Privacy in an Era of Rapid Change (2010 FTC Report), available at http://www.ftc.gov/os/2010/12/101201privacyreport.pdf.
80. Id. at iii.
81. Id. at 41. See also Ira S. Rubinstein, *Regulating Privacy by Design*, 26 BERKELEY TECH. L.J. 1409 (2011).
82. Agreement Containing Consent Order, In the Matter of Google, Inc., FTC File No. 102 3136 (March 30, 2011), available at http://www.ftc.gov/sites/default/files/documents/cases/2011/03/110330googlebuzzagreeorder.pdf.
83. See e.g. Matter of Sears Holdings Mgmt. Corp., FTC Docket No. C-4264 (2009), http://www.ftc.gov/enforcement/cases-and-proceedings/cases/2009/09/sears-holdings-management-corporation-corporation (alleging "My SHC Community" application violated implied restriction to collecting data about 'online browsing' to collect data from sources including non-Internet-related activity); Matter of Chitika, Inc., FTC Docket No. C-4324 (2011), http://www.ftc.gov/enforcement/cases-and-proceedings/cases/2011/06/chitika-inc-matter (alleging 10-day expiration of 'opt-out' choice violated implied meaning of consumer choice to opt out); Matter of Scanscout, Inc., FTC Docket No. C-4344 (2011), http://www.ftc.gov/enforcement/cases-and-proceedings/cases/2011/12/scanscout-inc-matter (alleging use of Flash objects to circumvent users' cookie deletion violated implications of privacy policy).
84. Private parties generally may not "intercept any wire . . . or electronic communication," 18 U.S.C. § 2511, "intentionally access without authorization a facility through which an electronic communication service is provided," 18 U.S.C § 2701, "intentionally exceed an authorization to access that facility and thereby obtain . . . a wire or electronic communication while it is in electronic storage in such system," 18 U.S.C. § 2701, or use electronic means to intercept routing information, such as telephone numbers dialed, 18 U.S.C § 3121.
85. 18 U.S.C. § 2702.
86. For recent discussions of these issues see e.g. Charles H. Kennedy, *An ECPA for the 21st Century: The Present Reform Efforts and Beyond*, 20 COMMLAW CONSPECTUS 129 (2011–12), and articles in the symposium Big Brother in the 21st Century? Reforming the Electronic Communications Privacy Act, 45 U.S.F. L. Rev. (2012).
87. For information about reform efforts, see e.g. digitaldueprocess.org.
88. 18 U.S.C. § 1030.
89. 15 U.S.C. § 6801.
90. See e.g. FTC Privacy of Consumer Financial Information Rule, http://www.business.ftc.gov/documents/bus67-how-comply-privacy-consumer-financial-information-rule-gramm-leach-bliley-act.
91. 15 U.S.C. § 1681 et seq.
92. 45 CFR § 160.102.

93. 45 CFR § 164.514.
94. See e.g. AMA Ethics Opinion 8.08, available at http://www.ama-assn.org//ama/pub/physician-resources/medical-ethics/code-medical-ethics/opinion808.page.
95. See http://www.ama-assn.org//ama/pub/physician-resources/legal-topics/patient-physician-relationship-topics/informed-consent.page.
96. See http://www.nlm.nih.gov/medlineplus/ency/patientinstructions/000445.htm.
97. 45 CFR § 164.502. Certain disclosures ancillary to patient care, such as to caregiving family members, require an opportunity for the patient to agree or object orally. 45 CFR § 164.510.
98. 45 CFR § 46.102.
99. 45 CFR § 46.111.
100. 45 CFR § 46.101(b).
101. 45 CFR § 164.512(i).
102. Professional ethics and institutional conflict of interest policies may provide additional constraints on marketing, but are not discussed here.
103. 45 CFR § 164.502(5).
104. 45 CFR § 164.508(a). There are limited exceptions to the authorization requirement for certain purposes related to public health or law enforcement.
105. 45 CFR § 164.508(c).
106. 45 CFR § 164.508(b)(5).
107. 45 CFR § 164.508(3) & (4).
108. 2010 FTC Report at 76–77.
109. See http://www.google.com/policies/privacy/.
110. Katherine J. Strandburg, *Free Fall: The Online Market's Consumer Preference Disconnect*, 2013 CHI. LEGAL FORUM 95.
111. See Omer Tene and Jules Polonetsky, *To Track or "Do Not Track": Advancing Transparency and Individual Control in Online Behavioral Advertising*, 13 MINN. J.L. SCI. & TECH. 281 (2012) ("Policymakers must engage with the underlying normative question" and "cannot continue to sidestep these questions in the hope that 'users will decide' for themselves"). But see Thierer, 36 HARV. J. L. & PUBLIC POL'Y at 437 (arguing instead for a "flexible framework" comprising "education, empowerment, and targeted enforcement of existing legal standards" to "help individuals cope with a world of rapidly evolving technological change and constantly shifting social and market norms as they pertain to information sharing").
112. See e.g. Dennis D. Hirsch, *Is Privacy Regulation the Environmental Law of the Information Age?* in Katherine J. Strandburg and Daniela S. Raicu, eds., PRIVACY AND TECHNOLOGIES OF IDENTITY: AN INTERDISCIPLINARY CONVERSATION (2005); Dennis D. Hirsch, *Achieving Global Privacy Rules through Sector-based Codes of Conduct*, 74 OHIO ST. L.J. 1029 (2013).
113. See e.g. Ira S. Rubinstein, *Regulating Privacy by Design*, 26 BERKELEY TECH. L.J. 1409 (2011) (suggesting regulatory mechanisms for encouraging and enforcing privacy by design); Kenneth A. Bamberger and Deirdre K. Mulligan, *Privacy on the Books and on the Ground*, 63 STAN. L. REV. 24 (2011) (discussing approaches

to managing privacy risks, including privacy audits, currently undertaken by private companies).

114. Michael Froomkin has suggested this type of approach. A. Michael Froomkin, *Privacy Impact Notices*, Privacy Law Scholars Conference 2013, abstract available at http://privacylaw.berkeleylawblogs.org/2013/05/24/a-michael-froomkin-privacy-impact-notices/. The 2010 FTC Report suggests that companies develop "comprehensive data management programs" and that "[w]here appropriate, the programs also should direct companies to assess the privacy impact of specific practices, products, and services to evaluate risks and ensure that the company follows appropriate procedures to mitigate those risks."
115. See e.g. Peter P. Swire, *Financial Privacy and the Theory of High-Tech Government Surveillance*, 77 WASH. U. L.Q. 461 (1999); Fred A. Cate, *Government Data Mining: The Need for a Legal Framework*, 43 HARV. C.R.-C.L. L. Rev. 435 (2008); Seth F. Kreimer, *Watching the Watchers: Surveillance, Transparency, and Political Freedom in the War on Terror*, 7 U. PA. J. CONST. L. 13 (2004). A discussion of the complex set of laws regulating surveillance and data acquisition by governments is beyond the scope of this chapter.
116. For a related suggestion, see Cate, 98 CAL. L. REV. at 1800 (suggesting that "personal information [] be provided to 'licensed' or 'registered' research facilities without any individual consent, but subject to strict privacy protections" and comparing this suggestion to the regime under FCRA, which "imposes strict requirements on 'Consumer Reporting Agencies' which are then allowed to collect consumer financial data without individual consent, but can only provide them to end users for 'permissible purposes' and subject to important restrictions on their disclosure and use").
117. 2010 FTC Report at 45.
118. Id. at 53–54.
119. Id. at 53–57.
120. The opportunity to opt in to first-party advertising can be provided concisely and with little burden on consumers. Many companies already provide such a process for users to consent to first-party advertising, such as discounts, newsletters, and 'special offers'.
121. *Added in second printing:* A somewhat similar conception of 'datafication' was introduced by Viktor Mayer-Schönberger and Kenneth Cukier in chapter 5 of *Big Data: A Revolution that Will Transform How We Live, Work and Think* (Eamon Dolan/Houghton Mifflin Harcourt, March 5, 2013). As they define it, "To datafy a phenomenon is to put it in a quantified format so it can be tabulated and analyzed." My meaning here is narrower. While both usages imply something 'different from digitization', theirs emphasizes quantification and includes pre-digital 'datafication', of which they give several interesting examples. As I use the term here, datafication signifies 'digitization plus', reflecting my belief that the combination of digitization with aggregation is the important feature for privacy law. Of course, many of the implications are the same for both usages. (The term has also been used in a completely distinct sense to describe the importance of data to business models. See e.g. J. Bertolucci, "Big Data's New Buzzword: Datafication," *Information Week*, Feb. 25, 2013.)

2 Big Data's End Run around Anonymity and Consent

Solon Barocas and Helen Nissenbaum

Introduction

Big data promises to deliver analytic insights that will add to the stock of scientific and social scientific knowledge, significantly improve decision making in both the public and private sector, and greatly enhance individual self-knowledge and understanding. They have already led to entirely new classes of goods and services, many of which have been embraced enthusiastically by institutions and individuals alike. And yet, where these data commit to record details about human behavior, they have been perceived as a threat to fundamental values, including everything from autonomy, to fairness, justice, due process, property, solidarity, and, perhaps most of all, privacy.[1] Given this apparent conflict, some have taken to calling for outright prohibitions on various big data practices, while others have found good reason to finally throw caution (and privacy) to the wind in the belief that big data will more than compensate for its potential costs. Still others, of course, are searching for a principled stance on privacy that offers the flexibility necessary for these promises to be realized while respecting the important values that privacy promotes.

This is a familiar situation because it rehearses many of the long-standing tensions that have characterized each successive wave of technological innovation over the past half-century and their inevitable disruption of constraints on information flows through which privacy had been assured. It should come as no surprise that attempts to deal with new threats draw from the toolbox assembled to address earlier upheavals. Ready-to-hand, anonymity and informed consent remain the most popular tools for relieving these tensions – tensions that we accept, from the outset, as genuine and, in many cases, acute. Taking as a given that big data implicates important ethical and political values,[2] we direct our focus instead on attempts to avoid or mitigate the conflicts that may arise. We do so because the familiar pair of anonymity and informed consent continues to strike many as the best and perhaps only way to escape the need to actually resolve these conflicts one way or the other.

Anonymity and informed consent emerged as panaceas because they presented ways to 'have it all'; they would open the data floodgates while ensuring that no one was unexpectedly swept up or away by the deluge. Now, as then, conscientious industry practitioners, policymakers, advocates, and researchers across the disciplines look to anonymity and informed consent as counters to the worrisome aspects of emerging applications of big data. We can see why anonymity and consent are attractive: anonymization seems to take data outside the scope of privacy, as it no longer maps onto identifiable subjects, while allowing information subjects to give or withhold consent maps onto the dominant conception of privacy as control over information about oneself. In practice, however, anonymity and consent have proven elusive, as time and again critics have revealed fundamental problems in implementing both.[3]

The argument that we develop in this chapter goes further. Those committed to anonymity and consent do not deny the practical challenges; their solution is to try harder, to be more creative, to utilize more sophisticated mathematical and statistical techniques, and to become astute to the cognitive and motivational contours of users. Although we accept that improvements can result and have resulted from these efforts (e.g. more digestible privacy policies, more robust guarantees of anonymity, more usable choice architectures, and more supple policy), the transition to big data has turned definitional and practical fault lines that have worried policymakers, pundits, practitioners, and scholars into impassable chasms. After tracing progressive difficulties for anonymity and informed consent, respectively, we reveal virtually intractable challenges to both. In the case of anonymity, where important work has already shown it to be rather elusive, we argue that, even where strong guarantees of anonymity can be achieved, common applications of big data undermine the values that anonymity traditionally had protected. Even when individuals are not 'identifiable', they may still be 'reachable', may still be comprehensibly represented in records that detail their attributes and activities, and may be subject to consequential inferences and predictions taken on that basis. In the case of consent, too, commonly perceived operational challenges have distracted from the ultimate inefficacy of consent as a matter of individual choice and the absurdity of believing that notice and consent can fully specify the terms of interaction between data collector and data subject. Both, we argue, lead to the inescapable conclusion that procedural approaches cannot replace policies based on substantive moral and political principles that serve specific contextual goals and values.

Definitions and Background Theory

Many of the terms in this chapter have ambiguous and often contested meanings. To avoid disagreements originating in terminological differences, we specify the interpretations of two key terms – big data and privacy – assumed throughout the rest of this chapter. We have reason to believe that these interpretations contribute positively to the substantive clarity, but, for the most part, we set these out as starting assumptions.

Big Data

Taking into consideration wide-ranging uses of 'big data' in public discussions, specialized applications,[4] government initiatives,[5] research agendas,[6] and diverse scientific,[7] critical,[8] and popular publications, we find that the term better reflects a paradigm than a particular technology, method, or practice. There are, of course, characteristic techniques and tools associated with it,[9] but, more than the sum of these parts, big data, the paradigm, is a way of thinking about knowledge through data and a framework for supporting decision making, rationalizing action, and guiding practice.[10] For better or worse, it is challenging entrenched epistemic and decision-making traditions across various domains, from climate science to medicine, from finance to marketing, from resource management to urban planning, and from security to governance.[11] Statistics, computer science, and information technology are crucial enablers and supporters of this paradigm,[12] but the ascent of big data involves, fundamentally, a belief in the power of finely observed patterns, structures, and models drawn inductively from massive datasets.[13]

Privacy as Contextual Integrity

There is some disagreement over how important privacy is among the various ethical and political issues raised by big data.[14] Downplaying privacy, the argument is that *real* problems include how we use the data, whether it is fair to treat people as part of a group, whether data is representative, whether we diminish the range of choices we make about their own lives and fates, whether data about us and the data that we generate belong to us, invoking thereby justice, fairness, autonomy, and property rights.

Revealing these wide-ranging ethical dimensions of big data is important, but an impoverished working conception of privacy can result in the failure to appreciate the crucial ways that these other values and privacy interact.

The conception we adopt here gives privacy a wider berth. To begin, we take privacy to be the requirement that information about people ('personal information') flows appropriately, where appropriateness means in accordance with informational norms. According to the theory of contextual integrity, from which this conception is drawn, informational norms prescribe information flows according to key actors, types of information, and constraints under which flow occurs ('transmission principles'). Key actors include recipients, information subjects, and senders, where the last two are often one and the same. Social contexts form the backdrop for this approach to privacy, accounting for the range over which the parameters of actors, information types, and transmission principles vary. Put more concretely, informational norms for a health care context would govern flow between and about people in their context-specific capacities, such as physicians, patients, nurses, insurance companies, pharmacists, and so forth. Types of information would range over relevant fields, including, say, symptoms, diagnoses, prescriptions, as well as biographical information. And notable among transmission principles, confidentiality is likely to be a prominent constraint on the terms under which information types flow from, say, patients to physicians. In drawing comparisons between contextual integrity and other theories of privacy, one key difference is that control over information about oneself is merely one in an indefinitely large class of transmission principles, not presumed unless the other parameters – (context specific) actors and information types – warrant it.[15]

Contextual informational norms, like other social norms, generally, are not fixed and static, but may shift, fade, evolve, and even reverse at varying rates, slowly or suddenly, sometimes due to deliberate cultural, legal, and societal alterations and other times in response to contingencies beyond human or societal control. Science and technology is a significant agent of change; in particular, computing and information technologies have been radically disruptive, enabling information practices that frequently diverge from entrenched informational norms. To explain why such disruptions are morally problematic – or rather to distinguish between those that are and are not – a norm-based account of privacy, such as contextual integrity, must offer a basis for drawing such distinctions. This enables a systematic critical perspective on informational norms in flux. For the theory of contextual integrity, the touchstones of moral legitimacy include interests and general moral and political values (and associated rights), commonly cited in accounts of privacy. Beyond these, however, a further distinctive set of considerations are context-specific ends, purposes, and values. Although this is not the place to elaborate in detail, consider as a quick illustration the rules limiting access to results of an HIV test. Generally, we might consider embarrassment, job security, danger to sexual partners, autonomy, various freedoms, and so on. Beyond these, however, contextual integrity further considers how the shape of access rules may affect whether people choose to undergo testing at all. As such, access rules could influence how effectively the purposes and values of the health care context are achieved. Ideal norms, therefore, are those that promote relevant ends, purposes, and values. And since the world is a messy place, rife with conflict and uncertainty, it is usually on the basis of partial knowledge only that we seek to optimize on these factors. In concrete circumstances where science and technology enable disruptions of entrenched norms, a heuristic supported by contextual integrity sets entrenched norms as default but allows that if novel practices are more effective in promoting interests, general moral and political values, and context-specific ends, purposes, and values, they should be favored over the status quo.

Now we are ready to weave together the disparate threads thus far spun. Big data involves practices that have radically disrupted entrenched information flows. From modes of acquiring to aggregation, analysis, and application, these disruptions affect actors, information types, and transmission principles. Accordingly, privacy, understood as contextual integrity, is fundamentally part of the big data story for it immediately alerts us to the ways any practice conflicts with the expectations we may have based on entrenched information-flow norms. But that is merely the beginning. Evaluating disruptive practices means judging whether they move us closer or farther from ideal informational flows, that is, whether they are more or less effective in promoting interests, general moral and political values, and context-specific ends, purposes, and values. In other words, we proceed from observing disruptive flows to assessing their comparative impacts on ethical and political values, such as fairness, justice, freedom, autonomy, welfare, and others more specific to the context in question. Take, for example, an applicant who is denied admission to college based on predictive analytics performed on a dataset aggregated from diverse sources, including many that have not traditionally featured into admissions decisions. Imagine further that these additional sources allowed the college to discriminate – perhaps unwittingly – against applicants on the basis of criteria that happen to correlate with socioeconomic status and thus with the likely need for financial aid.[16] While the outcome of such decisions may be judged unfair for many reasons worth discussing, it is the role of privacy – the role of disruptive informational flow – that we wish to note in this case.

Why, one may ask, insist on the centrality of privacy? First, doing so deepens our understanding of privacy and its instrumental value and at the same time highlights the distinctive ways that other ethical values are impinged and sustained, specifically, by the ways information does and does not flow. Privacy is important, in part, because it implicates these other values. Second, doing so also allows us to better formulate interventions, regulations, or remediation for the sake of these values. By keeping in view connections with specific information flows, certain options become salient that might otherwise not have been. Parsing cases in which big data gives rise to discrimination in terms of contextual integrity forces us to be much more specific about the source of that unfairness because it compels us to account for the disruption that made such discrimination possible.[17] And it likewise allows us to ask if anonymity and informed consent limit or mitigate the potential consequences of such disruptions – that is, whether they actually protect the values at stake when novel applications of big data (threaten to) violate contextual integrity.

Anonymity

Anonymity obliterates the link between data and a specific person not so much to protect privacy but, in a sense, to bypass it entirely.[18] Anonymity is an attractive solution to challenges big data poses to privacy when identities associated with information in a dataset are not necessary for the analysis to proceed. For those in search of group-level regularities, anonymity may allow for relatively unfettered access to databases. The greatest consensus around the utility of anonymization seems to have emerged in the sciences, including medicine, public and population health, urban planning, and education, to name a few, with exciting prospects for advancing knowledge, diminishing risk, and improving decision making.[19] But incumbents in many other sectors have begun to stake out this moral high ground by claiming that their analytics apply only to anonymized datasets, particularly those in marketing and other commercial sectors.[20]

As we well know, however, anonymity is not unassailable. One of the earliest public demonstrations of its limits came with AOL's release of a large set of anonymized search queries with the stated purpose of facilitating academic research. This well-intended act backfired when a pair of enterprising news reporters identified a number of individuals based on the content of searches.[21] Following these revelations, efforts to anonymize search query data, which were not particularly persuasive,[22] have more or less fizzled out. The promise of anonymization was further chipped away by rigorous demonstrations by Sweeney, joint work by Narayanan and Shmatikov, and ongoing efforts by Dwork,[23] with implications further drawn by Ohm and others in areas of law and policy, where debates rage on.[24]

It is impossible, within the scope of this article, to render anything close to a thorough account of the contemporary debate around anonymity; we merely mention key positions on threats to anonymity and attempts to defend it that are relevant to the general argument that we wish to develop. According to the literature, the promise of anonymity is impossible to fulfill if individual records happen to contain information – information that falls outside the scope of the commonly defined set of personally identifiable information – that nevertheless uniquely distinguishes a person enough to associate those records to a specific individual. So-called 'vanity searches' are an obvious example of this problem,[25] as AOL discovered,[26] but so, too, are records that contain extremely rich (e.g. location) data that necessarily map onto specific individuals.[27] The literature has also demonstrated many less obvious ways in which anonymity cannot be guaranteed due to the threat of so-called re-identification attacks.[28] These attacks depend on a variety of methods: overlaying an anonymized dataset with a separate dataset that includes identifying information, looking for areas of overlap (commonly described as a linkage attack)[29] or performing a sequence of queries on an anonymized dataset that allow the attacker to deduce that a specific person must be in the dataset because only one person has *all* of the queried attributes (differencing attack).[30] Responding to these challenges, computer scientists have developed a number of approaches to limit, if not eliminate, the chances of deducing identity, such as k-anonymity[31] and differential privacy,[32] which work in certain settings by abstracting or perturbing data to a level or degree set by data controllers. At the time of writing, this area of research is burgeoning, even though few real-world applications have been successfully implemented.

Let us review the main threads of this argument: anonymity is an attractive solution to challenges big data poses to privacy when identities associated with information in a dataset are not necessary for the analysis to proceed. Scientific and policy debates have swirled around whether robust anonymization is possible and whether the impact of intractable challenges is a fringe phenomenon of little practical importance (and thus merely of academic interest) or fatal to the entire enterprise. *The concerns we have are neither about whether anonymization is possible nor about how serious a problem it poses for practical purposes; they are whether, in the first place, anonymization addresses privacy and related ethical issues of big data.* In so saying, we wish to shift the locus of attention away from the usual debates – conceding, at the same time, that they are extremely important and significant – to a different set of questions, where, for the sake of argument, we assume that the problem of anonymization, classically speaking, has been solved.

In order to see why anonymity does not solve ethical problems relating to privacy in a big data age, we should ask why we believe it does. And to do that, we need to ask not only whether in this age we are able to preserve the present-day equivalent of a traditional understanding of anonymity as namelessness, but whether this equivalent preserves what is at stake in protecting anonymity. In short, we need to ask whether it is worthwhile to protect whatever is being protected when, today, we turn to anonymity to avoid the ethical concerns raised by the big data paradigm.

Scholarship, judicial opinions, and legislative arguments have articulated the importance of anonymity in preserving and promoting liberal democratic values. We summarized these in earlier work, where we wrote that anonymity

> offers a safe way for people to act, transact, and participate without accountability, without others 'getting at' them, tracking them down, or even punishing them. [As such, it] may encourage freedom of thought and expression by promising a possibility to express opinions, and develop arguments, about positions that for fear of reprisal or ridicule they would not or dare not do otherwise. Anonymity may enable people to reach out for help, especially for socially stigmatized problems like domestic violence, fear of HIV or other sexually transmitted infection, emotional problems, suicidal thoughts. It offers the possibility of a protective cloak for children, enabling them to engage in internet communication without fear of social predation or – perhaps less ominous but nevertheless unwanted – overtures from commercial marketers. Anonymity may also provide respite to adults from commercial and other solicitations. It supports socially valuable institutions like peer review, whistle-blowing and voting.[33]

In this work, we argued that the value of anonymity inheres not in namelessness, and not even in the extension of the previous value of namelessness to all uniquely identifying information, but instead to something we called 'reachability', the possibility of knocking on your door, hauling you out of bed, calling your phone number, threatening you with sanction, holding you accountable – with or without access to identifying information.[34]

These are problematic because they may curtail basic ethical and political rights and liberties. But also at stake are contextual ends and values such as intellectual exploration, wholehearted engagement in social and economic life, social trust, and the like. The big data paradigm raises the stakes even further (to a point anonymity simply cannot extend and the concept of reachability did not locate) for a number of related reasons.

'Anonymous Identifiers'

First and perhaps foremost, many of anonymity's proponents have different meanings in mind, few of which describe practices that achieve unreachability. For example, when commercial actors claim that they only maintain anonymous records, they do not mean that they have no way to distinguish a specific person – or his browser, computer, network equipment, or phone – from others. Nor do they mean that they have no way to recognize him as the same person with whom they have interacted previously. They simply mean that they rely on unique persistent identifiers that differ from those in common and everyday use (i.e. a name and other so-called personally identifiable information (PII)). Hence the seemingly oxymoronic notion of an 'anonymous identifier', the description offered by, among others, Google for its forthcoming AdID,[35] an alternative to the cookie-based tracking essential for targeted advertising.[36] If its very purpose is to enable Google to identify (i.e. recognize) the same person on an ongoing basis, to associate observed behaviors with the record assigned to that person, and to tailor its content and services accordingly, AdID is anonymous only insofar as it does not depend on traditional categories of identity (i.e. names and other PII). As such, the identifier on offer does nothing to alleviate worries individuals might have in the universe of applications that rely on it. This understanding of anonymity instead assumes that the real – and only – issue at stake is how easily the records legitimately amassed by one institution can be associated with those held by *other* institutions, namely an association that would reveal the person's legal or real-world identity.[37]

The reasons for adopting this peculiar perspective on anonymity becomes clear when we explore why names, in particular, tend to generate such anxiety. As a persistent and common identifier, names have long seemed uniquely worrisome because they hold the potential to act as an obvious basis for seeking out *additional* information that refers to the same person by allowing institutions to match records keyed to the same name. Indeed, this is the very business of commercial data brokers: "Acxiom and other database marketing companies sell services that let retailers simply type in a customer's name and zip code and append all the additional profile information that retailers might want".[38] But this is highly misleading because, as scholars have long argued, a given name and address is just one of many possible ways to recognize and associate data with a specific person.[39] Indeed, *any* unique identifier or sufficiently unique pattern can serve as the basis for recognizing the same person in and across multiple databases.[40]

The history of the Social Security Number is highly instructive here: as a unique number assigned to all citizens, the number served as a convenient identifier that *other* institutions could adopt for their own administrative purposes. Indeed, large institutions were often attracted to the Social Security Number because it was necessarily more unique than given names, the more common of which (e.g. John Smith) could easily recur multiple times in the same database. The fact that people had existing reasons to commit this number to memory also explains why other institutions would seize upon it. In so doing, however, these institutions turned the Social Security Number, issued by the government for administering its own welfare programs, into a *common* unique identifier that applied across multiple silos of information. A Social Security Number is now perceived as sensitive, not because of any quality inherent to the number itself, but rather because it serves as one of the few common unique identifiers that enable the straightforward matching of the disparate and detailed records held by many important institutions.

The history of the Social Security Number makes clear that any random string that acts as a unique persistent identifier should be understood as a pseudonym rather than an 'anonymous identifier',[41] that pseudonyms place no inherent restrictions on the matching of records, and that the protective value of pseudonyms decreases as they are adopted by or shared with additional institutions.[42] This is evident in the more recent and rather elaborate process that Facebook has adopted to facilitate the matching of its records with those maintained by outside advertisers while ensuring the putative anonymity of the people to whom those records refer:

> A website uses a formula to turn its users' email addresses into jumbled strings of numbers and letters. An advertiser does the same with its customer email lists. Both then send their jumbled lists to a third company that looks for matches. When two match, the website can show an ad targeted to a specific person, but no real email addresses changed hands.[43]

While there might be some merit to the argument, advanced by a representative of the Interactive Advertising Bureau, that such methods demonstrate that online marketers are not in the business of trying "to get people's names and hound them",[44] they certainly fall short of any common understanding of the value of anonymity. They place no inherent limits on an institution's ability to recognize the same person in subsequent encounters, to associate, amass, and aggregate facts on that basis, and to draw on these facts in choosing if and how to act on that person. The question is whether, in the big data era, this still constitutes a meaningful form of unreachability.

Comprehensiveness

A further worry is that the comprehensiveness of the records maintained by especially large institutions – records that contain no identifying information – may become so rich that they subvert the very meaning of anonymity.[45] Turow, for instance, has asked, "[i]f a company knows 100 data points about me in the digital environment, and that affects how that company treats me in the digital world, what's the difference if they know my name or not?"[46] The answer from industry is that it seems to matter very little indeed: "The beauty of what we do is we don't know who you are [. . .] We don't want to know anybody's name. We don't want to know anything recognizable about them. All we want to do is [. . .] have these attributes associated with them."[47] This better accounts for the common refrain that companies have no particular interest in who someone is because their ability to tailor their offerings and services to individuals is in no way limited by the absence of such information. And it helps to explain the otherwise bizarre statement by Facebook's Chief Privacy Officer that they "serve ads to you based on your identity [. . .] but that doesn't mean you're identifiable."[48] On this account, your legal or real-world identity is of no significance. What matters are the properties and behaviors that your identity comprises – the kinds of details that can be associated with a pseudonym assigned to you without revealing your actual identity. Where these details are sufficiently extensive, as is the case with platforms that deal in big data, and where all of these details can be brought to bear in deciding how to treat people, the protections offered by 'anonymity' or 'pseudonymity' may amount to very little.[49] They may enable holders of large datasets to act on individuals, under the cover of anonymity, in precisely the ways anonymity has long promised to defend against. And to the extent that results in differential treatment that limits available choices and interferes with identity construction, it threatens individual autonomy and social justice. For these reasons, Serge Gutwirth and Paul Hert have warned that if it is "possible to control and steer individuals without the need to identify them, the time has probably come to explore the possibility of a shift from personal data protection to data protection tout court."[50] In other words, we can no longer turn to anonymity (or, more accurately, pseudonymity) to pull datasets outside the remit of privacy regulations and debate.

Inference

But even this fails to appreciate the novel ways in which big data may subvert the promise of such protections: inference. However troubling the various demonstrations by computer scientists about the challenge of ensuring anonymity, there is perhaps more to fear in the expanding range of facts that institutions can infer and upon which they have become increasingly willing to act. As Brian Dalessandro has explained, "a lot can be predicted about a person's actions without knowing anything personal about them."[51] This is a subtle but crucially important point: insights drawn from big data can furnish additional facts about an individual (in excess of those that reside in the database) without any knowledge of their specific identity or any identifying information. Data mining breaks the basic intuition that identity is the greatest source of potential harm because it substitutes inference for using identifying information as a bridge to get at additional facts. Rather than matching records keyed to the same name (or other PII) in different datasets, data mining derives insights that simply allow firms to guess at these qualities instead. In fact, data mining opens people up to entirely new kinds of assessments because it can extend the range of inferable qualities far beyond whatever information happens to reside in records elsewhere. And as Dalessandro again explains, firms that adopt these tactics may submit to few, if any, constraints, because "PII isn't really that useful for a lot of predictive modeling tasks."[52] This explains a recent anecdote relayed by Hardy: "Some years ago an engineer at Google told me why Google wasn't collecting information linked to people's names. 'We don't want the name. The name is noise.' There was enough information in Google's large database of search queries, location, and online behavior, he said, that you could tell a lot about somebody through indirect means."[53] These indirect means may allow data collectors to draw inferences about precisely those qualities that have long seemed unknowable in the absence of identifying information. Rather than attempt to de-anonymize medical records, for instance, an attacker (or commercial actor) might instead infer a rule that relates a string of more easily observable or accessible indicators to a specific medical condition,[54] rendering large populations vulnerable to such inferences even in the absence of PII. Ironically, this is often the very thing about big data that generates the most excitement: the capacity to detect subtle correlations and draw actionable inferences. But it is this very same feature that renders the traditional protections afforded by anonymity (again, more accurately, pseudonymity) much less effective.

Research Underwritten by Anonymity

The very robustness of the new guarantees of anonymity promised by emerging scholarship may have perverse effects if findings from the research that they underwrite provide institutions with new paths by which to infer precisely those attributes that were previously impossible to associate with specific individuals in the absence of identifying information. Ironically, this is the very purpose of differential privacy, which attempts to permit useful analysis of datasets while providing research subjects with certain guarantees of anonymity.[55] *However much these protect volunteers, such techniques may license research studies that result in findings that non-volunteers perceive as menacing because they make certain facts newly inferable that anonymity once promised to keep beyond reach.*

A recent study demonstrating that students suffering from depression could be identified by their Internet traffic patterns alone was met with such a reaction.[56] Much of this seemed to stem from one of the applications that the researchers envisioned for their results: "[p]roactively discovering depressive symptoms from passive and unobtrusive Internet usage monitoring."[57] The study is noteworthy for our purposes for having taken a number of steps to ensure the anonymity and privacy of its research subjects while simultaneously – if unintentionally – demonstrating the limits of those very same protections for anyone who might be subject to the resulting model. The point is not to pick on these or other academic researchers; rather, it is to show that anonymity is not an escape from the ethical debates that researchers should be having about their obligations not only to their data subjects, but also to others who might be affected by their studies for precisely the reasons they have chosen to anonymize their data subjects.

Informed Consent

Informed consent is believed to be an effective means of respecting individuals as autonomous decision makers with rights of self-determination, including rights to make choices, take or avoid risks, express preferences, and, perhaps most importantly, resist exploitation. Of course, the act of consenting, by itself, does not protect and support autonomy; individuals must first understand how their assent plays out in terms of specific commitments, beliefs, needs, goals, and desires. Thus, where anonymity is unachievable or simply does not make sense, informed consent often is the mechanism sought out by conscientious collectors and users of personal information.

Understood as a crucial mechanism for ensuring privacy, informed consent is a natural corollary of the idea that privacy means control over information about oneself. For some, these are the roots of privacy that must be respected in all environments and against all threats. Its central place in the regulation of privacy, however, was solidified with the articulation and spread of the Fair Information Practice Principles (FIPPs) in the domains of privacy law and countless data protection and privacy regulation schemes around the world. These principles, in broad brushstrokes, demand that data subjects be given notice, that is to say, informed who is collecting, what is being collected, how information is being used and shared, and whether information collection is voluntary or required.[58]

The Internet challenged the 'level playing field' embodied in FIPPS.[59] It opened unprecedented modalities for collecting, disseminating, and using personal information, serving and inspiring a diverse array of interests. Mobile devices, location-based services, the Internet of things, and ubiquitous sensors have expanded the scope even more. For many, the need to protect privacy meant and continues to mean finding a way to support notice and choice without bringing this vibrant ecology to a grinding halt. This need has long been answered by online privacy policies offered to individuals as unilateral terms-of-service contracts (often dubbed 'transparency and choice' or 'notice and consent'). In so doing, privacy questions have been turned into practical matters of implementation. As in the arena of human subjects research, the practical challenge has been how to design protocols for embedding informed consent into interactions of data subjects and research subjects with online actors and researchers, respectively. In both cases, the challenge is to come up with protocols that appropriately model both notice and consent. What has emerged online are privacy policies similar to those already practiced in hard copy by actors in the financial sector, following the Gramm-Leach-Bliley privacy rules.[60]

Over the course of roughly a decade and a half, privacy policies have remained the linchpin of privacy protection online, despite overwhelming evidence that most of us neither read nor understand them.[61] Sensitive to this reality, regulatory agencies, such as the Federal Trade Commission, have demanded improvements focusing attention on (1) ways privacy policies are expressed and communicated so that they furnish more effective notice and (2) mechanisms that more meaningfully model consent, reviving the never-ending stalemate over opt-in versus opt-out.[62] While the idea that informed consent *itself* may no longer be a match for challenges posed by big data has been floated by scholars, practitioners, advocates, and even some regulators,[63] such thinking has not entered the mainstream. As before, the challenge continues to be perceived as purely operational, as a more urgent need for new and inventive approaches to informing and consenting that truly map onto the states of understanding and assenting that give moral legitimacy to the practices in question.

In this chapter, we take a different path. We accept that informed consent is a useful privacy measure in certain circumstances and against certain threats and that existing mechanisms can and should be improved, but, against the challenges of big data, consent, by itself, has little traction. After briefly reviewing some of the better-known challenges to existing models of informed consent, we explore those we consider insurmountable.

The Transparency Paradox

There is little value in a protocol for informed consent that does not meaningfully model choice and, in turn, autonomy. The ideal offers data or human subjects true freedom of choice based on a sound and sufficient understanding of what the choice entails. Community best practices provide standards that best approximate the ideal, which, because only an approximation, remains a subject of philosophical and practical debate.[64] Online tracking has been one such highly contentious debate[65] – one in which corporate actors have glommed onto the idea of plain language, simple-to-understand privacy policies, and plain-to-see boxes where people can indicate their assent or consent. A number of scholars continue to hold out hopes for this approach,[66] as do regulators, such as the FTC, who continues to issue guiding principles that reflect such commitments.[67] But situations involving complex data flows and diverse institutional structures representing disparate interests are likely to confront a challenge we have called 'the transparency paradox',[68] meaning that simplicity and clarity unavoidably results in losses of fidelity. Typical of the big data age is the business of targeted advertising, with its complex ecology of back-end ad networks and their many and diverse adjuncts. For individuals to make considered decisions about privacy in this environment, they need to be informed about the types of information being collected, with whom it is shared, under what constraints, and for what purposes. Anything less than this requires a leap of faith. Simplified, plain-language notices cannot provide information that people need to make such decisions. The detail that would allow for this would overwhelm even savvy users because the practices themselves are volatile and indeterminate as new parties come on board and new practices, squeezing out more value from other sources of information (e.g. social graphs), are constantly augmenting existing flows. Empirical evidence is incontrovertible: the very few people who read privacy policies do not understand them.[69] But the paradox identified above suggests that even when people understand the text of plain-language notices, they still will not – indeed cannot – be informed in ways relevant to their decisions whether to consent.

Indeterminate, Unending, Unpredictable

What we have said, thus far, emerges from a discussion of notice and choice applied to online behavioral advertising, but with clear parallels for the big data paradigm generally. Consider typical points of contact for data gathering: signing up for a smart utility meter, joining an online social network, joining a frequent flier program, buying goods and services, enrolling in a MOOC, enrolling in a health self-tracking program, traveling, participating in a medical trial, signing up for a supermarket loyalty card, clicking on an online ad, commenting on a book, a movie, or a product, applying for insurance, a job, a rental apartment, or a credit card. Because these mundane activities may yield raw material for subsequent analysis, they offer a potential juncture for obtaining consent, raising the natural question of how to describe information practices in ways that are relevant to privacy so that individuals meaningfully grant or withhold consent. The machinations of big data make this difficult because data moves from place to place and recipient to recipient in unpredictable ways. Further, because its value is not always recognized at collection time, it is difficult to predict how much it will travel, how much it will be in demand, and whether and how much it may be worth. In the language of contextual integrity, unless recipients and transmission principles are specified, the requirements of big data are for a blank check.

While questions of information type and use might, at first, seem straightforward, they are extremely difficult when considered in detail: it may be reasonably easy for a utility company to explain to customers that, with smart meters, it can monitor usage at a fine grain, can derive aggregate patterns within and across customers, and can use these as a basis for important decisions about allocation of resources and for targeted advisement about individual customers' energy usage. It may clearly explain who will be receiving what information and to what end. With notice such as this, consent is meaningful. However, big data analytics typically do not stop here; an enterprising company may attempt to figure out how many people are associated with a given account, what appliances they own, their routines (work, bedtime, and vacations). It may fold other information associated with the account into the analysis and other information beyond the account – personal or environmental, such as weather. The company may extract further value from the information by collaborating with third parties to introduce further data fields. Not anomalous, practices such as these are the life blood of the big data enterprise for massive corporate data brokers and federal, state, and local government actors. How can they be represented to data subjects as the basis for meaningful consent?

Let us consider the challenges. The chain of senders and recipients is mazelike and potentially indefinite, incorporating institutions whose roles and responsibilities are not circumscribed or well understood. The constraints under which handoffs take place are equally obscure, including payments, reciprocity, obligation, and more. What can it mean to an ordinary person that the information will be shared with Axciom or Choicepoint, let alone the NSA? Characterizing the type of information is even tougher. Is it sufficient for the utility company to inform customers that it is collecting smart meter readings? The case is strong for arguing that notice should cover not only this information but, further, information that can be directly derived from it and even information that more sophisticated analysis might yield, including that which follows from aggregations of smart meter readings with information about other matters, personal or contextual. Intuitions on this matter are challenging, almost by definition, because the value of big data lies in the unexpectedness of the insights that it can reveal.

Even if we knew what it meant to provide adequate notice to ensure meaningful consent, we would still not have confronted the deepest challenges. One is the possibility of detecting surprising regularities across an entire dataset that reveal actionable correlations defying intuition and even understanding. With the best of intentions, holders of large datasets willing to submit them to analyses unguided by explicit hypotheses may discover correlations that they had not sought in advance or anticipated. A lot hangs on what informed consent means in such cases.[70] Does the data controller's obligation end with informing subjects about data that is explicitly recorded, or must the data controller adopt a more encompassing approach, explaining what further information the institution may be able to glean?[71] If the more encompassing approach is taken, how does the data controller explain that it is impossible to know in advance what further information might be discoverable? These factors diminish the value of informed consent because they seem to require notice that does not delimit future uses of data and the possible consequences of such uses. As many have now argued, consent under those conditions is not meaningful.[72]

The Tyranny of the Minority

But big data troubles the long-standing focus on individual choice in a slightly more roundabout way because, as discussed earlier, the willingness of a few individuals to disclose certain information implicates everyone else who happens to share the more easily observable traits that correlate with the revealed trait. This is the tyranny of the minority: the volunteered information of the few can unlock the same information about the many. This differs markedly from the suggestion that individuals are ill equipped to make choices that serve their actual interests; rather, even if we accept that individuals can make informed, rational decisions concerning their own privacy, these decisions nonetheless affect what institutions (to whom these individuals have disclosed information) can now know (i.e. infer) about others.[73]

Such inferences can be drawn in a number of ways. In registering some kind of connection to another person through the formal process of 'friending' on a social networking site, we signal that this is a person with whom we share certain interests, affinities, and history. In associating with this person, we open ourselves up to inferences that peg us as people who share certain qualities with this other person. This is the familiar trope about 'the company I keep': what my friends say and do – or rather, what they are willing to say and do on social networking sites – will affect what others think of me. Hence danah boyd's point that "[i]t's no longer about what you do that will go down on your permanent record. Everything that everyone else does that concerns you, implicates you, or might influence you will go down on your permanent record."[74]

Computer scientists have turned this into a formal problem, asking whether techniques drawing from social network analysis and data mining can be used to infer undisclosed attributes of a user based on the disclosed attributes of the user's friends on social networking sites. And indeed a recent study has demonstrated that, where a certain portion of their friends disclose such facts, social networking sites may be able to infer users' undisclosed major, graduation year, and dorm.[75] Other – more widely reported – research has also shown that homosexuality can be inferred with some reliability from the fact that a user holds a number of relationships and interacts with an otherwise disproportionate number of 'out' users.[76] Yet another study, building on this earlier work, has even shown that it is possible to make inferences about people who are not even a part of an online social network (i.e. to learn things about obviously absent *non*members).[77]

These demonstrations have tended to focus on cases of explicit association and the drawing of inferences based on confirmed relations, but, when we move away from discussions of online social networking, we find that no such explicit associations are necessary to engage in this same kind of guesswork. More significantly, similar inferences can be made about an entire population even if only a small fraction of people who share no ties are willing to disclose. This describes the dynamics of the Target pregnancy prediction score.[78] In this case, Target did not infer the likelihood of a woman giving birth by looking at her group of friends; rather, the company looked over the records from its baby shower registry to find women who had actively disclosed the fact that they had given birth and then went about trying to figure out if these women's shopping habits, leading up to the baby shower, seemed to differ from other customers' habits such that Target could then recognize the telltale signs in the future shopping habits of *other* women.[79] Which is to say that Target was able to infer a rule about the relationship between purchases and pregnancy from what must have been a tiny proportion of all its customers who actually decided to tell the company that they recently had a baby. Not only is this the tyranny of the minority, it is a choice forced upon the majority by a minority with whom they have no meaningful or recognized relations.[80]

Computer science researchers are tackling this question head-on: what proportion of people need to disclose that they possess a certain attribute for an adversary to then be able to identify all the other members in the population who also have this attribute? The findings from Mislove et al.'s study are rather startling: "multiple attributes can be inferred globally when as few as 20% of the users reveal their attribute information."[81] Of course, reaching this minimum threshold is really just a matter of arriving at a sufficiently representative sample whose analysis generates findings that are generalizable to an entire population. As such, the value of any particular individual's withheld consent diminishes incrementally the closer the dataset of those who granted consent approaches representativeness – a point beyond which companies may have no further reason to pass. So long as a data collector can overcome sampling bias with a relatively small proportion of the consenting population,[82] this minority will determine the range of what can be inferred for the majority and it will discourage firms from investing their resources in procedures that help garner the willing consent of more than the bare minimum number of people. In other words, once a critical threshold has been reached, data collectors can rely on more easily observable information to situate all individuals according to these patterns, rendering irrelevant whether or not those individuals have consented to allowing access to the critical information in question. Withholding consent will make no difference to how they are treated!

Conclusion

Those swept up in the great excitement that has placed big data at the forefront of research investment and the national scientific policy agenda may take courage. For them, these findings, particularly those concerning consent, prove once and for all that privacy is an unsustainable constraint if we are to benefit, truly, from big data. Privacy and big data are simply incompatible and the time has come to reconfigure choices that we made decades ago to enforce certain constraints. The arguments presented here give further reason to dislodge privacy from its pedestal and allow the glorious potential of big data to be fulfilled.[83] We think these people are wrong in part because they adhere to a mistaken conception of privacy, often as control or as secrecy. Because they see privacy at odds with any distribution and use of data instead of focusing only on the inappropriate, they set up a false conflict from the start. They also may wrongly be conflating the *operationalization* of informed consent with informed consent *itself.*

Others say that we should remain concerned about ethical issues raised by big data, that, while privacy may be a lost cause, the real problems arise with use.[84] Those deserving urgent attention include unfair discrimination, being limited in one's life choices, being trapped inside stereotypes, being unable to delineate personal boundaries, being wrongly judged, embarrassed, or harassed.[85] Pursuing privacy as a way to address these issues is not only retrograde but a fool's errand, a conclusion reinforced by the arguments in our paper. Better, they would say, to route around privacy and pursue directly its ends. We agree that individual interests and ethical and, we would add, context-specific values are vitally important, but we think that it is reckless to sever, prematurely, the conceptual and practical ties between privacy and these moral and political ends. To fathom the ways that big data may threaten interests and values, we must distinguish among the origins and nature of threats to individual and social integrity, between, say, unfair discrimination originating in inappropriate information flows and unfair discrimination originating from other causes. For one thing, different sources may indicate different solutions.

We are not yet ready to give up on privacy, nor completely on anonymity and consent. The paradigm shift of big data calls for a paradigm shift in our responses and, though it may seem that the arguments of this chapter leave no place for anonymity and consent and, for some, therefore, no place for privacy, we reach different conclusions.

Let us begin with informed consent and imagine it foregrounded against a social landscape. In academic and regulatory circles, attention has focused on the foreground, suggesting ways to shape, tweak, and augment informed consent so that it covers everything important about the relationship between a data controller and a data subject. FIPPS and its innumerable descendants are a case in point. These efforts ensure that, in principle, nothing should go unremarked, unrevealed, unnoticed; in practice, informed consent has groaned under the weight of this burden with results – such as the transparency paradox – that have been noted here and elsewhere.

Informed consent also has a great legacy in the domain of human subjects research, where it remains the subject of ongoing deliberation, and has generated a mature philosophical literature. In *Rethinking Informed Consent in Bioethics*, philosophers Neil Manson and Onora O'Neill address a concern, analogous to the one confronted by privacy researchers and regulators, over how to communicate with human subjects to ensure that consent is meaningful. They observe that the transaction of informed consent in medical treatment and biomedical research can only be understood against a rich social backdrop, which integrates medical practice and research into the background fabric of social and political life. When individuals – human subjects – enter into a study or treatment regime, they engage not as tabula rasa in a vacuum expecting that the protocol of informed consent will specify fully what will happen and respective rights, obligations, and responsibilities. It does not and cannot constitute the complete relationship between the medical researcher or practitioner and the subject. Instead, the protocol is set against a rich background of social and professional roles, ethical standards, and legal and other obligations, which shape a subject's reasonable expectations. Notice generally only covers notable departures from these expectations and consent is a limited and selective waiver of rights that subjects normally would expect to be respected. In other words, individuals understand that

> obligations and expectations are presupposed by informed consent practices. When they are waived by giving consent, they are not discarded or marginalized: they are merely waived in limited ways, for a limited time, for a limited purpose. In consenting to an appendectomy I do not consent to other irrelevant incisions, or to incisions by persons other than the relevant surgeon. In consenting to take part in a clinical trial I do not consent to swallow other novel medicines, let alone medicines that are irrelevant to my condition. Informed consent matters because it offers a standard and controllable way of setting aside obligations and prohibitions for limited and specific purposes.[86]

According to O'Neill and Manson, consent is not required for acceptable, expected behaviors, but only for those that depart from it. The burden on notice, therefore, is to describe clearly the violations of norms, standards, and expectations for which a waiver is being asked and not to describe everything that will be done and not done in the course of treatment or research, which both the researcher and the subjects can safely presume. Manson and O'Neill decline to produce a general or universal list of legal and ethical claims that applies to all treatment and research scenarios because, while all would surely include a common set of obvious prohibitions on, say, killing, stealing, injury, torture, fraud, deception, manipulations, and so forth, each would further include prohibitions and prescriptions relevant to the particular treatment or study in which subjects are engaged. For example, subjects may reasonably expect physicians, researchers, and others to perform in accordance with the training and professional commitments required in their respective fields, for example, to prescribe only the treatment and medication they believe to be the best and necessary for a patient's condition.

It is not sufficient for researchers to provide assurances that subjects are given a choice to waive or not to waive; they must be able to justify "actions that otherwise violate important norms, standards or expectations."[87] According to O'Neill and Manson, "[a]ny justification of informed consent has therefore to start from a recognition of the underlying legal and ethical claims and legitimate expectations that are selectively waived by consent transactions, and the reasons individuals may have for waiving them in particular cases."[88] In other words, selective waivers may not be requested for just anything but are acceptable under two conditions, either concerning actions for which individuals are presumed to have reasons to waive rights and obligations, or concerning actions that promise significant benefits to others and to society at large. In other words, consent cannot exist as an excuse for anything, a limitation further emphasized by the second and third key principles of scientific integrity in the treatment of human subjects, namely, justice and beneficence (or non-maleficence.) Scientists requesting a limited waiver must ensure that subjects are well informed of departures from expected behaviors and they should ensure that the waiver they are requesting is consistent with the reasons their subjects have for waiving these rights. But informed consent is constrained in one further, crucial way – namely, by the requirements of beneficence, non-maleficence, and justice. These constrain what a subject can be asked to consent to.

When we understand informed consent as a limited waiver of rights and obligations, certain aspects of existing practices applied to privacy come to light. To begin, since FIPPs have served as a guide to law and policy, the focus has been on specifying the characteristics of notice and consent and very little on rights and obligations. Drawing on Manson and O'Neill, it is quite clear why this has not worked; it is impossible, even absurd to believe that notice and consent can fully specify the terms of interaction between data collector and data subject. The arguments in our paper attest to this. For too long, we have focused on the foreground, working at it from every angle. In good faith, we have crammed into the notice and consent protocol all our moral and political anxieties, believing that this is the way to achieve the level playing field,[89] to promote the autonomy of data subjects, to energize a competitive marketplace for good data practices, and more. In our view, this became a futile effort at some point along the way for reasons we and others have repeatedly offered. It is time to contextualize consent by bringing the landscape into focus. It is time for the background of rights, obligations, and legitimate expectations to be explored and enriched so that notice and consent can do the work for which it is best suited.[90]

Until now, the greatest obligation of data gatherers was either to anonymize data and pull it outside various privacy requirements or to inform and obtain consent. After charting the increasing difficulty of fulfilling these obligations in the face of big data, we presented the ultimate challenge: not of practical difficulty but of irrelevance. Where, for example, anonymizing data, adopting pseudonyms, or granting or withholding consent makes no difference to outcomes for an individual, we had better be sure that the outcomes in question can be defended as morally and politically legitimate. When anonymity and consent do make a difference, we learn from the domain of scientific integrity that simply because someone is anonymous or pseudonymous or has consented does not by itself legitimate the action in question. A burden is upon the collector and user of data to explain why a subject has good reason to consent, even if consenting to data practices that lie outside the norm. That, or there should be excellent reasons why social and contextual ends are served by these practices.

We have argued that background and context-driven rights and obligations have been neglected in favor of anonymity and consent to the detriment of individuals and social integrity. Although our chapter will be deeply vexing to those who have placed anonymization and consent at the foundation of privacy protection, we welcome the shift in focus to the purposes to which data practices are being put and how these comport with individual interests as well as ethical, political, and context-driven values.

Acknowledgements The authors gratefully acknowledge research support from Intel Science and Technology Center for Social Computing, DHHS Strategic Healthcare Information Technology Advanced Research Projects on Security (SHARPS), NSF Cyber-Trust Collaborative Research (CNS-0831124), and Lady Davis Trust, The Hebrew University of Jerusalem.

NOTES

1. For a wide-ranging set of opinions on these matters, see David Bollier, *The Promise and Peril of Big Data* (Washington, DC: The Aspen Institute, 2010) and Janna Anderson and Lee Rainie, *The Future of Big Data* (Washington, DC: Pew Research Center, July 20, 2012).
2. For a broad overview, see Solon Barocas, "Data Mining: An Annotated Bibliography," *Cyber-Surveillance in Everyday Life: An International Workshop* (Toronto, Canada: University of Toronto, 12–15 May 2011), http://www.digitallymediatedsurveillance.ca/wp-content/uploads/2011/04/Barocas_Data_Mining_Annotated_Bibliography.pdf.
3. See e.g. Latanya Sweeney, "K-Anonymity: A Model for Protecting Privacy," *International Journal of Uncertainty, Fuzziness and Knowledge-Based Systems* 10, no. 5 (October 2002): 557–570, doi:10.1142/S0218488502001648; Arvind Narayanan and Vitaly Shmatikov, "Robust De-Anonymization of Large Sparse Datasets" (presented at the 2008 IEEE Symposium on Security and Privacy, IEEE, 2008), 111–125, doi:10.1109/SP.2008.33; Paul Ohm, "Broken Promises of Privacy: Responding to the Surprising Failure of Anonymization," *UCLA Law Review* 57, no. 6 (August 2010): 1701–1777; Solon Barocas and Helen Nissenbaum, "On Notice: The Trouble with Notice and Consent" (presented at the Engaging Data: First International Forum on the Application and Management of Personal Electronic Information, Cambridge, MA, 2009); Lorrie Faith Cranor, "Necessary but

Not Sufficient: Standardized Mechanisms for Privacy Notice and Choice," *Journal on Telecommunications and High Technology Law* 10, no. 2 (Summer 2012): 273–445; Alessandro Acquisti and Jens Grossklags, "Privacy and Rationality in Individual Decision Making," *IEEE Security and Privacy Magazine* 3, no. 1 (January 2005): 26–33, doi:10.1109/MSP.2005.22; Daniel J. Solove, "Privacy Self-Management and the Consent Dilemma," *Harvard Law Review* 126, no. 7 (May 2013): 1880–1880.

4. James Manyika et al., *Big Data: The Next Frontier for Innovation, Competition, and Productivity* (McKinsey Global Institute, 2011).
5. *Demystifying Big Data: A Practical Guide to Transforming the Business of Government* (Washington, DC: TechAmerica Foundation, 2012).
6. "NSF Advances National Efforts Enabling Data-Driven Discovery" (Washington, DC: National Science Foundation, November 12, 2013).
7. E.g. *Big Data* (http://www.liebertpub.com/big).
8. E.g. *Big Data and Society* (http://bigdatasoc.blogspot.com/p/big-data-and-society.html).
9. See Tony Hey, Stewart Tansley, and Kristin Tolle, eds., The Fourth Paradigm: Data-Intensive Scientific Discovery (Redmond, WA: Microsoft Research, 2009); *Frontiers in Massive Data Analysis* (Washington, DC: The National Academies Press, 2013); Pete Warden, *Big Data Glossary* (Sebastopol, CA: O'Reilly Media, 2011).
10. Mireille Hildebrandt, "Defining Profiling: A New Type of Knowledge?" in *Profiling the European Citizen: Cross-Disciplinary Perspectives*, ed. Mireille Hildebrandt and Serge Gutwirth (Dordrecht, Netherlands: Springer, 2008), 17–45, doi:10.1007/978-1-4020-6914-7_2; danah boyd and Kate Crawford, "Critical Questions for Big Data," *Information, Communication & Society* 15, no. 5 (June 2012): 662–679, doi:10.1080/1369118X.2012.678878; Christopher Steiner, *Automate This: How Algorithms Came to Rule Our World* (New York: Portfolio/Penguin, 2012); Viktor Mayer-Schönberger and Kenneth Cukier, *Big Data: A Revolution That Will Transform How We Live, Work, and Think* (New York: Houghton Mifflin Harcourt, 2013).
11. Manyika et al., *Big Data: The Next Frontier for Innovation, Competition, and Productivity*; Steiner, *Automate This: How Algorithms Came to Rule Our World*; Mayer-Schönberger and Cukier, *Big Data: A Revolution That Will Transform How We Live, Work, and Think.*
12. *Frontiers in Massive Data Analysis.*
13. See Usama Fayyad, "The Digital Physics of Data Mining," *Communications of the ACM* 44, no. 3 (March 1, 2001): 62–65, doi:10.1145/365181.365198; David Weinberger, "The Machine That Would Predict the Future," *Scientific American* 305, no. 6 (November 15, 2011): 52–57, doi:10.1038/scientificamerican1211-52; Foster Provost and Tom Fawcett, "Data Science and Its Relationship to Big Data and Data-Driven Decision Making," *Big Data* 1, no. 1 (March 2013): 51–59, doi:10.1089/big.2013.1508; Vasant Dhar, "Data Science and Prediction," *Communications of the ACM* 56, no. 12 (December 1, 2013): 64–73, doi:10.1145/2500499.
14. See e.g. Oscar H. Gandy Jr., "Consumer Protection in Cyberspace," *tripleC: Communication, Capitalism & Critique* 9, no. 2 (2011): 175–189; Cynthia Dwork and Deirdre K. Mulligan, "It's Not Privacy, and It's Not Fair," *Stanford Law*

Review Online 66 (September 3, 2013): 35–40; Omer Tene and Jules Polonetsky, "Judged by the Tin Man: Individual Rights in the Age of Big Data," *Journal on Telecommunications and High Technology Law*, August 15, 2013; Jonas Lerman, "Big Data and Its Exclusions," *Stanford Law Review Online* 66 (September 3, 2013): 55–63; Kate Crawford and Jason Schultz, "Big Data and Due Process: Toward a Framework to Redress Predictive Privacy Harms," *Boston College Law Review* 55, no. 1 (2014).

15. For a more detailed account, see Helen Nissenbaum, *Privacy in Context: Technology, Policy, and the Integrity of Social Life* (Stanford, CA: Stanford University Press, 2010).
16. Applicants' ability to pay is already a controversial factor in the admissions decisions of many colleges; in locating less obvious correlates for the ability to pay, analytics may grant colleges the capacity to pursue similar ends without direct access to such information while also shielding such contentious practices from view.
17. Solon Barocas, "How Data Mining Discriminates," in *Data Mining: Episteme, Ethos, and Ethics*, PhD dissertation, New York University (Ann Arbor, MI: ProQuest Dissertations and Theses, 2014).
18. Such was the thinking in the so-called HEW report, where anonymized datasets were treated differently and separately under the heading of 'statistical databases'. Secretary's Advisory Committee on Automated Personal Data Systems, *Records, Computers and the Rights of Citizens* (U.S. Department of Health, Education, and Welfare, July 1973).
19. White House Office of Science and Technology Policy and the Networking and Information Technology R&D, *Data to Knowledge to Action*, Washington, DC, November 12, 2013, http://www.nitrd.gov/nitrdgroups/index.php?title=Data_to_Knowledge_to_Action.
20. Emily Steel and Julia Angwin, "On the Web's Cutting Edge, Anonymity in Name Only," *The Wall Street Journal*, August 4, 2010.
21. Michael Barbaro and Tom Zeller, "A Face Is Exposed for AOL Searcher No. 4417749," *The New York Times*, August 9, 2006.
22. Vincent Toubiana and Helen Nissenbaum, "An Analysis of Google Logs Retention Policies," *Journal of Privacy and Confidentiality* 3, no. 1 (2011): 2.
23. Sweeney, "K-Anonymity: A Model for Protecting Privacy;" Narayanan and Shmatikov, "Robust De-Anonymization of Large Sparse Datasets"; Cynthia Dwork, "Differential Privacy" (presented at the ICALP'06 Proceedings of the 33rd International Conference on Automata, Languages and Programming, Berlin: Springer, 2006), 1–12, doi:10.1007/11787006_1; Cynthia Dwork, "A Firm Foundation for Private Data Analysis," *Communications of the ACM* 54, no. 1 (January 1, 2011): 86, doi:10.1145/1866739.1866758.
24. Ohm, "Broken Promises of Privacy: Responding to the Surprising Failure of Anonymization"; Jane Yakowitz, "Tragedy of the Data Commons," *Harvard Journal of Law & Technology* 25, no. 1 (Autumn 2012): 1–67; Felix T Wu, "Defining Privacy and Utility in Data Sets," *University of Colorado Law Review* 84, no. 4 (2013): 1117–1177; Jane Bambauer, Krishnamurty Muralidhar, and Rathindra Sarathy, "Fool's Gold: An Illustrated Critique of Differential Privacy," *Vanderbilt Journal of Entertainment & Technology Law* 16 (2014).

25. Christopher Soghoian, "The Problem of Anonymous Vanity Searches," *I/S: A Journal of Law and Policy for the Information Society* 3, no. 2 (2007).
26. Barbaro and Zeller, "A Face Is Exposed for AOL Searcher No. 4417749."
27. Yves-Alexandre de Montjoye et al., "Unique in the Crowd: The Privacy Bounds of Human Mobility," *Scientific Reports* 3 (2013): 1376, doi:10.1038/srep01376.
28. Khaled El Emam et al., "A Systematic Review of Re-Identification Attacks on Health Data," ed. Roberta W Scherer, *PLoS ONE* 6, no. 12 (December 2, 2011): e28071, doi:10.1371/journal.pone.0028071.s001.
29. Narayanan and Shmatikov, "Robust De-Anonymization of Large Sparse Datasets."
30. Dwork, "A Firm Foundation for Private Data Analysis."
31. Sweeney, "K-Anonymity: a Model for Protecting Privacy."
32. Dwork, "Differential Privacy."
33. Helen Nissenbaum, "The Meaning of Anonymity in an Information Age," *The Information Society* 15, no. 2 (May 1999): 142, doi:10.1080/019722499128592.
34. Of course, this is why anonymity, in certain contexts and under certain conditions, can be problematic.
35. Alistair Barr, "Google May Ditch 'Cookies' as Online Ad Tracker," *USA Today*, September 17, 2013.
36. Ashkan Soltani, "Questions on the Google AdID," *Ashkan Soltani*, September 19, 2013, http://ashkansoltani.org/2013/09/19/questions-on-the-google-adid/.
37. In practice, this has tended to refer to what we commonly conceive as 'contact information'.
38. Natasha Singer, "Acxiom, the Quiet Giant of Consumer Database Marketing," *The New York Times*, June 16, 2012.
39. Gary T. Marx, "What's in a Name? Some Reflections on the Sociology of Anonymity," *The Information Society* 15, no. 2 (May 1999): 99–112, doi:10.1080/019722499128565.
40. Arvind Narayanan and Vitaly Shmatikov, "Myths and Fallacies of 'Personally Identifiable Information'," *Communications of the ACM* 53, no. 6 (June 1, 2010): 24–26, doi:10.1145/1743558.
41. This explains the ongoing attempt, as part of European Data Protection reform, to broaden the definition of 'personal data' to cover any such data that allows for the "singling out" of individuals, whether or not they can be identified as traditionally understood. The Article 29 Working Party, for instance, has advised "that a natural person can be considered identifiable when, within a group of persons, (s)he can be distinguished from others and consequently be treated differently. This means that the notion of identifiability includes singling out." *Statement of the Working Party on Current Discussions Regarding the Data Protection Reform Package* (European Commission, February 27, 2013). For a more detailed discussion of the contours of this debate in the context of online behavioral advertising, see Frederik Zuiderveen Borgesius, "Behavioral Targeting: A European Legal Perspective," *IEEE Security and Privacy Magazine* 11, no. 1 (January 2013): 82–85, doi:10.1109/MSP.2013.5.
42. This is not to suggest that pseudonyms are valueless. Most immediately, pseudonyms limit the potential to infer gender, race, national origin, religion,

or class position from names that possess any such obvious associations. *One-off* pseudonyms (i.e., unique identifiers that are *not* common to multiple databases) also do not lend themselves to the kind of straightforward matching of records facilitated by traditional categories of identity. In principle, only the institution that assigns a one-off pseudonym to a specific person can recognize that person *according* to that pseudonym. And where this pseudonym has been abandoned or replaced (e.g. by expiring or deleting a cookie), even the institution that assigned it to a specific individual will no longer be able to recognize or associate prior observations with that person.

43. Jennifer Valentino-Devries and Jeremy Singer-Vine, "They Know What You're Shopping for," *The Wall Street Journal*, December 7, 2012.
44. "'Drinking from a Fire Hose': Has Consumer Data Mining Gone Too Far?," *Knowledge@Wharton*, November 22, 2011, http://knowledge.wharton.upenn.edu/article.cfm?articleid=2886.
45. Steel and Angwin, "On the Web's Cutting Edge, Anonymity in Name Only."
46. "'Drinking From a Fire Hose': Has Consumer Data Mining Gone Too Far?"
47. Cindy Waxer, "Big Data Blues: The Dangers of Data Mining," *Computerworld*, November 4, 2013, http://www.computerworld.com/s/article/print/9243719/Big_data_blues_The_dangers_of_data_mining.
48. Valentino-Devries and Singer-Vine, "They Know What You're Shopping For."
49. This, too, explains the dispute over an article in the current draft of the proposed revision of the European Data Protection laws that stipulates that profiling "based solely on the processing of pseudonymous data should be presumed not to significantly affect the interests, rights or freedoms of the data subject." For more details about this point of debate, see Monika Ermert, "EU Data Protection: Bumpy Piece of Road Ahead," *Internet Policy Review*, October 24, 2013, http://policyreview.info/articles/news/eu-data-protection-bumpy-piece-road-ahead/209.
50. Serge Gutwirth and Paul Hert, "Regulating Profiling in a Democratic Constitutional State," in *Profiling the European Citizen: Cross-Disciplinary Perspectives*, ed. Mireille Hildebrandt and Serge Gutwirth (Dordrecht, Netherlands: Springer, 2008), 289, doi:10.1007/978-1-4020-6914-7_14.
51. Brian Dalessandro, "The Science of Privacy," *Ad:Tech*, July 30, 2013, http://blog.ad-tech.com/the-science-of-privacy/.
52. Dalessandro, "The Science of Privacy."
53. Quentin Hardy, "Rethinking Privacy in an Era of Big Data," *The New York Times*, June 4, 2012, http://bits.blogs.nytimes.com/2012/06/04/rethinking-privacy-in-an-era-of-big-data/.
54. Solon Barocas, "Extending the Frontier of the Inferable: Proxies, Proximity, and Privacy," in *Data Mining: Episteme, Ethos, and Ethics*.
55. Dwork, "A Firm Foundation for Private Data Analysis."
56. The authors of the study, somewhat surprised by the fierce reaction to news of their findings, assembled and responded to various criticisms: Frances H. Montgomery et al., "Monitoring Student Internet Patterns: Big Brother or Promoting Mental Health?" *Journal of Technology in Human Services* 31, no. 1 (January 2013): 61–70, doi:10.1080/15228835.2012.756600.

57. Raghavendra Katikalapudi et al., "Associating Internet Usage with Depressive Behavior among College Students," *IEEE Technology and Society Magazine* 31, no. 4 (Winter 2012): 73–80, doi:10.1109/MTS.2012.2225462.
58. Among other things, FIPPs also require that adequate steps be taken to secure the information.
59. OECD, *The Evolving Privacy Landscape: 30 Years after the OECD Privacy Guidelines*, vol. 176, April 6, 2011, doi:10.1787/5kgf09z90c31-en.
60. Gramm-Leach-Bliley Act, 15 USC, § 6801-6809. See also Chapters 1 and 7 in this volume.
61. Yannis Bakos, Florencia Marotta-Wurgler, and David R. Trossen, "Does Anyone Read the Fine Print? Testing a Law and Economics Approach to Standard Form Contracts," *SSRN Electronic Journal* (2009), doi:10.2139/ssrn.1443256; Aleecia M. McDonald and Lorrie Faith Cranor, "The Cost of Reading Privacy Policies," *I/S: a Journal of Law and Policy for the Information Society* 4, no. 3 (2008): 540–565; Aleecia M. McDonald et al., "A Comparative Study of Online Privacy Policies and Formats" (presented at the PETS 2009, Springer, 2009), 37–55. Also discussed in Helen Nissenbaum, "A Contextual Approach to Privacy Online," *Daedalus*, 140, no. 4 (Fall 2011): 32–48.
62. Julie Brill, "Reclaim Your Name: Privacy in the Age of Big Data" (presented at the Sloan Cyber Security Lecture, Brooklyn, NY, 2013).
63. In 2012, Microsoft hosted a series of 'dialogs' in which scholars. executives, advocates, and regulators discussed the future of notice and consent in the wake of big data, many of whom expressed this sentiment. For a summary of the full range of opinions at these events, see Fred H. Cate and Viktor Mayer-Schönberger, "Notice and Consent in a World of Big Data," *International Data Privacy Law* 3, no. 2 (May 20, 2013): 67–73, doi:10.1093/idpl/ipt005. Similar arguments have been advanced in Mireille Hildebrandt, "Who Is Profiling Who? Invisible Visibility," in *Reinventing Data Protection?* ed. Serge Gutwirth et al. (Dordrecht, Netherlands: Springer, 2009), 239–252, doi:10.1007/978-1-4020-9498-9_14; Christopher Kuner et al., "The Challenge of 'Big Data' for Data Protection," *International Data Privacy Law* 2, no. 2 (April 23, 2012): 47–49, doi:10.1093/idpl/ips003; *Big Data and Analytics: Seeking Foundations for Effective Privacy Guidance* (Washington, DC: The Centre for Information Policy Leadership, February 28, 2013); Omer Tene and Jules Polonetsky, "Big Data for All: Privacy and User Control in the Age of Analytics," *Northwestern Journal of Technology and Intellectual Property* 11, no. 5 (April 2013): 239–272; Ira Rubinstein, "Big Data: The End of Privacy or a New Beginning?" *International Data Privacy Law* 3, no. 2 (May 20, 2013): 74–87, doi:10.1093/idpl/ips036.
64. Nir Eyal, "Informed Consent," in *The Stanford Encyclopedia of Philosophy* (Fall 2012 Edition), ed. Edward N. Zalta (ed.), http://plato.standford.edu/archives/fall2012/entries/informed-consent/.
65. Lorrie Faith Cranor, "Can Users Control Online Behavioral Advertising Effectively?" *IEEE Security and Privacy Magazine* 10, no. 2 (n.d.): 93–96, doi:10.1109/MSP.2012.32; Pedro Giovanni Leon et al., "What Do Online Behavioral Advertising Privacy Disclosures Communicate to Users?" (presented at the WPES '12 Proceedings of the 2012 ACM workshop on Privacy in the electronic society, New York, NY: ACM Press, 2012), 19–30,

doi:10.1145/2381966.2381970; Omer Tene and Jules Polonetsky, "To Track or 'Do Not Track': Advancing Transparency and Individual Control in Online Behavioral Advertising," *Minnesota Journal of Law, Science & Technology* 13, no. 1 (Winter 2012): 281–357; Frederik J. Zuiderveen Borgesius, "Consent to Behavioural Targeting in European Law – What Are the Policy Implications of Insights from Behavioural Economics?" *SSRN Electronic Journal* (2013), doi:10.2139/ssrn.2300969; Joseph Turow, "Self-Regulation and the Construction of Media Harms: Notes on the Battle over Digital 'Privacy'," in *Routledge Handbook of Media Law*, ed. Monroe E Price, Stefaan Verhulst, and Libby Morgan (New York, NY: Routledge, 2013).

66. "Carnegie Mellon Leads NSF Project to Help People Understand Web Privacy Policies," *Carnegie Mellon News* (Pittsburgh, PA: Carnegie Mellon University, August 20, 2013). See Usable Privacy Policy Project: http://www.usableprivacy.org/.
67. *Protecting Consumer Privacy in an Era of Rapid Change* (Washington, DC: Federal Trade Commission, March 2012).
68. Helen Nissenbaum, "A Contextual Approach to Privacy Online," *Daedalus* 140, no. 4 (October 2011): 32–48, doi:10.1162/DAED_a_00113.
69. For a summary of the relevant research, see Solove, "Privacy Self-Management and the Consent Dilemma."
70. Mireille Hildebrandt, "Profiling and the Rule of Law," *Identity in the Information Society* 1, no. 1 (December 19, 2008): 55–70, doi:10.1007/s12394-008-0003-1.
71. Data miners were concerned with the principle of consent for this very reason from early on in the field's history; see e.g. Daniel E. O'Leary, "Some Privacy Issues in Knowledge Discovery: The OECD Personal Privacy Guidelines," *IEEE Expert: Intelligent Systems and Their Applications* 10, no. 2 (1995): 48–59.
72. Herman T. Tavani, "KDD, Data Mining, and the Challenge for Normative Privacy," *Ethics and Information Technology* 1, no. 4 (1999): 265–273, doi:10.1023/A:1010051717305; Mireille Hildebrandt, "Who Is Profiling Who?"; Mireille Hildebrandt, "Profiling and AmI," in *The Future of Identity in the Information Society*, ed. Kai Rannenberg, Denis Royer, and André Deuker (Berlin: Springer, 2009), 273–310, doi:10.1007/978-3-642-01820-6_7; Serge Gutwirth and Mireille Hildebrandt, "Some Caveats on Profiling," in *Data Protection in a Profiled World*, ed. Serge Gutwirth, Yves Poullet, and Paul De Hert (Dordrecht, Netherlands: Springer, 2010), 31–41, doi:10.1007/978-90-481-8865-9_2; Cate and Mayer-Schönberger, "Notice and Consent in a World of Big Data"; *Big Data and Analytics: Seeking Foundations for Effective Privacy Guidance*; Tene and Polonetsky, "Big Data for All: Privacy and User Control in the Age of Analytics"; Rubinstein, "Big Data: The End of Privacy or a New Beginning?"
73. We should stress that this is not the same argument, advanced by a number of scholars, that the capacity to assess any particular individual depends on the willingness (or unwillingness, as the case may be) of other individuals to reveal data about themselves. The common example in these arguments is that *I* am a good customer because *you* are less profitable. Taking the example of car insurance, Tverdek explains that what constitutes a 'good' driver is a statistical artifact that can only be made in contrast to a statistically 'reckless' driver. But the phenomena that we are describing here, to stick with the same example, is the capacity for

insurers to predict that *I* am bad driver because I share certain qualities with the limited number of other bad drivers who chose to report their accidents. Edward Tverdek, "Data Mining and the Privatization of Accountability," *Public Affairs Quarterly* 20, no. 1 (2006): 67–94. See also Scott R. Peppet, "Unraveling Privacy: The Personal Prospectus and the Threat of a Full Disclosure Future," *Northwestern University Law Review* 105, no. 3 (2011): 1153–1203, and, more recently, Evgeny Morozov, "The Real Privacy Problem," *MIT Technology Review*, October 22, 2013.

74. danah boyd, "Networked Privacy" (presented at the Personal Democracy Forum 2011, New York, NY, 2011).
75. Alan Mislove et al., "You Are Who You Know: Inferring User Profiles in Online Social Networks" (presented at the WSDM '10 Proceedings of the Third ACM International Conference on Web Search and Data Mining, New York, NY: ACM Press, 2010), 251–260, doi:10.1145/1718487.1718519.
76. Carter Jernigan and Behram F. T. Mistree, "Gaydar: Facebook Friendships Expose Sexual Orientation," *First Monday* 14, no. 10 (September 25, 2009), doi:10.5210/fm.v14i10.2611.
77. Emöke-Ágnes Horvát et al., "One Plus One Makes Three (for Social Networks)," ed. Sergio Gómez, *PLoS ONE* 7, no. 4 (April 6, 2012): e34740, doi:10.1371/journal.pone.0034740.s011.
78. Charles Duhigg, "How Companies Learn Your Secrets," *The New York Times Magazine*, February 16, 2012.
79. Rachel Nolan, "Behind the Cover Story: How Much Does Target Know?" *The New York Times*, February 21, 2012, http://6thfloor.blogs.nytimes.com/2012/02/21/behind-the-cover-story-how-much-does-target-know/.
80. Scholars have described this as the problem of 'categorical privacy',[80] whereby an individual's apparent membership in a group reveals more about them than can be observed directly (i.e., inferring that they likely possess the same traits as other group members). But the focus of this line of thinking has been on the impossibility of individuals foreseeing these potential inferences, the problem of inaccurate stereotyping, and the absence of associational ties, rather than the limited amount of examples that would be necessary to induce the rule and then apply it to others. See, in particular, Anton Vedder, "KDD: the Challenge to Individualism," *Ethics and Information Technology* 1, no. 4 (1999): 275–281, doi:10.1023/A:1010016102284.
81. Mislove et al., "You Are Who You Know: Inferring User Profiles in Online Social Networks," 255. We should put this result into context to ensure that we do not overstate its significance: Mislove et al. were looking at the relatively innocuous details posted by college students on Facebook, specifically their major, year of expected graduation, and college (at this particular university, students are assigned to a residential college where they tend to remain for the duration of their college career). The stakes are not especially high in this case. That said, these inferences were only based on the very limited set of variables (in fact, they only looked at the same attributes that they hoped to infer), rather than the far richer data that social networking sites accrue about their users. The authors speculate that inferences about far more sensitive attributes would be possible if the analysis were to consider a larger set of criteria that might prove statistically relevant.

82. Scholars have already proposed such a method: Christina Aperjis and Bernardo A. Huberman, "A Market for Unbiased Private Data: Paying Individuals According to Their Privacy Attitudes," *First Monday* 17, no. 5 (May 4, 2012), doi:10.5210/fm.v17i5.4013.
83. Yakowitz, "Tragedy of the Data Commons."
84. Tal Z. Zarsky, "Desperately Seeking Solutions Using Implementation-Based Solutions for the Troubles of Information Privacy in the Age of Data Mining and the Internet Society," Maine Law Review 56, no. 1 (2004): 14–59; Cate and Mayer-Schönberger, "Notice and Consent in a World of Big Data"; *Big Data and Analytics: Seeking Foundations for Effective Privacy Guidance*; Tene and Polonetsky, "Big Data for All: Privacy and User Control in the Age of Analytics"; Rubinstein, "Big Data: the End of Privacy or a New Beginning?"
85. Oscar H. Gandy, *Coming to Terms with Chance: Engaging Rational Discrimination and Cumulative Disadvantage* (Burlington, VT: Ashgate, 2009); Dwork and Mulligan, "It's Not Privacy, and It's Not Fair"; Tene and Polonetsky, "Judged by the Tin Man"; Omer Tene and Jules Polonetsky, "A Theory of Creepy: Technology, Privacy and Shifting Social Norms," *Journal on Telecommunications and High Technology Law*, September 16, 2013; Gutwirth and Hildebrandt, "Some Caveats on Profiling"; Tal Z. Zarsky, "'Mine Your Own Business!': Making the Case for the Implications of the Data Mining of Personal Information in the Forum of Public Opinion," *Yale Journal of Law & Technology* 5 (2004): 1–57.
86. Neil C. Manson and Onora O'Neill, *Rethinking Informed Consent in Bioethics* (New York: Cambridge University Press, 2012), 73.
87. Ibid., 75.
88. Ibid., 73.
89. Secretary's Advisory Committee on Automated Personal Data Systems, *Records, Computers and the Rights of Citizens*.
90. This idea has been picked up by the Federal Trade Commission, which, as one of the key recommendations of its 2012 report, suggested that "companies do not need to provide choice before collecting and using consumers' data for practices that are consistent with the context of the transaction, consistent with the company's relationship with the consumer," *Protecting Consumer Privacy in an Era of Rapid Change*, vii.

3 The Economics and Behavioral Economics of Privacy

Alessandro Acquisti

Introduction

Imagine a world in which consumers' preferences can be so precisely estimated by observing their online behavior that firms are able to anticipate consumers' needs, offering the right product at exactly the right time. Imagine the same world, but now consider that extensive knowledge of consumers' preferences also allows precise inferences about their reservation prices (the maximum price each consumer is willing to pay for a good), so that firms can charge different prices for the same product to each of their buyers and absorb the entire surplus arising from an economic transaction.

Imagine a world in which the collection and analysis of individual health data allow researchers to discover the causes of rare diseases and the cures for common ones. Now, consider the same world, but imagine that employers are able to predict job candidates' future health conditions from a few data points extracted from the candidates' social network profiles – and then, imagine those employers making hiring decisions based on those predictions, without any candidate's consent or even awareness.

The economics of privacy attempts to study the costs and benefits associated with personal information – for the data subject, the data holder, and society as a whole. As a field of research, it has been active for some decades. Advances in data mining, business analytics, and so-called big data have the potential to magnify the size and augment the scope of economic benefits and dangers alike. This chapter surveys the growing body of theoretical and empirical research on the economics and behavioral economics of privacy, and discusses how these streams of research can be applied to the investigation of the implications of consumer data mining and business analytics. Among the many possible interpretations of privacy, this investigation focuses on its informational aspects: the trade-offs arising from the protection or disclosure of personal data.

Since the second half of the last century, progress in information technology and the transformation of advanced economies into service economies have made it possible for organizations to monitor, collect, store, and analyze increasing amounts of individual data. These changes have also raised significant, and in some cases novel, privacy concerns. Attempting to analyze privacy in the age of big data from an economic perspective does not imply that we assume all modern privacy issues have explicit monetary dimensions. Rather, this type of analysis stems from the realization that, with or without individuals' awareness, decisions that data subjects and data holders make about personal data often carry complex trade-offs. The mining of personal data can help increase welfare, lower search costs, and reduce economic inefficiencies; at the same time, it can be a source of losses, economic inequalities, and power imbalances between those who hold the data and those whose data is controlled. For instance, a firm may reduce its inventory costs by mining and analyzing the behavior of many individual consumers; however, the infrastructure needed to carry out such analysis may require substantial investment, and if the analysis is conducted in a way that raises consumers' privacy concerns, that investment may backfire. Likewise, a consumer may benefit from contributing her data to a vast database of individuals' preferences (for instance, by sharing music interests with an online vendor and receiving, in return, targeted recommendations for new music to listen to); that same consumer, having lost control over that data, may end up suffering from identity theft, price discrimination, or stigma associated with the information unintended parties can acquire about her.

The microeconomic theory of privacy has brought forward arguments supporting both the view that privacy protection may increase economic efficiency in a marketplace and the view that protection decreases efficiency: personal information, when shared, can become a public good whose analysis reduces inefficiencies and increases economic welfare; when abused, it can lead to transfer of economic wealth from data subjects to data holders. Similarly, empirical evidence has been offered of both the benefits and costs, for data subjects and data holders alike, of privacy protection. It is unlikely that economics can answer questions such as what is the 'optimal' amount of privacy and disclosure for an individual and for society – but it can help us think about the trade-offs associated with personal information.

We first provide a brief summary of some relevant results from the microeconomic theory of privacy, then describe the potential trade-offs associated with privacy and disclosure in the age of big data, consumers' privacy valuations, and their behaviors. We conclude by discussing the role of privacy-enhancing technologies and market forces in balancing the value of data and the value of privacy.

Privacy and Economic Theory

Among the many heterogeneous dimensions of privacy (Solove 2006), formal economic analysis has predominantly (albeit not solely) focused on privacy as concealment of personal information – a form of information asymmetry (Akerlof 1970). For instance, before a consumer interacts with a seller, the seller may not have knowledge of the consumer's 'type' (such as her preferences, or her reservation price for a product). After the consumer has interacted with the seller (for instance, she has completed a purchase of a certain product at a certain price), it is the consumer who may not know how the seller is going to use the information it acquired through the transaction. The former is a case of hidden information; the latter, of hidden action.

Some of the earliest explicit economic discussions of privacy appeared in the literature near the end of the 1970s and the beginning of the 1980s – thanks in particular to the work of scholars belonging to the so-called Chicago School. Among them, Stigler (1980) argued that the protection of privacy may lower the quality of information about economic agents available in the marketplace. Hence, excessive protection of privacy rights may end up being economically inefficient and redistributive, as it may deny to the market the signals needed to allocate, compensate, and efficiently price productive factors. Similarly, Posner (1981) argued that concealing personal information may transfer costs from one party to another: for instance, the employer who cannot fully scrutinize the job candidate may end up paying the price of hiring an unsuitable employee. According to this view, legislative initiatives that favor privacy protection by restricting the activities of companies are likely to create inefficiencies, raise firm costs, and ultimately decrease economic welfare.

Hirshleifer (1980), however, took positions that may be considered at variance with Stigler and Posner. He noted that economic studies based on the assumption of neo-classically rational economic agents may not adequately capture the nuances of transactions that occur outside the logic of the market, such as those involving privacy. Earlier, Hirshleifer (1971) had also noted that investment in private information gathering may be inefficient: using private information may have redistributive effects, which leads to overinvestment in information gathering. A similar conclusion is reached by Taylor (2004b), who finds that market forces alone may not guarantee efficient economic outcomes (under competition, firms have a private incentive to invest more than is socially optimal in collecting consumer data).

The contraposition of the results found by Stigler (1980) or Posner (1981) and those by Hirshleifer (1971) or Taylor (2004b) highlights a common theme in the economic literature on privacy: privacy costs and privacy benefits are inextricably related. Work by Varian (1996), Noam (1996), Taylor (2004a), and Acquisti and Varian (2005) offers further examples.

Varian (1996) noted that a general ban on the dissemination of personal data would not be in the interest of the consumer herself, or that of the firms she interacts with. A consumer may naturally be interested in disclosing certain personal traits to some firms (for example, her preferences and tastes) so as to receive services. However, the same consumer may have an interest in keeping other types of information private (for example, her reservation price for a particular good). Noam (1996), applying Ronald Coase's 'theorem' to the study of privacy, argued that in absence of transaction costs, the interaction between consumers interested in the protection of their data and firms interested in accessing it will, under free market exchanges, lead to an equilibrium in which the agent with the greatest interest in either accessing the data or protecting it from access will be the one to actually achieve its goal – independent of the initial assignments of privacy rights or rights over access to consumer data. However, transaction costs and uncertainties regarding the initial assignment of rights over personal information are likely to be substantial in the interaction between consumers and firms, in which case it is no longer guaranteed that market forces alone will produce the most efficient privacy outcomes. Similarly, both Taylor (2004a) and Acquisti and Varian (2005) studied the economic impact of tracking technologies that make customer identification possible. In these types of models (which typically analyze intertemporal interactions between consumers and merchants, and focus on consumers' reservation prices as private information the merchant is interested in inferring), when consumers are rational decision makers, a regulatory regime for privacy protection turns out not to be necessary. For instance, in Acquisti and Varian (2005), consumers who expect to be tracked can engage in strategic behaviors that render tracking counterproductive; to avoid this, firms must use consumer information to offer personalized services that consumers will value.

This series of microeconomic results suggests that not only does privacy protection (or lack thereof) carry both potential costs and potential benefits for data subjects and data holders alike, but also that economic theory should not be expected answer the question "what is the economic impact of privacy (or lack thereof) on consumer and aggregate welfare?" in an unambiguous, unequivocal manner. Economic analysis certainly can help us to carefully investigate local trade-offs associated with privacy, but the economic consequences of privacy are nuanced, and the evolution of technologies for data mining and business intelligence are more likely to emphasize, rather than resolve, those nuances, as we highlight in the next section.

Markets for Privacy

Due to the concurrent evolution of Internet technologies, online business models, and data mining and business analytics tools, economic transactions of privacy relevance occur nowadays in different types of markets. We distinguish three markets for privacy in this section.

The first type of transaction that has privacy relevance actually occurs in the market for ordinary, non-privacy goods: in the process of exchanging or acquiring products or services, individuals often reveal personal information, which may be collected, analyzed, and then used by the counterpart in the transaction in a variety of ways. In this situation, the exchange of personal data, and the privacy implications of such exchange, is a secondary aspect of a primary transaction involving a good which is not, per se, privacy related. An example of a transaction happening in this market may be the purchase of a book completed on an online merchant's site.

The second type of privacy-related transaction occurs in what may be called the market for personal data. This market itself includes a variety of exchanges. One form of exchange involves 'infomediaries' that trade consumer data among themselves or with other data-holding firms. For instance, firms such as Acxiom or credit reporting agencies such as TransUnion both acquire, and sell, consumer data by interacting with other consumer-facing firms. The data subjects are not generally active agents in these transactions. A second form of exchange involves so-called free products or services provided to consumers in exchange for their data. This market includes search engines and online social networks. In these exchanges, consumers are directly involved in the transaction, although the exchange of their personal information is not always a *visible*, explicit component of the transaction: while the price for services in this type of exchange may be nominally zero, the customer is effectively purchasing the service by selling her data.

A third form of privacy-related transaction occurs in what may be called the market for privacy. In this market, consumers explicitly seek products and services to manage and protect their personal information. For instance, they may acquire a privacy-enhancing technology to protect their communications or hide their browsing behavior. The business models associated with providing consumers with more protection for their data have evolved rapidly, also due to the attention paid to the potential costs and benefits associated with the sharing and mining of consumer data. Indeed, some business models become a bridge between the market for privacy and the market for personal data, in that they aim at giving consumers more 'ownership' over (exchanges involving) their personal information, including – sometimes – the potential ability to monetize it.

Privacy Trade-offs

The evolution and success of data mining and business analytics tools is and will keep affecting the markets described in the previous sections and their emerging trade-offs, especially in the form of both positive and negative externalities that arise when a consumer's data is aggregated and analyzed together with the data of many other consumers.[1] As anticipated in our discussion of microeconomic theory, the resulting costs and benefits for data subjects, data holders, and society at large are complex and nuanced. On the one hand, expected benefits can emerge from disclosed data for both data holders and data subjects (as well as opportunity costs when information is not shared or collected), together with the expected costs of the investments necessary to collect and process that data. On the other hand, expected benefits can arise from *protecting* data and expected costs can arise from privacy intrusions; however, costs are also associated with the protection of personal data. While a complete analysis of such dual benefits and costs associated with either sharing or protecting data are outside the scope of this chapter, in this section we provide a few key examples that are especially relevant to the context of data mining and business analytics.

We first consider some of the benefits of data sharing, as well as some of the costs associated with data protection.

Firms can capitalize in various ways on the data of current and potential customers. Detailed knowledge of a consumer's preferences and behavior can help firms better target their products or ads, lowering advertising costs (Blattberg and Deighton 1991), providing enhanced, personalized services (Acquisti and Varian 2005), increasing consumer retention and loyalty, but also enforcing profit-enhancing price discrimination (Varian 1985) (although the last effect may not always hold in presence of competition; Fudenberg and Tirole 2000). For instance, the granular targetability made possible by online advertising may increase revenues for marketers and merchants (according to Beales, 2010, the price of behaviorally targeted advertising is almost three times higher than the price of untargeted advertising). Similarly, by aggregating consumer data, firms can forecast trends and predict individual preferences, giving them the ability to provide valuable product recommendations (Bennett and Lanning 2007), or improve or redesign services based on observed behavior. Furthermore, revenues from targeted consumers may allow firms to provide lower-cost versions of a product to other consumers or to support services provided to consumers at a price of zero – but in exchange for their data.

Indeed, some of the benefits data holders gain from data may be passed on to, or shared with, data subjects themselves (Lenard and Rubin 2009; Goldfarb and Tucker 2010), in the form of free content or services (made possible by advertising or personal data trades), personalized services, reduced search costs, or more efficient interactions with merchants or their sites. Positive externalities may also materialize. For instance, better consumer data may allow firms to target the right consumers, reducing the amount of marketing investment that is wasted on consumers uninterested in the product and potentially leading to lower product prices (see also Blattberg and Deighton 1991), or aggregation of web searches of many individuals could help detect disease outbreaks (Wilson and Brownstein 2009), or the aggregation of location data could be used to improve traffic conditions and reduce road congestion. In other words, the aggregation of private data can create a public good, with societal benefits accruing from big data.

One should note, however, that many of the benefits consumers can enjoy from data sharing may also be obtained without the disclosure of *personally identified* data. In other words, reaping the benefits of big data while protecting privacy may not necessarily be contradictory goals: as further discussed in the final section of this chapter, advances in privacy-enhancing technologies promise many shades of gray between the polar extremes of absolute sharing and complete protection of personal data; rather, it may be possible to selectively protect or disclose different types of personal information, and modulate their identifiability, in order to optimize privacy trade-offs for individuals and society as a whole. As a result, benefits from data may be gained while data is also protected, and the actual societal costs of privacy protection may prove to be limited. For instance, the privacy-enhancing provisions of the Fair Credit Reporting Act did not raise the significant barriers to profitable uses of consumer data that critics of the act feared before its passage (Gellman 2002); the possible reduction of ads' effectiveness caused by regulations limiting behavioral targeting may simply be offset by using ads on sites with specific content, larger ads, or ads with interactive, video, or audio features (Goldfarb and Tucker 2010); and certain types of privacy regulation in health care may actually foster innovation in the form of higher probability of success of 'health information exchanges' (Adjerid et al. 2013).

As for the costs that come from privacy violations or disclosed data, they can be both tangible and intangible for data holders and data subjects alike. From the perspective of the data subject, Calo (2011) distinguishes between subjective and objective privacy harms: the former derive from unwanted perceptions of observation; the latter consist of the unanticipated or coerced use of information concerning a person against that person. Hence, the former relate to the anticipation, and the latter to the consequences, of losing control over personal information. Subjective harms include anxiety, embarrassment, or fear; the psychological discomfort associated with feeling surveilled; the embarrassment associated with public exposure of sensitive information; or the chilling effects of fearing one's personal life will be intruded upon. These harms may be hard to capture and evaluate in economic terms, and usually are not recognized by U.S. courts as *actual* damage (Romanosky and Acquisti 2009). Objective harms can be immediate and tangible, or indirect and intangible. They include the damage caused by identity theft; the effort spent deleting (or avoiding) junk mail; the time spent dealing with annoying telemarketing; the higher prices paid due to (adverse) price discrimination; but also the effects of profiling, segmentation, and discrimination. For instance, profiling could be used to nudge consumers toward products that do not enhance their well-being,[2] and information revealed on a social network may lead to job market discrimination (Acquisti and Fong 2012). In more general terms, as an individual's data is shared with other parties, those parties may gain a bargaining advantage in future transactions with that individual. For instance, while a consumer, thanks to behavioral advertising, may receive targeted ads for products she is actually interested in, other entities (such as marketers and merchants) will accumulate data about the consumer that may permit the creation of a detailed dossier of her preferences and tastes, and the prediction of her future behavior. As we noted, microeconomic models predict that, in the presence of myopic customers, this information will affect the allocation of the surplus of future transactions, increasing the share of the data holder at the expense of the data subject. Ultimately, the disclosure of personal data affects the balance of power between the data subject and the data holder.

Data holders can also, under certain conditions, bear costs from the misuse of consumer data. For instance, Romanosky and Acquisti (2009) note that, following a data breach, firms suffer in terms of consumer notification costs, fines, settlement costs, stock market losses, or loss of consumers' trust. However, a firm can often externalize the privacy costs of using consumer data, while internalizing the gain (Swire and Litan 1998).

Often, objective privacy harms are merely probabilistic: when data is revealed or breached, it may or may not lead to the actual negative consequences we have just described. For instance, poor data-handling practices by a credit reporting firm may later cause a consumer's mortgage request to be wrongfully denied; or the breach of a database containing consumers' credit card information may later lead to identity theft (Camp 2007). The metaphor of a 'blank check' has been used to describe the uncertainty associated with privacy costs: disclosing personal information is like signing a blank check, which may never be cashed in – or perhaps cashed in at some unpredictable moment in time with an indeterminably low, or high, amount to pay. In economic terms, the damage from disclosed data is, as Knight (1921) says, *ambiguous* and, up to a point, unknowable.[3] Consider, in fact, that some privacy costs are high-probability events with negligible individual impact (e.g. spam); other costs are low-probability events with very significant adverse consequences (e.g. some of the costs associated with the more pernicious forms of identity theft). Because of this and their often intangible dimensions, privacy costs may be hard to assess and therefore also to act upon. Either because the likelihood of occurrence is low, or the perceived magnitude of damage is limited, privacy costs may therefore be dismissed as unimportant at the individual level – even when, in the aggregate, they may cause significant societal damage, or a significant transfer of wealth from data subjects to others (including data holders).

Do Consumers Value Privacy?

Farrell (2012) notes that privacy is both a final and an intermediate good: "[c]onsumers care about privacy in part for its own sake: many of us at least sometimes feel it's just icky to be watched and tracked. [. . .] Consumers also care about privacy in a more instrumental way. For instance, loss of privacy could identify a consumer as having a high willingness to pay for something, which can lead to being charged higher prices if the competitive and other conditions for price discrimination are present." In this section, we summarize a number of empirical investigations of consumers' privacy valuations. In the next section, we examine the hurdles that consumers face in making privacy decisions consistent with those valuations.

Numerous factors influence individuals' privacy concerns (Milberg et al. 1995), and therefore the mental 'privacy calculus' that individuals make when deciding whether to protect or disclose personal information (Laufer and Wolfe 1977; Culnan and Armstrong 1999; Dinev and Hart 2006). Researchers from such diverse disciplines as economics, marketing, information systems, and computer science have attempted to estimate empirically the value that, in this calculus, individuals assign to privacy and their personal data. The results suggest that privacy valuations are highly context dependent. Furthermore, willingness to pay, or reservation prices, may not adequately capture the value of privacy for those individuals who simply do not feel they should have to pay to protect their privacy.

Huberman et al. (2005) used a second-price auction to estimate the price at which individuals were willing to publicly reveal personal information such as their weight. Individuals whose weight was more deviant from the perceived norm for the rest of the group were more likely to exhibit higher valuations. Wathieu and Friedman (2005) found that survey participants were more acceptive of an organization sharing their personal information after the economic benefits of doing so were explained to them. Cvrček et al. (2006) reported large differentials across EU countries in the price EU citizens would accept to share mobile phone location data. Hann et al. (2007) focused on online privacy and, using a conjoint analysis, found that protection against errors, improper access, and secondary use of personal information was worth US$30.49–$44.62 among U.S. subjects. Both Varian et al. (2005) and Png (2007) estimated U.S. consumers' implicit valuation of protection from telemarketers using data about Do Not Call list adoptions. They found widely differing values, from a few cents to as much as $30. Tsai et al. (2011) found that, when information about various merchants' privacy policies was made available to them in a compact and salient manner, subjects in an experiment were more likely to pay premia of roughly 50 cents to purchase products from more privacy-protective merchants.

At the same time, various studies have highlighted a dichotomy between self-professed privacy attitudes and actual self-revelatory behavior. Tedeschi (2002) reported on a Jupiter Research study in which the overwhelming majority of surveyed online shoppers would give personal data to new shopping sites for the chance to win $100. Spiekermann et al. (2001) found that even participants in an experiment who could be classified as privacy conscious and concerned were willing to trade privacy for convenience and discounts: differences across individuals in terms of reported concerns did not predict differences in self-revelatory behavior. Similar findings were obtained in different settings by Acquisti and Grossklags (2005) and Acquisti and Gross (2006). Coupled with the observation that businesses focused on providing privacy-enhancing applications have met difficulties in the marketplace (Brunk 2002), these results suggest a potential privacy paradox: people want privacy, but do not want to pay for it, and in fact are willing to disclose sensitive information for even small rewards (for an overview of this area, see Acquisti, 2004, and Acquisti and Grossklags, 2007). In fact, Acquisti et al. (2013b) have recently presented an application of the endowment effect to the privacy domain: subjects who started an experiment from positions of greater privacy protection were found to be five times more likely than other subjects (who did not start with that protection) to forgo money to preserve their privacy. These results illustrate the challenges with pinpointing exact valuations of personal data: consumers' privacy valuations are not only context dependent, but affected by numerous heuristics and biases (see the next section), and so are individuals' decisions to share or to protect personal information (John et al. 2011). In addition, awareness of privacy risks (and potential solutions to privacy threats) may also significantly affect consumers' privacy choices – which is why revealed preference arguments (which rely on observing consumers' choices in the marketplace – for instance, in the case of privacy, their propensity to share data online or to use protecting technology) do not necessarily provide the fuller or clearer picture of what privacy is ultimately worth to individuals. We discuss some of the hurdles that affect privacy decision making and valuations in the next section.

Hurdles in Privacy Behavior

A stream of research investigating the so-called privacy paradox has focused on the hurdles that hamper individuals' privacy-sensitive decision making. If consumers act myopically, or not fully rationally (in the neo-classical economic sense of utility-maximizing, Bayesian-updating agents who make use of all information available to them), then market equilibria may not in fact guarantee privacy protection. For instance, in the absence of regulatory protection for consumers' data, firms will tend to extract the surplus generated from transactions with consumers, by using consumers' data for price discrimination (Acquisti and Varian 2005; Taylor 2004a).

There is, indeed, evidence that consumers face known decision-making hurdles when making privacy trade-offs, such as (a) incomplete information, (b) bounded cognitive ability to process available information, and (c) a number of heuristics and cognitive or behavioral biases, which lead to systematic deviations from theoretically rational decision making – sometimes, various combinations of these factors affect consumer decision making at the same time.

Consider, first, the problem of incomplete information. In many scenarios – such as those associated with behavioral monitoring and targeting – a consumer may not even realize the extent to which her behavior is being monitored and exploited. Furthermore, after an individual has released control of her personal information, she is in a position of information asymmetry with respect to the party with whom she is transacting. In particular, the subject might not know if, when, or how often the information she has provided will be used. For example, a customer might not know how the merchant will use the information that she has just provided through a website.

Furthermore, the 'value' itself of an individual's information might be highly uncertain and variable. The subject and the parties she interacts with may evaluate differently the same piece of information, and the specific environmental conditions or the nature of the transaction may affect the value of information in unpredictable ways. For example, a customer might not know what damage she will incur because of her personal information becoming known, she might not know how much profit others will make thanks to that information, and she might not know the benefits she will forgo if her data is not shared. To what, then, is the subject supposed to anchor the valuation of her personal data and its protection?

Second, findings from behavioral economics document consumers' inability to exhaustively consider the possible outcomes and risks of data disclosures, due to bounded rationality. Furthermore, an individual will often find herself in a weaker bargaining position than the parties she is interacting with (e.g. merchants). In many transactions, an individual is unable to negotiate a desired level of information protection; she rather faces take-it-or-leave-it offers of service in exchange for personal data.

Third, even if a consumer has access to complete information about all trade-offs associated with data sharing and data protection, she will suffer from cognitive and behavioral biases that are more intense in scenarios where preferences are more likely to be uncertain. One such example is that, if the expected negative payoff from privacy invasions could be estimated, some individuals might seek immediate gratification, discounting hyperbolically (Rabin and O'Donoghue 2000) future risks (for example, of being subject to identity theft), and choosing to ignore the danger. Hence, because of asymmetric information, self-gratification bias, overconfidence, or various other forms of misrepresentation studied in the behavioral economics literature, individuals might choose not to protect their privacy *possibly* against their own best interest. They might act *myopically* when it comes to protecting their privacy even when they might act *strategically* (as rational agents) when bargaining for short-term advantages such as discounts (Acquisti 2004).

Consider, for instance, the case of data breaches. As discussed in Romanosky and Acquisti (2009), after being notified of a breach of her financial information, a consumer may not be able to identify the right course of action: should she, for instance, punish the financial firm that due to faulty security controls compromised her data, by changing to a competitor? While this may appear to be a risk-reducing behavior, by doing so the consumer discloses her personal information to another firm – and actually materially increases the probability that a future breach will involve her data. Furthermore, the cost of acting may be significant: calling the breached firm to obtain details about the breach and its consequences, notifying financial institutions of the occurred breach and of potentially compromised accounts, and subscribing to credit alert and insurance services are all actions which carry perceived cognitive, transaction, and actual costs. Such costs may appear greater to the consumer than the perceived benefit from action. It could also be that, because of psychological habituation from repeated media reports of data breaches, the consumer becomes desensitized to their effects – which counters the desired impact of notifications. Ultimately, the consumer may 'rationally' decide to remain 'ignorant' (following the Choicepoint breach, fewer than 10% of affected individuals availed themselves of the free credit protection and monitoring tools offered by Choicepoint; Romanosky and Acquisti 2009). This example suggests how nuanced and full of obstacles is the path that leads from consumer notification of a privacy problem to her actually taking action to solve that problem.

Based on these hurdles, recent behavioral privacy research has questioned the validity and effectiveness of regimes based on transparency and control mechanisms (also known as choice and notification; Brandimarte et al. 2013; Adjerid et al. 2013a, 2013b; Acquisti et al. 2013a).[4]

An improved understanding of cognitive and behavioral biases that hamper privacy (and security) decision making, however, could also be exploited for normative purposes. Specifically, knowledge of those biases could be used to design technologies and policies that anticipate and counter those very biases (Acquisti 2009). Such technologies and policies would be informed by the growing body of research in behavioral economics on soft or asymmetric paternalism (Loewenstein and Haisley 2008) as well as research on privacy and security usability. They may help consumers and societies achieve their desired balance between information protection and information sharing.

Technology, Regulation, and Market Forces

As noted in Acquisti (2010), advances in computer science, statistics, and data mining have not only produced potentially privacy-eroding business analytic tools and big data technologies, they have also led to the development, over the past few decades, of privacy-enhancing technologies that allow the protection of (certain) individual data simultaneously with the sharing, or analysis, of aggregate, de-identified, or non-sensitive identified data. Online activities and transactions for which privacy-preserving versions exist include electronic payments (Chaum 1983), online communications (Chaum 1985), Internet browsing (Dingledine et al. 2004), credentials (Camenisch and Lysyanskaya 2001), and even online recommendations (Canny 2002). One of the most interesting directions of research relates to executing calculations in encrypted spaces (Gentry 2009), and investigating whether these types of computations will make it possible to have both privacy *and* big data, confidentiality *and* analytics. In the best case, the deployment of privacy-enhancing technologies results in a win–win for data holders and data subjects: certain data is protected (thereby avoiding the costs associated with certain privacy intrusions), whereas other data is shared, analyzed, and used (thereby enjoying the benefits and the value of data, big or small). Alternatively, the old economic adage that there is no free lunch may apply: whenever protection is applied to a dataset, the utility of that dataset is decreased (Duncan et al. 2001). The interesting economic question then becomes, whose utility will be adversely affected – or, in other words, who will bear the costs if privacy-enhancing technologies become more popular in the age of big data: data subjects (whose benefits from business analytics and big data would shrink if they restrict the amount of information they share), data holders (who would face increasing costs associated with collecting and handling consumer data), or both?

This remains an open research question. An additional and related open question is whether, even if privacy-enhancing technologies were found to increase data subjects' welfare more than they would adversely affect data holders' welfare, market forces alone would lead to the deployment and success of those technologies. While there is no lack of evidence online of both disclosure/publicity-seeking and privacy-seeking behavior, privacy-enhancing technologies (as opposed to security technologies such as anti-virus software or firewalls) have not gained widespread adoption. This situation has several possible explanations. On the consumer side, a first obvious explanation is low consumer demand for privacy; however, other, more nuanced (and non–mutually exclusive) explanations include users' difficulties and costs in using privacy technologies (see Whitten and Tygar 1999), switching costs, as well as biases such as immediate gratification, which reduce demand for those products even by privacy-sensitive consumers. On the data holder side, in absence of regulatory intervention, or of clear evidence that privacy protection can act as a distinctive source of competitive advantage for a firm, it is unlikely that firms will incur costs to transition to technologies that may, in the short run, limit their access to consumer data relative to their competitors.

The debate over the comparative economic advantages of regulation and self-regulation of privacy remains intense to this date. On the one hand, Gellman (2002) challenges the view that unrestricted trafficking in personal information always benefits the consumer, and that privacy trade-offs may be evaluated merely on the basis of monetary costs and benefits. He concludes that an unregulated, privacy-invasive market in personal data can be costly for consumers. Cate (2002), Cate et al. (2003), Rubin and Lenard (2001), and Lenard and Rubin (2009), on the other hand, claim that legislative initiatives that restrict the amount of personal information available to business would actually penalize the consumers themselves: regulation should be undertaken only when a given market for data is not functioning properly, and when the benefits of new measures outweigh their costs.

It may not be possible to resolve this debate using purely economic tools. Economic theory, as we have discussed, has brought forward arguments supporting both the views that privacy protection *increases* economic efficiency, and that it *decreases* it. Empirically, the costs and benefits associated with the protection and revelation of consumers' data have not proven amenable to aggregation. First, as soon as one attempts an aggregate evaluation of the impact of privacy regulation, one faces the challenge of delimiting the problem: data breaches, identity theft, spam, profiling, and price discrimination are all examples of privacy problems, yet they comprise very different expected benefits and costs for the parties involved. Second, even within each scenario, it may be hard to measure statically the aggregate costs and benefits of data protection and data sharing, since the benefits and costs of privacy happen over time (for instance, data revealed today may only damage the individual years from now). And third, in addition to measurable outcomes (such as the financial losses due to identity theft or the opportunity costs of spam), privacy invasions require an estimation of consumers' valuations of privacy. Furthermore, as we have noted elsewhere in this chapter, many of the benefits associated with data disclosure may, in fact, still be gained when data is protected. Evaluations and conclusions regarding the economic value of privacy and the optimal balance between disclosure and protection are, therefore, far from simple.

Acknowledgement The author would like to thank the editors, the anonymous reviewers, Veronica Marotta, Laura Brandimarte, and Sonam Samat for particularly insightful comments and suggestions. This chapter is partly based on previous work by the author, including Acquisti (2010) and Brandimarte and Acquisti (2012).

NOTES

1. This section is based on discussion in Acquisti (2010, sec. 3).
2. Some consumer data firms advertise databases with contacts to individuals suffering from various types of addiction, such as gambling (see http://www.dmnews.com/media-one-gamblers-database/article/164172/).
3. Knight (1921) distinguished between *risk* (the random outcomes of an event can be described with a known probability distribution) and *ambiguity* (those probabilities are unknown).
4. See also Chapter 2 of this volume, by Barocas and Nissenbaum.

REFERENCES

Acquisti, A. 2004. Privacy in electronic commerce and the economics of immediate gratification. In *Proc. ACM Conference on Electronic Commerce (EC '04)*, 21–29.

Acquisti, A. 2009. Nudging privacy: The behavioral economics of personal information. *IEEE Security & Privacy* 7(6): 82–85.

Acquisti, A. 2010. The economics of personal data and the economics of privacy. Background paper for OECD Joint WPISP-WPIE Roundtable, December 1.

Acquisti, A., I. Adjerid, and L. Brandimarte. 2013a. Gone in 15 seconds: The limits of privacy transparency and control. *IEEE Security & Privacy* 11(4): 72–74.

Acquisti, A., and C. Fong. 2012. An experiment in hiring discrimination via online social networks. In *5th Privacy Law Scholars Conference (PLSC)*, June 7–8.

Acquisti, A., and R. Gross. 2006. Imagined communities: Awareness, information sharing, and privacy on the Facebook. In *Workshop on Privacy Enhancing Technologies (PET '06)*, 36–58.

Acquisti, A., and J. Grossklags. 2005. Privacy and rationality in individual decision making. *IEEE Security & Privacy* 3(1): 24–30.

Acquisti, A., and J. Grossklags. 2007. What can behavioral economics teach us about privacy? In *Digital Privacy: Theory, Technologies and Practices*, ed. S. G. C. L. Alessandro Acquisti and Sabrina De Capitani di Vimercati, 363–377. Boca Raton, FL: Auerbach Publications.

Acquisti, A., L. K. John, and G. Loewenstein. 2013b. What is privacy worth? *Journal of Legal Studies* 42(2): 249–274.

Acquisti, A., and H. R. Varian. 2005. Conditioning prices on purchase history. *Marketing Science* 24(3): 367–381.

Adjerid, I., A. Acquisti, L. Brandimarte, and G. Loewenstein. 2013a. Sleights of privacy: Framing, disclosures, and the limits of transparency. In *Symposium on Usable Privacy and Security (SOUPS)*, 9.

Adjerid, I., A. Acquisti, R. Padman, R. Telang, and J. Adler-Milstein. 2013b. The impact of privacy regulation and informed consent on technology adoption: The case of health information exchanges. Economics of Information Technology and Digitization Workshop, NBER Summer Institute, 2013, http://users.nber.org/~confer/2013/SI2013/PRIT/pritprg.html.

Akerlof, G. A. 1970. The market for 'lemons': Quality uncertainty and the market mechanism. *Quarterly Journal of Economics* 84(3): 488–500.

Beales, H. 2010. The value of behavioral targeting. Network Advertising Initiative.

Bennett, J., and S. Lanning. 2007. The Netflix prize. In *Proc. KDD Cup and Workshop*.

Blattberg, R.C., and J. Deighton. 1991. Interactive marketing: Exploiting the age of addressability. *Sloan Management Review* 33(1): 5–14.

Brandimarte, L., and A. Acquisti. 2012. The economics of privacy. In *Handbook of the Digital Economy*, ed. M. Peitz and J. Waldfogel. New York: Oxford University Press.

Brandimarte, L., A. Acquisti, and G. Loewenstein. 2013. Misplaced confidences privacy and the control paradox. *Social Psychological and Personality Science* 4(3): 340–347.

Brunk, B. D. 2002. Understanding the privacy space. *First Monday* 7(10), http://dx.doi.org/10.5210%2Ffm.v7i10.991.

Calo, R. 2011. The boundaries of privacy harm. *Indiana Law Journal* 86:1131–1162.

Camenisch, J., and A. Lysyanskaya. 2001. An efficient system for non-transferable anonymous credentials with optional anonymity revocation. In *Advances in Cryptology – EUROCRYPT '01*, LNCS 2045, 93–118. Heidelberg: Springer.

Camp, L. J. 2007. *Economics of Identity Theft: Avoidance, Causes and Possible Cures*. New York: Springer.

Canny, J.F. 2002. Collaborative filtering with privacy. In *IEEE Symposium on Security and Privacy*, 45–57.

Cate, F. H. 2002. Principles for protecting privacy. *Cato Journal* 22(1): 33–57.

Cate, F. H., R. E. Litan, M. Staten, and P. Wallison (2003). Financial privacy, consumer prosperity, and the public good: Maintaining the balance. Federal Trade

Commission Workshop on Information Flows: The costs and benefits to consumers and businesses of the collection and use of consumer information.

Chaum, D. 1983. Blind signatures for untraceable payments. In *Advances in Cryptology*, 199–203. New York: Plenum Press.

Chaum, D. 1985. Security without identification: Transaction systems to make big brother obsolete. *Communications of the ACM* 28(10): 1030–1044.

Culnan, M. J., and P. K. Armstrong. 1999. Information privacy concerns, procedural fairness, and impersonal trust: An empirical investigation. *Organization Science* 10(1): 104–115.

Cvrcek, D., M. Kumpost, V. Matyas, and G. Danezis. 2006. A study on the value of location privacy. In *ACM Workshop on Privacy in the Electronic Society (WPES)*, 109–118.

Dinev, T., and P. Hart. 2006. An extended privacy calculus model for e-commerce transactions. *Information Systems Research* 17(1): 61–80.

Dingledine, R., N. Mathewson, and P. Syverson. 2004. Tor: The second-generation onion router. In *Proc. 13th Conference on USENIX Security Symposium*, 13:21.

Duncan, G. T., S. A. Keller-McNulty, and S. L. Stokes. 2001. Disclosure risk vs. data utility: The R-U confidentiality map. Technical Report No. 121. Research Triangle Park, NC: NISS.

Farrell, J. 2012. Can privacy be just another good? *Journal on Telecommunications and High Technology Law* 10:251–445.

Fudenberg, D., and J. Tirole. 2000. Customer poaching and brand switching. *RAND Journal of Economics* 31(4): 634–657.

Gellman, R. 2002. Privacy, consumers, and costs – how the lack of privacy costs consumers and why business studies of privacy costs are biased and incomplete. March.

Gentry, C. 2009. Fully homomorphic encryption using ideal lattices. In *Proc. 41st Annual ACM Symposium on Theory of Computing*, 169–178.

Goldfarb, A., and C. Tucker. 2010. Privacy regulation and online advertising. Available at SSRN: http://ssrn.com/abstract=1600259.

Hann, I.-H., K.-L. Hui, T. S. Lee, and I. P. Png. 2007. Overcoming online information privacy concerns: An information processing theory approach. *Journal of Management Information Systems* 42(2): 13–42.

Hirshleifer, J. 1971. The private and social value of information and the reward to inventive activity. *American Economic Review* 61(4): 561–574.

Hirshleifer, J. 1980. Privacy: Its origins, function and future. *Journal of Legal Studies* 9, no. 4 (December): 649–664.

Huberman, B. A., E. Adar, and L. R. Fine. 2005. Valuating privacy. *IEEE Security & Privacy* 3:22–25.

John, L. K., A. Acquisti, and G. Loewenstein. 2011. Strangers on a plane: Context-dependent willingness to divulge sensitive information. *Journal of Consumer Research* 37(5): 858–873.

Knight, F. 1921. *Risk, Uncertainty and Profit.* Boston: Hart, Schaffner & Marx; Houghton Mifflin.

Laufer, R. S., and M. Wolfe. 1977. Privacy as a concept and a social issue: A multidimensional developmental theory. *Journal of Social Issues* 33(3): 22–42.

Lenard, T. M., and P. H. Rubin. 2009. In defense of data: Information and the costs of privacy. Washington, DC: Technology Policy Institute, May.

Loewenstein, G., and E. Haisley. 2008. The economist as therapist: Methodological issues raised by light paternalism. In *Perspectives on the Future of Economics: Positive and Normative Foundations*, ed. A. Caplin and A. Schotter. New York: Oxford University Press.

Milberg, S. J., S. J. Burke, H. J. Smith, and E. A. Kallman. 1995. Values, personal information privacy, and regulatory approaches. *Communications of the ACM* 38(12): 65–74.

Noam, E. M. 1996. Privacy and self-regulation: Markets for electronic privacy. In Chapter 1 of *Privacy and Self-Regulation in the Information Age*. Washington, DC: National Telecommunications and Information Administration.

Png, I. 2007. On the value of privacy from telemarketing: Evidence from the 'Do Not Call' registry. Working paper, National University of Singapore.

Posner, R. A. 1981. The economics of privacy. *American Economic Review* 71, no. 2 (May): 405–409.

Rabin, M., and T. O'Donoghue. 2000. The economics of immediate gratification. *Journal of Behavioral Decision Making* 13(2): 233–250.

Romanosky, S., and A. Acquisti. 2009. Privacy costs and personal data protection: Economic and legal perspectives. *Berkeley Technology Law Journal* 24(3): 1060–1100.

Rubin, P. H., and T. M. Lenard. 2001. *Privacy and the Commercial Use of Personal Information*. Boston: Kluwer Academic Publishers.

Solove, D. J. 2006. A taxonomy of privacy. *University of Pennsylvania Law Review* 154(3): 477.

Spiekermann, S., J. Grossklags, and B. Berendt. 2001. E-privacy in 2nd generation e-commerce: Privacy preferences versus actual behavior. In *Proc. 3rd ACM Conference on Electronic Commerce*, 38–47.

Stigler, G. J. 1980. An introduction to privacy in economics and politics. *Journal of Legal Studies* 9, no. 4 (December): 623–44.

Swire, P. P., and R. E. Litan. 1998. *None of Your Business – World Data Flows, Electronic Commerce, and the European Privacy Directive*. Washington, DC: Brookings Institution Press.

Taylor, C. R. 2004a. Consumer privacy and the market for customer information. *RAND Journal of Economics* 35(4): 631–651.

Taylor, C. R. 2004b. Privacy and information acquisition in competitive markets. Technical Report 03-10. Durham, NC: Duke University, Economics Department.

Tedeschi, B. 2002. E-commerce report; everybody talks about online privacy, but few do anything about it. *The New York Times*, June 3.

Tsai, J. Y., S. Egelman, L. Cranor, and A. Acquisti. 2011. The effect of online privacy information on purchasing behavior: An experimental study. *Information Systems Research* 22(2): 254–268.

Varian, H. 1985. Price discrimination and social welfare. *American Economic Review* 75(4): 870–875.

Varian, H. R. 1996. Economic aspects of personal privacy. In Chapter1 in *Privacy and Self-Regulation in the Information Age*. Washington, DC: National Telecommunications and Information Administration.

Varian, H., F. Wallenberg, and G. Woroch. 2005. The demographics of the do-not-call list. *IEEE Security & Privacy* 3(1): 34–39.

Wathieu, L., and A. Friedman. 2005. An empirical approach to understanding privacy valuation. In *4th Workshop on the Economics of Information Security*.

Whitten, A., and J. D. Tygar. 1999. Why Johnny can't encrypt: A usability evaluation of PGP 5.0. In *Proc. 8th USENIX Security Symposium*, 14.

Wilson, K., and J. Brownstein. 2009. Early detection of disease outbreaks using the Internet. *Canadian Medical Association Journal* 180(8): 829.

Changing the Rules: General Principles for Data Use and Analysis

Paul Ohm

Introduction

How do information privacy laws regulate the use of big data techniques, if at all? Do these laws strike an appropriate balance between allowing the benefits of big data and protecting individual privacy? If not, how might we amend or extend laws to better strike this balance?

This chapter attempts to answer questions like these. It builds on Chapter 1 of this volume, by Strandburg, which focused primarily on legal rules governing the *collection* of data. This chapter will focus primarily on the law of the United States, although it will make comparisons to the laws of other jurisdictions, especially the European Union, which is well covered in Chapter 8 of this volume.

Most information privacy law focuses on collection or disclosure and not use. Once data has been legitimately obtained, few laws dictate what may be done with the information. The exceptions to this general pattern receive attention below; laws that govern use tend to focus on particular types of users, especially users that lawmakers have deemed owe obligations of confidentiality to data subjects. For example, law regulating the health and financial industries, industries that historically have evolved obligations of confidentiality, constrain not only collection and disclosure but also use.

This chapter argues that our current information privacy laws are failing to protect individuals from harm. The discussion focuses primarily on shortcomings in the law that relate to specific features of big data, although it also describes a few shortcomings that relate only tangentially to these features. All of these shortcomings expose some individuals to the risk of harm in certain circumstances. We need to develop ways to amend the laws to recalibrate the balance between analytics and risk of harm. Ultimately, the chapter proposes five general approaches for change.

Current State of the Law Governing the Use of Information

Privacy law tends to divide the activities of data analysis (or, to use the more recent term, data science) into three steps, each subject to its own rules with specific, tailored levels of coverage and burden. These steps are collection, use, and disclosure. Chapter 1 focused mostly on collection and some on disclosure; this chapter will tackle use.

Simply put, most uses of data are not regulated. Many data analysis practices fall wholly outside any privacy law, at least in the United States. Those that are nominally governed by law tend to be restricted lightly, if at all. The few use restrictions that exist tend to build upon the Fair Information Practice Principles (FIPPs). This section offers a thumbnail sketch of a few example laws, but even as to these laws, the analysis is necessarily brief and incomplete. Space does not allow for a thorough, much less complete, survey of the law. The purpose of this chapter is not to offer legal advice, but rather to critique and consider the need to expand the law.

The Fair Information Practice Principles

The FIPPs were promulgated first in an influential report issued by an Advisory Committee to the Secretary of Health, Education, and Welfare (HEW) which suggested in 1973 five rules for protecting the privacy of individuals in record-keeping systems:

1. There must be no personal data record-keeping systems whose very existence is secret.
2. There must be a way for a person to find out what information about the person is in a record and how it is used.
3. There must be a way for a person to prevent information about the person that was obtained for one purpose from being used or made available for other purposes without the person's consent.
4. There must be a way for a person to correct or amend a record of identifiable information about the person.
5. Any organization creating, maintaining, using, or disseminating records of identifiable personal data must assure the reliability of the data for their intended use and must take precautions to prevent misuses of the data.[1]

The FIPPs have been embraced by many scholars and policymakers, and form the basis of numerous government regulations. Each regulation, however, seems to embrace a different form of the FIPPs, with some that look very different from this original set.

For example, the HEW FIPPs greatly influenced the OECD's *Guidelines on the Protection of Privacy and Transborder Flows of Personal Data*, an influential set of voluntary standards developed by members of the OECD, which specify eight principles: collection limitation, data quality, purpose specification, use limitation, security safeguards, openness, individual participation, and accountability.[2] The Federal Trade Commission (FTC) has for many years focused on its own version of the FIPPs as an organizing principle for its privacy mandates. The FTC's version proposes five requirements: notice, consent, access, data integrity, and enforcement.[3]

Legal scholar Fred Cate has noted that, at least since the OECD guidelines, every example of the FIPPs has "reflect[ed] a distinct goal of data protection as empowering consumers to control information about themselves, as opposed to protecting individuals from uses of information about them that are unfair or harmful."[4] This manifests in particular on a focus on notice and choice, which Chapter 2 argues persuasively is insufficient for protecting privacy in big data contexts. According to Cate, "in the past two decades most FIPPS have been applied in practice to require primarily notice and, in some instances, choice," citing FTC pronouncements and the HIPAA rules as examples of how completely notice and choice has taken hold in the United States.[5]

But notice and choice is not only an American phenomenon, argues Cate, as even many of the provisions of the EU Data Protection Directive "can be waived with consent," notwithstanding assertions by EU officials that the "directive is not concerned with notice and consent."[6]

Sectoral Privacy Protection

Privacy law in the United States is often referred to as 'sectoral'.[7] No privacy law sweeps broadly across many different industries in the way the EU Data Protection Directive does in Europe. Rather, privacy laws in this country tend to focus on particular sectors – usually differentiated by industry segments or the type of information held. Thus, the Health Insurance Portability and Accountability Act (HIPAA) regulates the privacy of information generated and stored by the health industry,[8] the Family Educational Rights and Privacy Act (FERPA) protects the privacy of information in schools,[9] and the Gramm-Leach-Bliley Act provides some protection for information stored by financial institutions.[10]

For the most part, privacy law in the United States focuses on the collection or disclosure of information, not on its use. Most U.S. privacy laws do not implement a strong form of the FIPP of purpose limitation – information may be collected (or not) and shared (or not) according to these rules, but once information has been legitimately obtained consistent with these laws, it can then be used for any purpose by any person associated with the organization possessing the information.

There are some exceptions to this general trend. Some privacy laws do limit the uses to which legitimately held information can be put. In the discussion that follows, the attention placed on these relative outliers should not obscure the broader point: most privacy laws do not contain these types of constraints. Once legitimately held, all information is fair game. Consider three examples of laws that regulate use: the Privacy Act, the Fair Credit Reporting Act (FCRA), and HIPAA.

The sweeping name given to the Privacy Act may mislead given the law's relatively narrow and specialized scope. It protects the privacy of information maintained in large government databases. Enacted in 1974 – a time of general distrust in government and emerging fear about the impact of computerization on liberty and privacy – it represents the strong influence of the FIPPs. Like most FIPPs-based approaches, it focuses much more on collection and disclosure and notice and choice than on dictating narrowly circumscribed rules for use. Information collected must be associated with a purpose, and that information can be disclosed within an organization only "to those officers and employees of the agency . . . who have a need for the record in the performance of their duties."[11] In addition, the act allows the maintenance of "only such information about an individual as is relevant and necessary to accomplish a purpose of the agency required to be accomplished by statute or by executive order of the President."[12] Thus, government agencies are not supposed to collect information for which there is no clearly specified purpose.

The FCRA covers the activities of consumer reporting agencies, primarily the 'Big Three' credit reporting agencies, Experian, TransUnion, and Equifax, but also any entity that "furnish[es] consumer reports to third parties" used for credit, insurance, or employment purposes.[13] The Federal Trade Commission, which shares authority with many other state and federal agencies under the FCRA to enforce the law, has argued that the law applies broadly beyond the Big Three. For example, in early 2012, the FTC sent letters to some of those who marketed mobile 'background screening' applications, warning them of their likely FCRA obligations.[14] The law limits the uses of consumer reports by consumer reporting agencies to activities enumerated in a long 'white list', such as issuing credit or evaluating a potential employee.[15] It forbids uses for other purposes.

HIPAA applies only to 'covered entities.' Many researchers who handle health information are not governed by HIPAA because they work for organizations that are not covered. For covered entities, such as many health care providers, HIPAA limits the use of information to a long list of permitted uses, such as law enforcement purposes or to facilitate the donation of cadaveric organs.[16] For research purposes, HIPAA generally requires the prior authorization of the subject of the data, with three limited exceptions: authorization is not required with permission of an Institutional Review Board or Privacy Board, for a purpose preparatory to research such as to assist preparing a research protocol, or involving only the information of decedents.[17] HIPAA also allows research outside these exceptions on datasets that have been de-identified according to a very specific and stringent standard.[18]

Gaps in the Law

None of the laws cited above focuses on the special attributes of big data. Instead, they regulate all uses of data, regardless of the size of the dataset or the techniques used. Generally speaking, this probably makes sense. We should not create laws that treat big data research as a wholly different and independent endeavor. We should instead create privacy laws that cover specific contexts of appropriate size and scope.

But big data techniques place special pressures on current privacy law, pressures that will lead some to call for a narrow revision or maybe even wholesale retrenchment of some privacy laws. Consider two. First, big data thrives on surprising correlations and produces inferences and predictions that defy human understanding. These characteristics call seriously into question laws that rely on providing notice to and receiving consent from data subjects. How can you provide notice about the unpredictable and unexplainable? This will lead some to call for us to abandon the FIPPs of use specification and purpose limitation.

Second, big data techniques resist attempts to reduce privacy risks by adding noise to personal information. Laws, such as HIPAA, reward anonymization and other methods of obscuring data, but big data techniques can often restore that which is removed or obscured. Consider both of these pressures in greater depth.

Because big data techniques produce surprising correlations, we increasingly will find it difficult to know that we are in the presence of risky data, the kind of data that will likely lead to privacy harm. Thinking about it from the regulator's point of view, big data's tendency to generate surprising correlations attacks at their core laws developed upon preconceived intuitions, for example privacy laws structured around the specification of 'bad data lists', lists of the types or categories of information that require special handling. Bad data lists require a nearly omniscient regulator capable of distinguishing, ex ante and from the regulator's typical remove, good data from bad. Big data undermines this approach. Mathematician Rebecca Goldin has said something similar about what big data might do to laws meant to prevent race discrimination. As described by *The Economist:*

> [R]acial discrimination against an applicant for a bank loan is illegal. But what if a computer model factors in the educational level of the applicant's mother, which in America is strongly correlated with race? And what if computers, just as they can predict an individual's susceptibility to a disease from other bits of information, can predict his predisposition to committing a crime?[19]

Although we have banned discrimination based on race, big data helps companies find a reasonable proxy for race. Following a similar mechanism, if a behavioral advertising law regulates only databases containing personally identifiable information, no matter how this term is defined, clever statisticians will find a way to infer the very same information they have been denied from seemingly unrelated data.

For example, suppose a study determines tomorrow that height and weight retain a surprising amount of identifying information. It would be a short-sighted response to add height and weight to the lists of types of information that had to be kept from researchers. Every list that prohibits the collection of A will be defeated once data owners realize that unregulated B and C can be used to derive A.

Similarly, the notice and choice at the heart of FIPPs cannot do enough to protect privacy in the age of big data. Big data succeeds by drawing inferences that confound expectations. A data scientist who does no more than confirm prior intuitions will soon be out of work. The best data scientists find results that are not only counterintuitive but also sometimes governed by mysterious, opaque mechanisms. Big data empowers through surprise. Thus, a regime which depends solely on limited purpose, notice, and choice cannot do enough to protect against the unpredictability that tomorrow promises.

In the same way, legal limits on purpose and use will have trouble addressing the problems of big data. An important principle found in many FIPPs is the principle of limited purpose. The HEW report demands that "[t]here must be a way for a person to prevent information about the person that was obtained for one purpose from being used or made available for other purposes without the person's consent."[20] The OECD requires both a 'Purpose Specification' and a 'Use Limitation.'[21] The EU Data Protection Directive similarly requires a 'purpose limitation.'[22]

It is difficult to square a purpose limitation (or even an obligation merely to specify a purpose) with big data. A big data analyst often cannot specify a purpose except at a very high level of abstraction – to "learn something about our customers," or to "find the patterns hiding in the data." This is why so many big data practitioners are loathe to delete old data; you never know what use we will find for it tomorrow! The very idea of big data seems anathema to a purpose limitation which would require that, using the language of the European Union's Article 29 Working Group, "data should be processed for a specific purpose and subsequently used or further communicated only insofar as this is not incompatible with the purpose of the transfer."[23]

Toward a New Conceptualization of Using Law to Protect Privacy

If big data puts new pressures on information privacy, we may need to change information privacy law. I believe that change is urgently needed. Too many people are considering forms of research that escape regulatory scrutiny, and too much of this research might subject individuals and society to jarring shifts from the historical status quo and wreak devastating harm to individuals. In this section, I sketch the impending harm, and I propose some prescriptions to help us restore what we might lose.

Big Data Research Needs to Be Regulated More

We need to fill significant gaps in our privacy laws for data practices generally and for big data practices in particular. The current regulatory landscape exposes too many people to too many unjustifiable risks of significant harm. These harms impact not only individuals but also groups and some of these harms extend across the entire society. The next, last section of this chapter will offer a road map for ways we might extend our present regulatory obligations, but first, consider the argument that change is necessary.

A touchstone for this work is Helen Nissenbaum's idea of privacy as contextual integrity.[24] Contextual integrity requires us, as the name suggests, to frame privacy problems and solutions in appropriately sized contexts as opposed to searching for comprehensive privacy solutions that cover all of society. The book in which this chapter appears focuses on a few narrow contexts: big data analytics, primarily (although not exclusively) conducted in traditional research institutions, with a special focus on the study of cities. Accordingly, I will focus my analysis on these contexts too, even though much of what I say will apply more broadly, for example to big data research conducted inside private corporations.

My argument for change builds also on two strands of the work of Danielle Citron. Citron has argued that individuals can suffer significant harm from the mishandling of information about them stored in massive databases.[25] She has also shined a spotlight on the way official decision making, particularly by the government, is increasingly becoming algorithmic.[26] In this second strand, she has argued that we face threats from the complexity and opacity of this shift, chronicling tales of people trapped with Kafka-esque bureaucracies abetted by the rise of new technology.[27]

In my recent work, I have built upon Citron's arguments. I have highlighted the risk of significant harm inherent in massive databases of information, something I have called the 'database of ruin.'[28] I have discussed a method for assessing this risk, one borrowed from computer security, of building threat models designed to reveal the kind and intensity of harm people might be exposed to, risks not only of identity theft and invidious discrimination, but also traditional privacy harms such as stalking, harassment, and blackmail.[29] These threat models suggest the need to focus not only on external risks such as hackers and government surveillance but also internal risks such as graduate students 'peeking' at sensitive information to satisfy idle curiosity or worse.[30] Internal threats like these are much more difficult to guard against, according to computer scientists.[31]

While Citron and I have focused primarily on 'traditional' and individualistic harms of privacy, there is a rich body of literature focused on broader harms to both the individual and society. This work has been labeled the 'New Privacy' movement by some.[32] There are many strands to this work and not enough space to elaborate all of them fully. New Privacy scholars argue that society suffers under the panoptic effect of surveillance. Julie Cohen focuses on how people will be chilled from experimentation and play, and as a result stunted in their development as emergent individuals.[33] Neil Richards focuses in particular on the intellectual development of individuals, worrying most about data involving evidence of the intellectual products we consume: what we read, listen to, and watch.[34] Paul Schwartz argues that society suffers when it is not afforded private spaces to engage in a deliberative democracy.[35] Others see privacy as essential to preventing imbalances in power.[36]

Finally, I draw from the work of scholars and other commentators who worry about the dehumanizing effect of treating individuals merely as data. Daniel Solove contends that "the information in databases often fails to capture the texture of our lives. Rather than provide a nuanced portrait of our personalities, they capture the stereotypes and the brute facts of what we do without the reasons."[37] Jaron Lanier decries what he calls "antihuman software design."[38] Danielle Citron writes about the mistakes that come from substituting a human role in decision making.[39]

Prescriptions

I offer five proposals for reform, at varying levels of generality. First, any rules we create or expand should be calibrated to the sensitivity of the information stored in the database, meaning according to the risk of significant harm to individuals or groups. Certain types of databases are riskier than others, and I offer some first-cut rules for making these distinctions. Second, previous rules based on distinctions between PII and non-PII no longer make sense and need to be abandoned. Third, we should build and police walls or gaps between different datasets. Fourth, we should remind data analysts repeatedly that the numbers they analyze represent the lives of people. These reminders should scale up with the sensitivity of information, meaning those who work with the most sensitive information should face constant, even uncomfortable, reminders. Fifth, and finally, researchers must constantly assess the ethics of new practices. These ethical conversations need to be imposed or at least incentivized by rule or law. The need for conversations about ethics are particularly required at the interface of traditional and new modes of research funding and organization, to counteract the likely jealousy that will be felt within traditional research centers (such as universities) for the relatively thin set of rules governing research being done in private corporations. Consider each of these prescriptions in greater depth in turn.

Sensitive Information Not all information can be used to cause harm. Over hundreds of years of legal evolution, we have begun to recognize some categories as especially likely to cause harm. These include health, sex, financial, and educational information, to name only a few. In other work, I have examined closely what makes a category of information sensitive, and I have concluded that there are at least four things.[40]

First, sensitive information can be used to cause harm. Second, this risk of harm is substantial, not improbable nor speculative. Third, categories of sensitive information often involve relationships we consider confidential, such as doctor–patient or bank–customer. Finally, categories of sensitive information tend to reflect majoritarian sentiments, protecting us from harms suffered by many.[41]

As we extend privacy law to cover currently unregulated forms of data analysis, we should focus on contexts involving sensitive information, including new categories of information not yet deemed sensitive but deserving of the label. I have pointed to three new candidates: precise geolocation, genomic information, and biometric information. We should write new privacy laws covering these (and other) non-regulated forms of information.[42]

Ending the PII/Non-PII Distinction Every privacy law that has ever been written has rewarded the anonymization of data. In fact, most privacy laws do not apply at all to data that has been stripped of identifiers. In the typical parlance, privacy law usually applies only to data containing personally identifiable information, or PII. We should abandon this mode of regulation, recognizing instead that even apparently anonymized information often contains enough residual data to re-link to individuals. Privacy laws should continue to apply even to data that has been de-identified, at least for the most sensitive forms of data. But this is not to say that de-identification cannot continue to play a role, because sensible de-identification can greatly reduce the risk of troubling outcomes. Rather than treating all purportedly anonymized data as completely regulated, we might reconfigure our privacy laws to act more like sliding scales, reducing the requirements of the rules for sufficiently de-identified information, along lines suggested by Paul Schwartz and Dan Solove.[43]

Legislating Gaps We should use law to create gaps between contexts. Many theorists have written about the special problems of total surveillance. Julie Cohen in particular advocates the imposition of gaps between contexts, allowing people to carry out different experiments – to play – in order to learn and develop.[44]

One way to do this is to declare some modes or categories of information off-limits from data processing. In some sense, this is what we have done through what I have called 'protected channel laws.'[45] Protected channel laws deem some methods of surveillance specially protected. One example is the federal Wiretap Act, which declares it a felony to use a packet sniffer to acquire the contents of communications in transit across a telephone or computer network, absent some exception.[46]

In the context of big data and research, because of protected channel laws like the Wiretap Act, we probably understand less about what people say on the telephone than we do what they say in stored media such as social networks. A sociologist simply cannot obtain a massive database of the recorded content of telephone calls, even if he has a good reason to want it, and even if he can describe profound benefits he might develop from studying this kind of information. This inability should not be viewed as regrettable; it is instead a laudable example of using law to impose Cohen's gaps.

To some data scientists, gaps may seem anathema, contrary to core goals of big data, because they represent troubling blind spots. But if we believe in the importance of privacy gaps, we should try to change this attitude. A legally imposed privacy gap is a feature not a bug.

Even more importantly, we need to disabuse people of the idea that legally imposed gaps need to be highly tailored to prevent only narrowly defined, highly specific types of horrific privacy harm. As Cohen puts it, the goal of gap building is to enable unexpected advantage. "Evolving subjectivity, or the everyday practice of self, responds to the play-of-circumstances in unanticipated and fundamentally unpredictable ways. . . . [T]he play-of-circumstances operates as a potent engine of cultural dynamism, mediating both evolving subjectivity and evolving collectivity, and channeling them in unexpected ways."[47] This means that the gaps we define will not be the product of hyper-rationalized decision making, but instead that it will be based on inexact, human-scale approaches. As Cohen puts it, "privacy consists in setting of limits precisely where logic would object to drawing lines."[48] Or as I have said earlier, "we will be forced to carve out these gaps using machetes not scalpels."[49]

Reminding Researchers about Humanity Some of the greatest concerns about big data build upon fears about its dehumanizing effects. We worry that big data techniques will replace traditional, official modes of decision making about the lives of individuals. Even if these techniques lead to results that we may consider more efficient, we should worry that they allow too much distance between the decision makers – increasingly those who call themselves data scientists – and those whose lives they control. These fears are in part based on instrumental concerns: we worry about the unfairness of being trapped by the machine or in the machine.

We worry about being victims of officious bureaucrats. But the concerns are not merely about instrumental effects. Even if data science can lead to results that are efficient, viewed through a utilitarian lens, they yet may not be better if they result in treating people as dehumanized widgets.

A prophylactic step we might take is to design mechanisms to constantly remind those who operate on information about the people whose lives they assess, direct, or divulge. We should consider forcing data scientists to think about the lives of those analyzed. We should use techniques like 'visceral notice'[50] to send cues to data scientists that lives are being impacted. Perhaps more controversially, for studies on sensitive information, we should opt to make the very act of data analysis uncomfortable in subtle ways, to remind the analysts that they are operating on people.

I am informed in this admittedly unorthodox idea from my experience as a former employee of the federal government. Federal employees are subjected to stringent ethical rules, for example, rules limiting the type and size of gifts they may accept. The purposes of these rules are primarily instrumental: they are intended to prevent bribery, blackmail, and fraud infecting the decision making of the government. But, in my personal experience, they serve a secondary, possibly unintended but still laudable, purpose: they remind government workers that it is they who serve the citizenry, not the other way around. The life of a federal worker can sometimes seem practically ascetic next to his or her corporate colleagues; I am reminded of a time when I was forced to decline a voucher to buy a cafeteria lunch at a meeting held on the campus of a Silicon Valley tech giant, even though everybody else was allowed to accept. There was never the possibility that a $10 turkey sandwich would have caused me to abuse the public trust, but I was reminded when I was forced to refuse that my job carried special, important burdens that I would be unwise to forget.

I think it makes sense to try to replicate this experience, sometimes, for data scientists, particularly those working with sensitive information. Data scientists who access individual salary information, for example, or data containing past drug use or criminal convictions should be reminded by their work conditions that they wield great power. We should place them within bare walls and under flickering fluorescent lighting rather than ply them with foosball tables and free soda. In fact, there may be precedent for this idea. Some information is considered so sensitive, the only people permitted to analyze it are forced to come to the data rather than the other way around. One example occurs in the use of 'data rooms' to facilitate corporate mergers or acquisitions.[51] These are secure and highly monitored rooms that potential acquirers can visit to examine sensitive documents without being able to retain verbatim copies. Access to information stored by government agencies is often subjected to similarly strict controls. Even though the point of these measures is security rather than to remind the analyst of the humanity underneath the numbers, we might think of imposing conditions like these for this second reason too.

Developing Ethical Norms Finally, it is imperative that data scientists begin to debate, agree upon, and perhaps even codify ethical norms of behavior around responsible data science. Lawmakers and policymakers abhor an ethical vacuum, and if ever a serious, headline-grabbing violation of big data techniques should occur – the information equivalent of the Tuskegee syphilis study – the first question outsiders will ask is whether the practice violated the relevant community's internal ethical norms.

This prescription builds on the work of many others. Helen Nissenbaum situates norms as key to understanding contextual integrity (and determining when contextual integrity has been breached.)[52] Ryan Calo has called for learning from human subjects review in medical research for big data.[53]

This prescription can take one of many different forms. At the very least, it encourages informal deliberation within communities of researchers. Better yet, groups can try to identify people or institutions that can convene others to discuss norms formally.

Similar efforts have been undertaken before. Notably, the Belmont Report established the modern baseline for biomedical and behavioral research and served as a precursor to today's Common Rule and the rise of Institutional Review Boards.[54] Although not nearly as prominent, researchers in information and communication technology, under the auspices of the Department of Homeland Security, produced the Menlo Report, which attempted to provide ethical guidance for computer research involving human subjects.[55]

Conclusion

Today, it is fair to say that a wide swath of research being done or proposed involving big data techniques is entirely unregulated, and what little work is regulated is regulated only with respect to how the data is collected or subsequently disclosed. Although some might cheer this state of affairs, complimenting the government for staying out of the way of research, I do not. Data analysis is poised to become a central tool of science, and in some fields it already has, and the results produced will become the basis for decision making. As we expand the reach and power and influence of data science, we must take steps to prevent harm, to ensure that this remains always a humanistic endeavor, and to help people preserve their sense of power and autonomy.

NOTES

1. Secretary's Advisory Committee on Automated Personal Data Systems, *Records, Computers and the Rights of Citizens* (Washington, DC: U.S. Department of Health, Education and Welfare, 1973), 41–42.
2. Organisation for Economic Co-operation and Development, *OECD Guidelines on the Protection of Privacy and Transborder Flows of Personal Data* (September 23, 1980), http://www.oecd.org/internet/ieconomy/oecdguidelinesonthe protectionofprivacyandtransborderflowsofpersonaldata.htm.
3. Federal Trade Commission, *Privacy Online: Fair Information Practices in the Electronic Marketplace: A Federal Trade Commission Report to Congress*, Washington, DC, May 2000, http://www.ftc.gov/reports/privacy-online-fair-information-practices-electronic-marketplace-federal-trade-commission.
4. Fred Cate, "The Failure of the Fair Information Practice Principles," in *Consumer Protection in the Age of the Information Economy*, ed. Jane K. Winn (Surrey, UK: Ashgate, 2006), 356.
5. Ibid., 357.
6. Ibid., 360.
7. Paul M. Schwartz, "Preemption and Privacy," *Yale Law Journal* 118 (2009): 902.
8. Health Insurance Portability & Accountability Act (HIPAA), 45 C.F.R. §§ 164.501, et seq.
9. Family Educational Rights & Privacy Act, 20 U.S.C. § 1232g.
10. Gramm-Leach-Bliley Act, 15 U.S.C. § 6801, et seq.
11. 5 U.S.C. § 552a(b)(1).
12. 5 U.S.C. § 552a(e)(1).
13. 15 U.S.C. § 1681a(d).
14. See http://www.ftc.gov/opa/2012/02/mobileapps.shtm.
15. 15 U.S.C. § 1681b.
16. See http://privacyruleandresearch.nih.gov/pdf/HIPAA_Privacy_Rule_Booklet .pdf.
17. 45 C.F.R. § 164.512(i).
18. 45 C.F.R. § 164.514(e)(2).
19. "New Rules for Big Data," *The Economist,* February 25, 2010.
20. Secretary's Advisory Committee on Automated Personal Data Systems, *Records.*
21. Organisation for Economic Co-operation and Development, *OECD Guidelines.*
22. EU Data Protection Directive of 1995, Directive 95/46 EC of the European Parliament and the Council.
23. Working Party on the Protection of Individuals with Regard to the Processing of Personal Data, "Working Document on Transfers of Personal Data to Third Countries: Applying Articles 25 and 26 of the EU Data Protection Directive," DG XV D/5025/98 WP 12 (July 24, 1998).

24. Helen Nissenbaum, *Privacy in Context: Technology, Policy, and the Integrity of Social Life* (Stanford, CA: Stanford University Press, 2009); see also Chapter 2 in this volume, by Barocas and Nissenbaum.
25. Danielle Keats Citron, 'Reservoirs of Danger: The Evolution of Public and Private Law at the Dawn of the Information Age," *California Law Review* 80 (2007): 241.
26. Danielle Keats Citron, "Technological Due Process," *Washington University Law Review* 85 (2008): 1249.
27. Ibid. See also Daniel J. Solove, "Privacy and Power: Computer Databases and Metaphors for Information Privacy," *Stanford Law Review* 53 (2001): 1393.
28. Paul Ohm, "Broken Promises of Privacy: Responding to the Surprising Failure of Anonymization," *UCLA Law Review* 57 (2010): 1701.
29. Paul Ohm, "Sensitive Information?" forthcoming, manuscript on file.
30. Ibid.
31. S. Stolfo et al., eds., *Insider Attack and Computer Security: Beyond the Hacker* (New York: Springer, 2008).
32. See Paul M. Schwartz and William M. Treanor, "The New Privacy," *Michigan Law Review* 101 (2012): 2163–2181.
33. Julie Cohen, *Configuring the Networked Self: Law, Code, and the Play of Everyday Practice* (New Haven, CT: Yale University Press, 2012).
34. Neil Richards, "Intellectual Privacy," *Texas Law Review* 87 (2008): 387.
35. Paul Schwartz, "Internet Privacy and the State," *Connecticut Law Review* 32 (2000): 815.
36. Priscilla M. Regan, *Legislating Privacy: Technology, Social Values, and Public Policy* (Chapel Hill, NC: University of North Carolina Press, 1995).
37. Solove, "Privacy and Power."
38. Jaron Lanier, *You Are Not a Gadget: A Manifesto* (New York: Knopf, 2010), 193.
39. Citron, "Technological Due Process."
40. Ohm, "Sensitive Information?"
41. Ibid.
42. Ibid.
43. Paul M. Schwartz and Daniel J. Solove, "The PII Problem: Privacy and a New Concept of Personally Identifiable Information," *NYU Law Review* 86 (2011): 1814.
44. Cohen, "Configuring the Networked Self."
45. Ohm, "Sensitive Information?"
46. 18 U.S.C. § 2511.
47. Cohen, "Configuring the Networked Self."
48. Ibid.
49. Paul Ohm, "Mind the Gap," http://www.concurringopinions.com/archives/2012/03/mind-the-gap.html.
50. Ryan Calo, "Against Notice Skepticism in Privacy (and Elsewhere)," 87 Notre Dame Law Review 87 (2012): 1027.
51. See http://www.digitaldataroom.com/What-is-a-Data-Room.html.
52. Nissenbaum, *Privacy in Context.*
53. Ryan Calo, "Consumer Subject Review Boards: A Thought Experiment," 66 *Stanford Law Review Online* 66 (2013): 97.

54. "Belmont Report: Ethical Principals and Guidelines for the Protection of Human Subjects of Research," Report of the National Commission for the Protection of Human Subjects of Biomedical and Behavioral Research, 44 Fed. Reg. 23,192 (April 18, 1979).
55. "The Menlo Report: Ethical Principles Guiding Information and Communication Technology Research," Department of Homeland Security, 76 Fed. Reg. 81,517 (December 28, 2011).

5 Enabling Reproducibility in Big Data Research: Balancing Confidentiality and Scientific Transparency

Victoria Stodden

Introduction

The 21st century will be known as the century of *data*. Our society is making massive investments in data collection and storage, from sensors mounted on satellites down to detailed records of our most mundane supermarket purchases. Just as importantly, our reasoning about these data is recorded in software, in the scripts and code that analyze this digitally recorded world. The result is a deep digitization of scientific discovery and knowledge, and with the parallel development of the Internet as a pervasive digital communication mechanism we have powerful new ways of accessing and sharing this knowledge. The term *data* even has a new meaning. Gone are the days when scientific experiments were carefully planned prior to data collection. Now the abundance of readily available data creates an observational world in itself suggesting hypotheses and experiments to be carried out after collection, curation, and storage of the data has already occurred. We have departed from our old paradigm of data collection to resolve research questions – nowadays, we collect data simply *because we can*.

In this chapter I outline what this digitization means for the independent verification of scientific findings from these data, and how the current legal and regulatory structure helps and hinders the creation and communication of reliable scientific knowledge.[1] Federal mandates and laws regarding data disclosure, privacy, confidentiality, and ownership all influence the ability of researchers to produce openly available and reproducible research. Two guiding principles are suggested to accelerate research in the era of big data and bring the regulatory infrastructure in line with scientific norms: the Principle of Scientific Licensing and the Principle of Scientific Data and

Code Sharing. These principles are then applied to show how intellectual property and privacy tort laws could better enable the generation of verifiable knowledge, facilitate research collaboration with industry and other proprietary interests through standardized research dissemination agreements, and give rise to dual licensing structures that distinguish between software patenting and licensing for industry use and open availability for open research. Two examples are presented to give a flavor of how access to data and code might be managed in the context of such constraints, including the establishment of 'walled gardens' for the validation of results derived from confidential data, and early research agreements that could reconcile scientific and proprietary concerns in a research collaboration with industry partners.

Technological advances have complicated the data privacy discussion in at least two ways. First, when datasets are linked together, a richer set of information about a subject can result but so can an increased risk of a privacy violation. Linked data presents a challenging case for open scientific research, in that it may permit privacy violations from otherwise non-violating datasets. In this case privacy tort law is suggested as a viable remedy for privacy violations that arise from linking datasets.

Second, the subjects of studies are becoming more knowledgeable about privacy issues, and may wish to opt for a greater level of access to their contributed data than that established by traditional research infrastructures, such as Institutional Review Boards. For data collection and release that happens today, research subjects have very little say over the future openness of their data. A suggestion is made to permit individuals to share their own data with provisions regarding informed consent. For example, an enrollee in a clinical trial for a new Crohn's disease treatment may wish to permit other Crohn's researchers access to the data arising from her participation, perhaps in an effort to help research advance in an area about which she cares deeply. At the moment, this is not only nonstandard, but downstream data use is difficult for the participant to direct.

Ownership itself can be difficult to construe since many resources typically go into creating a useful dataset, from research scientists who design the experiment, to data collectors, to participants, to curators, to industry collaborators, to institutes and funding agencies that support the research, further complicating the discussion of data and code access. Data access becomes increasingly complex, underscoring the need for a broad understanding of the value of maximizing open access to research data and code.

Trust and Verify: Reliable Scientific Conclusions in the Era of Big Data

Scientific research is predicated on an understanding of scientific knowledge as a public good – this is the rationale underlying today's multibillion-dollar subsidies of scientific research through various federal and state agencies. The scientific view is not one of adding nuggets of truth to our collective understanding, but instead one of weighing evidence and assigning likelihoods to a finding's probability of being true. This creates a normative structure of skepticism among scientists: the burden is on the discovering scientist to convince others that what he or she has found is more likely to be correct than our previous understanding. The scientific method's central motivation is the *ubiquity of error* – the awareness that mistakes and self-delusion can creep in absolutely anywhere and that the scientist's effort is primarily expended in recognizing and rooting out error. As a result, standards of scientific communication evolved to incorporate full disclosure of the methods and reasoning used to arrive at the proffered result.

The case for openness in science stems from Robert Boyle's exhortations in the 1660s for standards in scientific communication. He argued that enough information should be provided to allow others in the field to independently reproduce the finding, creating both the greatest chance of the accurate transmission of the new discoveries and also maximizing the likelihood that errors in the reasoning would be identified. Today, communication is changing because of the pervasive use of digital technology in research. Digital scholarly objects such as data and code have become essential for the effective communication of computational findings. Computations are frequently of such a complexity that an explanation sufficiently detailed to enable others to replicate the results is not possible in a typical scientific publication. A solution to this problem is to accompany the publication with the code and data that generated the results and communicate a *research compendium*.[2] However, the scientific community has not yet reached a stage where the communication of research compendia is standard.[3] A number of delicate regulatory and policy changes are essential to catalyze both scientific advancement and the development of applications and discoveries outside academia by making the data and code associated with scientific discoveries broadly available.

Challenge 1: Intellectual Property Law and Access to Digital Scholarly Objects

The Intellectual Property Clause of the United States Constitution has been interpreted to confer two distinct powers, the first providing the basis for copyright law: Securing for a limited time a creator's exclusive right to their original work;[4] and the second giving the basis for patent law: Endowing an inventor with a limited-term exclusive right to use their discoveries in exchange for disclosure of the invention. In this section the barrier copyright creates to open reproducible research will be discussed first, then the role of patent law in potentially obfuscating computational science.

Creators do not have to apply for copyright protection, as it adheres automatically when the original expression of the idea is rendered in fixed form. Many standard scientific activities, such as writing a computer script to filter a dataset or fit a statistical model, will produce copyrighted output, in this case the code written to implement these tasks. Building a new dataset through the original selection and arrangement of data will generate ownership rights through copyright for the dataset creator, to give another example.[5] The default nature of copyright creates an *intellectual property* framework for scientific ideas at odds with longstanding scientific norms in two key ways.[6] First, by preventing copying of the research work it can create a barrier to the legal reproduction and verification of results.[7] Second, copyright also establishes rights for the author over the creation of derivative works. Such a derivative work might be something as scientifically productive as, say, the application of a software script for data filtering to a new dataset, or the adaptation of existing simulation codes to a new area of research.

As computation becomes central to scientific investigation, copyright on code and data become barriers to the advancement of science. There is a copyright exception, titled *fair use,* which applies to "teaching (including multiple copies for classroom use), scholarship, or research"[8] but this does not extend to the full research compendium including data, code, and research manuscript. In principle, a relatively straightforward solution to the barrier copyright imposes would be to broaden the fair use exception to include scientific research that takes place in research institutions such as universities or via federal research grants; however, this is extremely challenging in practice.[9] Distinguishing legal fair use is not a clear exercise in any event, and an extension to research more broadly may still not sufficiently clarify rights. A more practical mechanism for realigning intellectual property rights with scientific norms is the Reproducible Research Standard (RRS), applying appropriate open licenses to remove restrictions on copying and reuse of the scientific work, as well as possibly adding an attribution requirement to elements of the research compendium. Components of the research compendium have different features that necessitate different licensing approaches and a principle for licensing scientific digital objects can guide choices:

Principle of Scientific Licensing *Legal encumbrances to the dissemination, sharing, use, and reuse of scientific research compendia should be minimized, and require a strong and compelling rationale before their application.*[10]

For media components of scientific work, the Reproducible Research Standard suggests the Creative Commons attribution license (CC BY), which frees the work for replication and re-use without prior author approval, with the condition that attribution must accompany any downstream use of the work.

Many licenses exist that allow authors to set conditions of use for their code. In scientific research code can consist of scripts that are essentially stylized text files (such as python or R scripts) or the code can have both a compiled binary form and a source representation (such as code written in C). Use of the CC BY license for code is discouraged by Creative Commons.[11] The Reproducible Research Standard suggests the Modified Berkeley Software Distribution (BSD) license, the MIT license, or the Apache 2.0 license, which permit the downstream use, copying, and distribution of either unmodified or modified source code, as long as the license accompanies any distributed code and the previous authors' names are not used to promote modified downstream code.[12] The Modified BSD and MIT licenses differ in that the MIT license does not include a clause forbidding endorsement.[13] The Apache 2.0 license differs in that it permits the exercise of patent rights that would otherwise extend only to the original licensor, meaning that a patent license is granted for those patents needed for use of the code.[14] The license further stipulates that the right to use the work without patent infringement will be lost if the downstream user of the code sues the licensor for patent infringement.

Collecting, cleaning, and preparing data for analysis can be a significant component of empirical scientific research. Copyright law in the United States forbids the copyrighting of 'raw facts' but original products derived from those facts can be copyrightable. In *Feist Publications,*

Inc. v. Rural Telephone Service, the Court held that the *original* "selection and arrangement" of databases is copyrightable:[15] the component falling under copyright must be original in that "copyright protection extends only to those components of the work that are original to the author, not to the facts themselves."[16] Attaching an attribution license to the original "selection and arrangement" of a database may encourage scientists to release the datasets they have created by providing a legal framework for attribution and reuse of the original selection and arrangement aspect of their work.[17] Since the raw facts themselves are not copyrightable, such a license cannot be applied to the data themselves. The selection and arrangement may be implemented in code or described in a text file accompanying the dataset, either of which can be appropriately licensed. Data can however be released to the public domain by marking with the Creative Commons CC0 standard.[18]

This licensing structure that makes the total of the media, code, data components – the research compendium – available for reuse, in the public domain or with attribution, is labeled the *Reproducible Research Standard*.

Patent law is the second component of intellectual property law that affects the disclosure of scientific scholarly objects. In 1980 Congress enacted two laws, the Stevenson-Wydler Act and the Bayh-Dole Act, both intended to promote the commercial development of technologies arising from federally funded research. This was to be facilitated through licensing agreements between research entities, such as universities, and for-profit companies. The Bayh-Dole Act explicitly gave federal agency grantees and contractors, most notably universities and research institutions, title to government-funded inventions and charged them with using the patent system to disclose and commercialize inventions arising in their institution. In 2009 this author carried out a survey of computational scientists, in order to understand why they either shared or withheld the code and data associated with their published papers. In the survey one senior professor explained that he was not revealing his software because he was currently seeking a patent on the code.[19] In fact, 40% of respondents cited patent seeking or other intellectual property constraints as a reason they were not sharing the code associated with published scientific results.[20] Rates of software patenting by academic institutions have been increasing over the last decade, posing a potentially serious problem for scientific transparency and reproducibility.[21] Instead of ready access to the code that generated published results, a researcher may be required to license access to the software through a university's technology transfer office, likely being prohibitively expensive for an academic scientist in both time and money. In December of 1999, the National Institutes of Health stated that

> the use of patents and exclusive licenses is not the only, nor in some cases the most appropriate, means of implementing the [Bayh-Dole] Act. Where the subject invention is useful primarily as a research tool, inappropriate licensing practices are likely to thwart rather than promote utilization, commercialization, and public availability.[22]

The federal funding agencies are without authority to issue regulations regarding patentable inventions, and the NIH viewpoint above does not appear to have been adopted by technology transfer offices at the university and institutional research level. A typical interpretation is that of Columbia University, where this author is employed, which follows: "The University claims, as it may fairly and rightfully do, the commercial rights in conceptions that result primarily from the use of its facilities or from the activity of members of its faculty while engaged in its service."[23] Not all universities make such an a priori claim to determine the patenting and licensing fate of research inventions. For example, Stanford University's *Research Policy Handbook* says that as a researcher, "I am free to place my inventions in the public domain as long as in so doing neither I nor Stanford violates the terms of any agreements that governed the work done."[24] The Bayh-Dole Act also grants agencies 'march-in' rights to obtain intellectual property (presumably to grant nonexclusive licenses, but not necessarily), but the process is long with multiple appeal opportunities. In July of 2013, however, in a letter to Francis Collins, head of the NIH, Senator Leahy recommended the use of march-in rights on patented breast cancer genetic research "to ensure greater access to genetic testing for breast and ovarian cancer."[25]

Challenge 2: Scale, Confidentiality, and Proprietary Interests

Even without intellectual property law encumbrances to the dissemination of digital scholarly objects, other barriers can create obstacles to access. For example, the sheer size of many datasets may require specialized computational infrastructure to permit access, or scale itself can even prohibit access. For example, the July 2013 release of the Sloan Digital Sky Survey (SDSS) is 71.2 terabytes in size, making a conventional download of data to a personal laptop impossible.[26] The approach of the SDSS is to create different websites for different data types, and provide a variety of tools for access including SkyServer SQL search, CasJobs, and Schema Browser, each with a different purpose in mind.[27] This infrastructure permits search and user-directed access to significantly smaller subsets of the entire database.

In some fields however even 70 terabytes would not seem large. CERN director general Rolf Heuer said in 2008 that, "[t]en or 20 years ago we might have been able to repeat an experiment. They were simpler, cheaper and on a smaller scale. Today that is not the case. So if we need to re-evaluate the data we collect to test a new theory, or adjust it to a new development, we are going to have to be able reuse it. That means we are going to need to save it as open data."[28] In March of 2013, the CERN data center passed a storage milestone by exceeding 100 petabytes of data.[29] It is not clear how this can be made open data in the sense discussed in this chapter, as Director Heuer suggests. The traditional approaches to making data and code available seem intractable for such datasets at the present time. I use these examples to introduce a Principle of Scientific Data and Code Sharing:

Principle of Scientific Data and Code Sharing *Access to the data and methods associated with published scientific results should be maximized, only subject to clearly articulated restrictions such as: privacy or confidentiality concerns, legal barriers including intellectual property or HIPAA regulations, or technological or cost constraints.*

This principle can also be phrased as 'Default to Open', meaning that it takes compelling and convincing reasons, articulated in detail (i.e. the precise section of HIPAA that is restricting disclosure, or the part of intellectual property law that is creating a barrier) to close data and code from public access.[30,31] A careful defense of any deviation from full openness will have the effect of maximizing the availability of data and code. A corollary effect is an uncovering of the reasons for not sharing data and presumably a greater understanding of the precise nature of legal barriers to disclosure and their appropriateness given the nature of the underlying data and code.[32] Sequestering a dataset due to 'confidentiality', with no further justification, should no longer be acceptable practice.

The second corollary from the Principle of Scientific Data and Code Sharing is that it implies *levels* of access. Whether due to privacy concerns, technological barriers, or other sources, restrictions on data and code availability do not necessarily imply absolute barriers. In the case of CERN, internal research design mechanisms exist to make up for some of the shortcomings in openness of research data and the inability of independent groups to verify findings obtained from empirical data. Specifically, either independent research groups within CERN access the data from the collider and carry out the research in isolation from each other, or the same group will verify analyses using independent toolsets.[33] Of crucial importance, these internal groups have access to the infrastructure and technologies needed to understand and analyze the data. In this case, there has been some openness of the data and the use of independent parallel research increases the chances of catching errors, all improvements over the more commonly seen research context where the data are accessed only by the original researchers and analyzed without any reported validation or verification cross-checks.

A second illustrative example originates from the Wandell Lab in the Psychology Department at Stanford University. Brian Wandell, the Isaac and Madeline Stein Family Professor of Psychology, has an MRI machine for his lab research. For the lifetime of the machine, each image has been carefully stored in a database with metadata including equipment settings, time, date, resolution, and other pertinent details of the experimental setup. The output image data are, however, subject to HIPAA regulations in that each image is a scan of a subject's brain and therefore privacy restrictions prevent these from being made publicly available. The Wandell Lab belongs to a consortium with several other research groups at different universities in California. In order to permit some potential verification of results based upon these images, there is no legal barrier to giving researchers within these authorized groups access to the database, and thereby creating the possibility for independent cross-checking of findings inside this 'walled garden'. While this does not achieve the same potential for finding errors as open release would (more eyes making more bugs increasingly shallow), it satisfies the Principle of Scientific Data and Code Sharing by maximizing access subject to the inherent legal constraints with which the data are endowed. Although the implementation details may differ for different data, understanding and developing infrastructure to facilitate these middle-ground data access platforms or walled gardens, will be essential for the reliability of results derived from confidential data.[34] One could also cast the CERN approach as a type of walled garden since it is characterized by independent research on the same question on closed data, carried out by different internal groups.

Another potential barrier to data and code release derives from collaboration with partners who may be unwilling to release the data and software that arise from the project, and may not be academic researchers bound by the same notions of scientific transparency. For example, industry research partners do not necessarily have the goal of contributing their research findings to the public good, but are frequent research collaborators with academics. A conflict can ensue, for example, at the point of publication when the academic partner wishes to publish the work in a journal or a conference proceedings that requires data and code disclosure, or when the researcher simply wishes to practice really reproducible research and make the data and code openly available.[35] One possible solution is to offer template agreements for data and code disclosure at the beginning of the collaboration, possibly through the institution's technology transfer office or through funding agency access policy.[36] Unfortunately the issue of data and code access is often ignored until the point at which one party would like to make them available after the research has been completed.[37]

When a patent is being sought on the software associated with the research, broader access can be achieved by implementing patent licensing terms that distinguish between commercial and research applications, in order to permit reuse and verification by researchers, while maintaining the incentives for commercialization and technology transfer provided by the Bayh-Dole Act. The Stanford Natural Language Processing Group for example uses such a dual licensing strategy. Their code is available for download by researchers under an open license and groups that intend commercial reuse must pay licensing fees.[38]

Challenge 3: Linked Data and Privacy Tort Law

Access to datasets necessarily means data with common fields can and will be linked. This is very important for scientific discovery as it enriches subject-level knowledge and opens new fields of inquiry, but it comes with risks such as revealing private information about individuals that the datasets in their isolated, unlinked form would not reveal. As has been widely reported, data release is now mandated for many government agencies through Data.gov. In 2009 Vivek Kundra, then–federal chief information officer,[39] was explicit – saying that, "the dream here is that you have a grad student, sifting through these datasets at three in the morning, who finds, at the intersection of multiple datasets, insight that we may not have seen, or developed a solution that we may not have thought of." On February 22, 2013, the Office of Science and Technology Policy directed federal agencies with significant research budgets to remit plans to make data arising from this research openly available.[40] This includes academic research funded by the National Science Foundation and the National Institutes for Health, for example.

An instructive example about the privacy risks from data linking that Kundra describes comes from the release of genomic information. An individual's genomic information could be uncovered by linking their relatives' genomic information together, when this individual has not shared any of his or her genetic information directly. Recall, we carry 50% of the DNA from each of our parents and children, and an average of 50% from each of our siblings. Privacy risks could include, for example, an insurance company linking the genetic signature information to medical records data, possibly through a genetic diagnostic test that was performed, and then to other insurance claims, for individuals whose relatives had made their DNA available though they themselves did not.[41] A number of cities are releasing data, for example public school performance data, social service agency visits, crime reports, and other municipal data, and there has been controversy over appropriate privacy protection for some of these data.[42] Research that links these datasets may have laudable aims – better understanding the factors that help students succeed in their education – but the risks to linking datasets can include privacy violations for individuals.

Much of the policy literature around privacy in digitally networked environments refers to corporate or government collected data used for commercial or policy ends.[43] Insufficient attention has been paid to the compelling need for access to data for the purposes of verification of data-driven research findings. This chapter does not advocate that the need for reproducibility should trump privacy rights, but instead that scientific integrity should be part of the discussion of access to big data, including middle ground solutions such as those as discussed earlier in this chapter.

Traditional scientists in an academic setting are not the only ones making inferences from big data and linked data, as Chapters 6 and 7 in this volume show. The goal of better decision making is behind much of the current excitement surrounding big data, and supports the emergence of 'evidence-based' policy, medicine, practice, and management. For conclusions that enter the public sphere, it is not unreasonable to expect that the steps that generated the knowledge be disclosed to the maximal extent possible, including making the data they are based on available for inspection, and making the computer programs that carried out the data analysis available.

We cannot know how data released today, even data that all would agree carry no immediate privacy violations, could help bring about privacy violations when linked to other datasets in the future. These other datasets may not be released, or even imagined, today. It is impossible to guard completely against the risk of these types of future privacy violations. For this reason a tort-based approach to big data access and privacy is an important alternative to creating definitive guidelines to protect privacy in linked data. Perhaps not surprisingly, however, privacy tort law developed in the pre-digital age and is not a perfect fit with today's notions of privacy protection and big data access.

Much of the current scholarly literature frames the online privacy violation question as protection again defamation or the release of private information by others, and does not explicitly consider the case of linked data. For example, privacy torts are often seen as redress for information made available online, without considering the case of harm from new information derived from combination of non-private sources. This can happen in the case of data linking, as described above, but differs in that a privacy violation can be wholly inadvertent and unforeseen, and may not be individually felt but can affect an entire class of people (those in the dataset). This, along with persistence of privacy-violating information on the web, changes the traditional notion of an individual right to privacy.[44] In current privacy tort law one must establish that the offender intended to commit a privacy invasion,[45] that the conduct was "highly offensive to the reasonable person," and that the information revealed was sufficiently private.[46]

Current privacy tort law protects against emotional, reputational, and proprietary injuries caused by any of: a public disclosure of private facts; an intrusion on seclusion; a depiction of another in a false light, or an appropriation of another's image for commercial enrichment.[47] Articulating privacy rights in big data and linked data founders on accountability since it is unlike securing private (real) property or a landlord ensuring his or her building is secured.[48] Potential privacy violations deriving from linked data cannot always be foreseen at the time of data release. The Principle of Scientific Data and Code Sharing frames a possible way forward: research data that does not carry any immediate privacy violations should be released (and otherwise released in a way that makes the data maximally available for independent reproducibility purposes that safeguards privacy); linked datasets should either be released or the methods to link the datasets should be released with caveats to check for new privacy violations; and if privacy violations still arise, redress could be sought through the tort system. If tort law responds in a way that matches our normative expectations regarding privacy in data, this will permit a body of law to grow around big data that protects privacy. In order for this to be effective, a broadening of tort law beyond the four types of privacy-violating behaviors needs to occur. Harms arising from the release of private information derived from data, and from linked data, could be included in the taxonomy of privacy torts. These may not be intentioned or foreseeable harms, and may potentially be mass torts as datasets with confidentiality violations are likely to contain records on a large number of people. The issue of liability and responsibility for privacy violations becomes more complex than in the past, and there may be chilling effects on the part of institutions and funding agencies with regard to open data. Finally, making code and data available is not costless as databases and access software can cost a considerable amount of money, and innovative middle-ground solutions that may be project specific can add to that cost.[49]

Research data poses yet another unique challenge to privacy law. Many research collaborations exist across international boundaries, and it is common for some members of a research team to be more heavily involved with the associated data than other members. Access to data on the Internet is not generally restricted by country and enforcing privacy violations across international borders poses a considerable challenge for scientific research. Data and code must be made available to maximally permit verification, subject to privacy and other barriers, and these data may be accessible from anywhere in the world through the Internet. Privacy violations from linked data can thus occur in countries with more stringent privacy standards though the release of the data may occur in a country that does not have a mechanism for legal redress of privacy violations.

Challenge 4: Changing Notions of Data Ownership and Agency

The notion of a data owner is a rapidly changing concept as many entities contribute to dataset creation, increasing the complexity of the data-sharing issue. Data is collected both by people and by automated systems such as sensor arrays, and goes through myriad processing in the course of information extraction. Different entities may carry out data cleaning and filtering, data curation and warehousing, facilitation of data access, recombination of datasets to create novel databases, or preservation and provenance through repositories and institutions – each possibly creating intellectual property and ownership rights in the data. There is a similar story for research code, as it evolves through different applications and extensions by different people and becomes an amalgam of many contributions. The open release of data and code means untangling ownership and tracking contributions. Versions of code and data are vitally important for reproducibility – as code is modified, even as bugs are fixed, or data are extended, corrected, or combined, it is important to track which instantiation produced which scientific findings.

There is a new source of potential ownership as well. Subjects in a study can feel a sense of ownership over information about themselves, including medical descriptions or choices they have made. It is becoming increasingly the case that study participants wish to direct the level of access to data about themselves and traditional notions of privacy protection may not match their desires. Some data owners would prefer that data about themselves, that might traditionally be considered worthy of privacy protection such as medical data or data resulting from clinical trials participation, should be made more fully available.[50] As noted in a World Economic Forum Report, "[o]ne of the missing elements of the dialogue around personal data has been how to effectively engage the individual and give them a voice and tools to express choice and control over how data about them are used."[51,52] Traditional mechanisms, such as the Institutional Review Board or national laboratory policy, may be overprotecting individuals at the expense of research progress if they are not taking individual agency into account.

These changing notions of ownership can impede sharing, if permission from multiple parties is required to grant open access, or to relinquish data, or even to simply participate in the development of infrastructure to support access. A careful assessment of ownership and contributions to dataset development will inform liability, in the case of breaches of privacy. While some of this assessment and tracking is done today for some datasets, for the majority of datasets there is very little provenance available and little clarity regarding ownership rights.

Conclusion

The goal of this chapter is to bring the consideration of scientific research needs to the discussion around data disclosure and big data. These needs comprise a variety of issues, but a primary one is the need for independent verification of results, for reproducibility from the original data using the original code. This chapter asserts two principles to guide policy thinking in this area: the **Principle of Scientific Licensing**, that legal encumbrances to the dissemination, sharing, use, and re-use of scientific research compendia should be minimized, and require a strong and compelling rationale before their application; and the **Principle of Scientific Data and Code Sharing**, that access to the data and methods associated with published scientific results should be maximized, only subject to clearly articulated restrictions interpreted in the most minimally restrictive way, including intellectual property or HIPAA restrictions, or technological or cost constraints.

The chapter outlines intellectual property barriers to the open release of research data and code, and proposes open licensing solutions. Templated sharing agreements are suggested to guide data and code release at the beginning of collaboration with industry partners who may have a different agenda to the open sharing of data and code that arise from the research. The chapter also argues for dual licensing of patented research code: license fees for commercial reuse, and open availability for academic research purposes. To address privacy and confidentiality in sharing there must be a move to maximize openness in the face of these concerns. Sharing within a group of authorized researchers in the field, or with scientists who have sought permission, can create a 'walled garden' that, while inferior to open sharing, can still obtain some of the properties and benefits of independent verification that is possible from public access. 'Middle-ground' platforms such as walled gardens are possible solutions to maximize the reliability of scientific findings in the face of privacy and confidentiality concerns.

The linking of open data sets is framed as an open-ended threat to privacy. Individuals may be identified through the linking of otherwise non-identifiable data. Since these linkages cannot, by definition, be foreseen and are of enormous benefit to research and innovation, the use of privacy tort law is suggested both to remedy harm caused by such privacy violations and to craft a body of case law that follows norms around digital data sharing.

Finally, privacy can be an overly restrictive concept, both legally and as a guiding principle for policy. Data ownership can be difficult to construe since many resources can create a useful dataset, and individuals may prefer to release what might be considered private information by some. In the structure of data collection and release today, such individuals have very little say over the future openness of their data. A sense of agency should be actively restored to permit individuals to share data.

Some of the concern about open data stems from the potential promulgation of misinformation as well as perceived privacy risks. In previous work I have labeled that concern 'Taleb's Criticism'.[53] In a 2008 essay, Taleb worries about the dangers that can result from people using statistical methodology without having a clear understanding of the techniques.[54]

An example of Taleb's Criticism appeared on UCSF's EVA website, a repository of programs for automatic protein structure prediction.[55] The UCSF researchers refuse to release their code publicly because, as they state on their website, "[w]e are seriously concerned about the 'negative' aspect of the freedom of the Web being that any newcomer can spend a day and hack out a program that predicts 3D structure, put it on the web, and it will be used." However, an analogy can be made to early free speech discussions that encouraged open dialog. In a well-known quote Justice Brandeis elucidated this point in *Whitney v. California* (1927), writing that "If there be time to expose through discussion the falsehood and fallacies, to avert the evil by the processes of education, the remedy to be applied is more speech, not enforced silence." In the open data discussion this principle can be interpreted to favor a deepening of the dialog surrounding research, which is in keeping with scientific norms of skepticism and the identification of errors. In the case of the protein structure software, the code remains closed and a black box in the process of generating research results.[56]

Increasing the proportion of verifiable published computational science will stem from changes in four areas: funding agency policy, journal publication policies, institutional research policies, and the attitudes of scientific societies and researchers themselves. Although there have been significant recent advances from each of these four stakeholder groups, changing established scientific dissemination practices is a collective action problem. Data and code sharing places additional burdens on all these groups, from curation and preparation through to hosting and maintenance, which go largely unrewarded in scientific careers and advancement. These burdens can be substantial for all stakeholders in terms of cost, time, and resources. However, the stakes are high. Reliability of the results of our investments in scientific research, the acceleration of scientific progress, and the increased availability of scientific knowledge are some of the gains as we begin to recognize the importance of data and code access to computational science.

Acknowledgement I would like to thank two anonymous and extraordinarily helpful reviewers. This research was supported by Alfred P. Sloan Foundation award number PG004545 "Facilitating Transparency in Scientific Publishing" and NSF award number 1153384 "EAGER: Policy Design for Reproducibility and Data Sharing in Computational Science."

NOTES

1. Because of the wide scope of data considered in this article, the term *computational science* is used in a very broad sense, as any computational analysis of data.

See V. Stodden, "Resolving Irreproducibility in Empirical and Computational Research," *IMS Bulletin*, November 2013, for different interpretations of reproducibility for different types of scientific research.
2. R. Gentleman and D. Temple Lang, "Statistical Analyses and Reproducible Research," Bioconductor Working Series, 2004. Available at http://biostats.bepress.com/bioconductor/paper2/.
3. D. Donoho, A. Maleki, I. Ur Rahman, M. Shahram, and V. Stodden, "Reproducible Research in Computational Harmonic Analysis," *Computing in Science and Engineering* 11, no. 1 (2009): 8–18. Available at http://www.computer.org/csdl/mags/cs/2009/01/mcs2009010008-abs.html.
4. For a discussion of the Copyright Act of 1976 see e.g. Pam Samuelson, "Preliminary Thoughts on Copyright Reform Project," *Utah Law Review* 2007 (3): 551–571. Available at http://people.ischool.berkeley.edu/~pam/papers.html.
5. See Feist Publications Inc. v. Rural Tel. Service Co., 499 U.S. 340 (1991) at 363–364.
6. For a detailed discussion of copyright law and its impact on scientific innovation see V. Stodden, "Enabling Reproducible Research: Licensing for Scientific Innovation," *International Journal for Communications Law and Policy*, no. 13 (Winter 2008–09). Available at http://www.ijclp.net/issue_13.html.
7. See V. Stodden "The Legal Framework for Reproducible Scientific Research: Licensing and Copyright," *Computing in Science and Engineering* 11, no. 1 (2009): 35–40.
8. U.S. 17 Sec. 107.
9. This idea was suggested in P. David, "The Economic Logic of 'Open Science' and the Balance between Private Property Rights and the Public Domain in Scientific Data and Information: A Primer." Available at http://ideas.repec.org/p/wpa/wuwpdc/0502006.html. For an analysis of the difficulty of an expansion of the fair use exception to include digital scholarly objects such as data see J. H. Reichman and R. L. Okediji, "When Copyright Law and Science Collide: Empowering Digitally Integrated Research Methods on a Global Scale," *Minnesota Law Review* 96 (2012): 1362–1480. Available at http://scholarship.law.duke.edu/faculty_scholarship/267.
10. A research *compendium* refers to the triple of the research article, and the code and data that underlies its results. See Gentleman and Temple Lang, "Statistical Analyses and Reproducible Research."
11. See "Can I Apply a Creative Commons License to Software?" http://wiki.creativecommons.org/FAQ.
12. http://opensource.org/licenses/bsd-license.php.
13. http://opensource.org/licenses/mit-license.php.
14. http://www.apache.org/licenses/LICENSE-2.0.
15. Miriam Bitton, "A New Outlook on the Economic Dimension of the Database Protection Debate," *IDEA: The Journal of Law and Technology* 47, no. 2 (2006): 93–169. Available at http://ssrn.com/abstract=1802770.
16. Feist v. Rural, 340. The full quote is "Although a compilation of facts may possess the requisite originality because the author typically chooses which facts to include, in what order to place them, and how to arrange the data so that readers

may use them effectively, copyright protection extends only to those components of the work that are original to the author, not to the facts themselves. . . . As a constitutional matter, copyright protects only those elements of a work that possess more than de minimis quantum of creativity. Rural's white pages, limited to basic subscriber information and arranged alphabetically, fall short of the mark. As a statutory matter, 17 U.S.C. Sec. 101 does not afford protection from copying to a collection of facts that are selected, coordinated, and arranged in a way that utterly lacks originality. Given that some works must fail, we cannot imagine a more likely candidate. Indeed, were we to hold that Rural's white pages pass muster, it is hard to believe that any collection of facts could fail."

17. See A. Kamperman Sanders, "Limits to Database Protection: Fair Use and Scientific Research Exemptions," *Research Policy* 35 (July 2006): 859, for a discussion of the international and WIPO statements of the legal status of databases.
18. For details on the CC0 protocol see http://creativecommons.org/press-releases/entry/7919.
19. V. Stodden, "The Scientific Method in Practice: Reproducibility in the Computational Sciences," MIT Sloan School Working Paper 4773-10, 2010. Available at http://papers.ssrn.com/sol3/papers.cfm?abstract_id=1550193.
20. Ibid.
21. V. Stodden and I. Reich, "Software Patents as a Barrier to Scientific Transparency: An Unexpected Consequence of Bayh-Dole," Conference on Empirical Legal Studies, 2012. Available at http://papers.ssrn.com/sol3/papers.cfm?abstract_id=2149717.
22. See National Institutes of Health, "Principles for Recipients of NIH Research Grants and Contracts on Obtaining and Disseminating Biomedical Research Resources: Request for Comments." Available at http://www.ott.nih.gov/policy/rt_guide.html.
23. See Columbia University, *Faculty Handbook*, Appendix D: "Statement of Policy on Proprietary Rights in the Intellectual Products of Faculty Activity." Available at http://www.columbia.edu/cu/vpaa/handbook/appendixd.html (accessed August 21, 2013).
24. See Stanford University, *Research Policy Handbook*. Available at http://doresearch.stanford.edu/sites/default/files/documents/RPH%208.1_SU18_Patent%20and%20Copyright%20Agreement%20for%20Personnel%20at%20Stanford.pdf (accessed August 21, 2013).
25. "Leahy Urges Action to Ensure Access to Affordable Life-Saving Diagnostic Tests for Breast and Ovarian Cancer" (press release). See http://www.leahy.senate.gov/press/leahy-urges-action-to-ensure-access-to-affordable-life-saving-diagnostic-tests-for-breast-and-ovarian-cancer (accessed August 21, 2013).
26. See http://www.sdss3.org/dr10/.
27. See http://www.sdss3.org/dr10/data_access/, including http://skyserver.sdss3.org/dr10/en/help/docs/sql_help.aspx, http://skyserver.sdss3.org/CasJobs/, and http://skyserver.sdss3.org/dr10/en/help/browser/browser.aspx (accessed August 23, 2013).
28. "In Search of the Big Bang," *Computer Weekly*, August 2008. Available at http://www.computerweekly.com/feature/In-search-of-the-Big-Bang (accessed August 23, 2013).

29. "CERN Data Centre Passes 100 Petabytes," *CERN Courier*, March 28, 2013. Available at http://cerncourier.com/cws/article/cern/52730 (accessed August 23, 2013). 100 petabytes is about 100 million gigabytes or 100,000 terabytes of data. This is equivalent to approximately 1500 copies of the Sloan Digital Sky Survey.
30. See D. H. Bailey, J. Borwein, and V. Stodden, "Set the Default to 'Open'," *Notices of the American Mathematical Society*, June/July 2013, available at http://www.ams.org/notices/201306/rnoti-p679.pdf, and V. Stodden, J. Borwein, and D. H. Bailey, "'Setting the Default to Reproducible' in Computational Science Research," *SIAM News*, June 3, 2013, available at http://www.siam.org/news/news.php?id=2078.
31. For a complete discussion of HIPAA, see Chapters 1 (Strandburg) and 4 (Ohm) in this volume.
32. Some of these barriers were elucidated through a survey of the machine learning community in 2009. See Stodden, "The Scientific Method in Practice."
33. E.g. "All results quoted in this paper are validated by using two independent sets of software tools. . . . In addition, many cross checks were done between the independent combination tools of CMS and ATLAS in terms of reproducibility for a large set of test scenarios" (from http://cds.cern.ch/record/1376643/files/HIG-11-022-pas.pdf).
34. Reichman and Uhlir proposed that contractual rules governing data sharing, for example providing licensing terms or compensating creators, create a knowledge "semi-commons." A 'semi-commons' can exist through data pooling and thus sharing the burden of warehousing and supporting access infrastructure and tools, in exchange for increased access to the data. However, the concept of the 'walled garden' is slightly different in this example in that authorized independent researchers are given full access to the resources for verification and/or reuse purposes thereby mimicking open data as fully as possible under the privacy constraints inherent in the data. J. H. Reichman and Paul F. Uhlir, "A Contractually Reconstructed Research Commons for Scientific Data in a Highly Protectionist Intellectual Property Environment," *Law and Contemporary Problems* 66 (Winter 2003): 315–462. Available at http://scholarship.law.duke.edu/lcp/vol66/iss1/12.
35. For an assessment of the reach of data and code disclosure requirements by journals, see V. Stodden, P. Guo, and Z. Ma, "Toward Reproducible Computational Research: An Empirical Analysis of Data and Code Policy Adoption by Journals." *PLoS ONE* 8, no. 6 (2013). Available at http://www.plosone.org/article/info%3Adoi%2F10.1371%2Fjournal.pone.0067111.
36. For a further discussion of such template agreements, see V. Stodden, "Innovation and Growth through Open Access to Scientific Research: Three Ideas for High-Impact Rule Changes," in *Rules for Growth: Promoting Innovation and Growth through Legal Reform* (Kansas City, MO: Kauffman Foundation, 2011). Available at http://www.kauffman.org/~/media/kauffman_org/research%20reports%20and%20covers/2011/02/rulesforgrowth.pdf.
37. Of course, researchers in private sector for-profit firms are not the only potential collaborators who may have a different set of intentions regarding data and code

availability. Academic researchers themselves may wish to build a start-up around the scholarly objects deriving from their research, for example. In a survey conducted by the author in 2009, one senior academic wrote he would not share his code because he intended to start a company around it. See Stodden, "The Scientific Method in Practice."

38. See http://nlp.stanford.edu/software/.
39. See http://www.whitehouse.gov/the_press_office/President-Obama-Names-Vivek-Kundra-Chief-Information-Officer/ (accessed September 1, 2013).
40. See "Expanding Public Access to the Results of Federally Funded Research," http://www.whitehouse.gov/blog/2013/02/22/expanding-public-access-results-federally-funded-research (accessed September 1, 2013).
41. For other examples see e.g. J. E. Wiley and G. Mineau, "Biomedical Databases: Protecting Privacy and Promoting Research," *Trends in Biotechnology* 21, no. 3 (March 2003): 113–116. Available at http://www.sciencedirect.com/science/article/pii/S0167779902000392.
42. See e.g. https://data.cityofchicago.org/ and http://schoolcuts.org, http://nycopendata.socrata.com/ and https://data.ny.gov/ (state level), and https://data.sfgov.org/. The Family Educational Rights and Privacy Act (FERPA) attempts to address this with a notion of student privacy; see http://www.ed.gov/policy/gen/guid/fpco/ferpa/index.html. The State of Oklahoma recently passed a bill, the Student DATA Act, to protect student school performance data; see http://newsok.com/oklahoma-gov.-mary-fallin-signs-student-privacy-bill/article/3851642. In New York State a case was filed in 2013 to prevent a third party from accessing student data without parental consent; see http://online.wsj.com/article/AP0d716701df9f4c129986a28a15165b4d.html.
43. See e.g. the report from the World Economic Forum, "Unlocking the Value of Personal Data: From Collection to Usage," (Geneva, 2013). Available at http://www.weforum.org/reports/unlocking-value-personal-data-collection-usage.
44. See e.g. D. Citron, "Mainstreaming Privacy Torts," *California Law Review* 98 (2010): 1805–1852.
45. See e.g. McCormick v. Haley, 307 N.E.2d 34, 38 (Ohio Ct. App. 1973).
46. Restatement (Second) of Torts § 652B (1977).
47. See Citron, 1809.
48. Citron; e.g. Kline v. 1500 Massachusetts Ave. Apartment Corp., 439 F.2d 477, 480–81 (D.C. Cir. 1970), holding the landlord liable for a poorly secured building when tenants were physically beaten by criminals.
49. See e.g. F. Berman and V. Cerf, "Who Will Pay for Public Access to Research Data?" *Science* 341, no. 6146 (2013): 616–617. Available at http://www.sciencemag.org/content/341/6146/616.summary.
50. Individuals may direct their data to be used for research purposes only, or to be placed in the public domain for broad reuse, for example. See e.g. Consent to Research, http://weconsent.us, which supports data owner agency and informed consent for data sharing beyond traditional privacy protection.
51. World Economic Forum, 12.
52. Some restrictions on subject agency exist; see e.g. Moore v. Regents of University of California 51 Cal.3d 120 (Supreme Court of California July 9, 1990). This

case dealt with ownership over physical human tissue, and not digital data, but the tissue could be interpreted as providing data for scientific experiments and research, in a role similar to that of data. See also the National Institutes of Health efforts to continue research access to the Henrietta Lacks cell line, taking into account Lacks family privacy concerns. E. Callaway, "Deal Done over HeLa Cell Line," *Nature News*, August 7, 2013. Available at http://www.nature.com/news/deal-done-over-hela-cell-line-1.13511.

53. V. Stodden, "Optimal Information Disclosure Levels: Data.gov and 'Taleb's Criticism'," http://blog.stodden.net/2009/09/27/optimal-information-disclosure-levels-datagov-and-talebs-criticism/.
54. N. Taleb, "The Fourth Quadrant: A Map of the Limits of Statistics," http://www.edge.org/3rd_culture/taleb08/taleb08_index.html (accessed September 1, 2013).
55. See http://eva.compbio.ucsf.edu/~eva/doc/concept.html (accessed September 1, 2013).
56. See e.g. A. Morin, J. Urban, P. D. Adams, I. Foster, A. Sali, D. Baker, and P. Sliz, "Shining Light into Black Boxes," *Science* 336, no. 6078 (2012): 159–160. Available at http://www.sciencemag.org/content/336/6078/159.summary.

Part II Practical Framework

The essays in this part of the book make powerful arguments for the value of data in the public sector. We are all aware of their value to the private sector; indeed, the market advantage of many of large companies in the United States, such as Google, Facebook, and Yahoo, lies in their access to large datasets on individual behavior, and their ability to turn data into privately held information. Yet the experience of the authors demonstrates that the gap between vision and reality in the public sector is large, for many reasons. The authors identify new approaches that can enable public sector custodians to combine and use data, and new approaches to enable researcher access so that data can be turned into publicly held information. A major leitmotif in each chapter is, of course, trust.

What is the vision? An illustrative, but not exhaustive, list identified by the authors of the potential and actual value of big data range from simply better, more targeted city management to reduced taxpayer cost and burden, from great transparency and less corruption to greater economic growth, and indeed to addressing problems of epidemics, climate change, and pollution. Interestingly, as Elias points out, the European Commission recognized the value of data as far back as 1950, when the European Convention on Human Rights noted, "There shall be no interference by a public authority with the exercise of this right [**to privacy**] except **such as in accordance with the law and is necessary in a democratic society in the interest of national security, public safety or the economic well-being of the country, for the protection of disorder or crime, for the protection of health or morals, or the protection of the rights and freedoms of others**" (emphasis added). Indeed, as Greenwood et al. point out in an evocative phrase, data can be seen as the oil in the new economy and we should work to provide the appropriate business, legal, and technical (BLT) infrastructure to facilitate its use.

The reality is quite different. In practice, as Goerge points out, there are many practical decisions to be made, including what data to access, how to build capacity to use and present the data, and how to keep data secure. And the challenges include the fact that public attorneys and data custodians are often reluctant to provide data access because of the unclear legal framework and the downside risk associated with privacy breaches. Many of the problems could be addressed with sufficient funding, but the primary challenge identified by both Goerge and Koonin and Holland is building the trust necessary to provide access.

Closing the gap between the vision and the reality is the practical thrust of most of the chapters. The most important task is building trust. Elias notes that a very useful UK survey provides a roadmap. It found that the key elements in building trust include identifying the legal status of those bodies holding data; developing agreed and common standards covering data security and the authentication of potential research users; developing public support for the use for research of de-identified personal information; and creating a coordinated governance structure for all activities associated with access, linking, and sharing personal information.

A logical set of next steps, then, is to move from artisanal approaches to protecting privacy to a much more systematic approach. The authors provide a set of illustrative suggestions that very much mirror the UK survey. Greenwood et al. propose a set of regulatory standards and financial incentives to entice owners to share data (very much in the spirit of the Acquisti chapter in Part I). They explicitly discuss BLT rules that can be developed by companies and governments. They propose using big data itself to keep track of user permissions for each piece of data and act as a legal contract. Most specifically, they propose building an open Personal Data Store (openPDS) personal cloud trust network and 'living informed consent', whereby the user is entitled to know what data is being collected by what entities and is put in charge of sharing authorizations. Landwehr is less sanguine about the adoption of such technologies. He argues for performing analysis on encrypted files or building systems in which information flow, rather than access control, is used to enforce policies. Wilbanks proposes a portable legal consent framework, which is a commons-based approach that can be used to recruit individuals who understand the risks and benefits of data analysis and use.

All of these approaches must be built to be scalable for big data. If the public is to know what is being done to their data, and users are to know the analytical properties of the data, it is critically important to track data provenance – and even more important, information flows. This is very difficult territory indeed, as Landwehr points out. Provenance information has been characterized formally as an acyclic directed graph; such graphs get complex very fast, yet tracing changes is necessary to both replication and validation of scientific results. Most applications in the public sector are not designed to assure users or data providers that big datasets are accessed according to prescribed policies; hence in the near future, unless a sustained effort is put in place to build applications to code, the only approaches are likely to be manual and individuals will need to trust researchers. This, in turn, raises a major problem, because, as noted by Wilbanks, the reality is that informed consent terms and procedures are written by non-experts, with fields of study very different from re-identification.

In sum, the vision of big data for the public good can be achieved – the authors provide evidence from cities from Chicago to New York and in areas from health and the environment to public safety. But if the vision is to be delivered in a large-scale fashion, the authors in this chapter also make it clear that the public sector must make substantial investments in building the necessary infrastructure. If big data are the oil of the new economy, we must build the data equivalent of interstate highways.

6 The Value of Big Data for Urban Science

Steven E. Koonin and Michael J. Holland

Introduction

The past two decades have seen rapid advances in sensors, database technologies, search engines, data mining, machine learning, statistics, distributed computing, visualization, and modeling and simulation. These technologies, which collectively underpin 'big data', are allowing organizations to acquire, transmit, store, and analyze all manner of data in greater volume, with greater velocity, and of greater variety. Cisco, the multinational manufacturer of networking equipment, estimates that by 2017 there will be three networked devices for every person on the globe.[1] The 'instrumenting of society' that is taking place as these technologies are widely deployed is producing data streams of unprecedented granularity, coverage, and timeliness.

The tsunami of data is increasingly impacting the commercial and academic spheres. A decade ago, it was news that Walmart was using predictive analytics to anticipate inventory needs in the face of upcoming severe weather events.[2] Today, retail (inventory management), advertising (online recommendation engines), insurance (improved stratification of risk), finance (investment strategy, fraud detection), real estate, entertainment, and political campaigns routinely acquire, integrate, and analyze large amounts of societal data to improve their performance. Scientific research is also seeing the rise of big data technologies. Large federated databases are now an important asset in physics, astronomy, the earth sciences, and biology. The social sciences are beginning to grapple with the implications of this transformation.[3] The traditional data paradigm of social science relies upon surveys and experiments, both qualitative and quantitative, as well as exploitation of administrative records created for non-research purposes. Well-designed surveys generate representative data from comparatively small samples, and the best administrative datasets provide high-quality data covering a total population of interest. The opportunity now presents to understand how these traditional tools can be complemented by large volumes of 'organic' data that are being generated as a natural part of a modern, technologically advanced society.[4] Depending upon how sampling errors, coverage errors, and biases are accounted for, we believe the combination can yield new insights into human behavior and social norms.

Governments too are exploring whether making their data more open can help them to become more participatory, decentralized, and agile institutions able to address problems faster and more successfully on behalf of their citizens. As a result, open government data portals are becoming common in the United States at the federal, state and local levels.[5] To seize these opportunities, agencies will need to build their own internal capacity for data analytics as well as make judicious use of the expertise of their vendor communities if they are to deliver services more efficiently, increase the precision and accuracy of enforcement actions, set more informed policies, or more effectively plan infrastructure improvements. Not only can administrative, regulatory, and enforcement agencies benefit from improved data analytics, but statistical agencies are looking for additional tools to help them fulfill their obligation to produce accurate national, state, or local statistics (while cautious given that harms resulting from disclosures can seriously impact participants' willingness to participate in surveys).[6] Citizens too are interested in urban data to ensure government transparency and accountability as well as to enhance their local government's opportunities to improve urban living.[7]

Recognizing the economic value of government data beyond the usual arguments for increasing government transparency and efficiency, the Obama administration issued in May 2013 an executive order to make information generated and stored by the federal government more open and accessible with an explicit goal of fueling entrepreneurship, innovation, and scientific discovery.[8] Putting urban data in the hands of citizens has the potential to improve governance and participation, but data in the hands of entrepreneurs and corporations can also stimulate the development of new products and services. Climate Corporation, a start-up that was acquired in late 2013 by Monsanto for about $1 billion, combines 30 years of weather data, 60 years of crop yield data, and 14 terabytes of soil data – all from government agencies – for such uses as research and pricing crop insurance.[9] A recent study by the Knight Foundation, which has supported activities at the nexus of technology, civic innovation, open government, and citizen engagement, found that these 'civic tech' firms in the U.S. garnered more than $430 million from private sector investors and foundations between January 2011 and May 2013.[10] Start-ups facilitating peer-to-peer sharing attracted the most private sector investment, while start-ups facilitating access, transparency, and usability of government data attracted the most foundation investment. In an analysis of the worldwide value of open data for both government and industry, the McKinsey Global Institute estimates that open data could enable more than $3 trillion in additional value annually across the seven domains it analyzed: education, transportation, consumer products, electricity, oil and gas, health care, and consumer finance;[11] that $3 trillion is 3.4% of the estimated 2013 gross world product of US$85 trillion.[12]

At the Center for Urban Science & Progress, our goal is to collect and analyze data that will allow us to characterize and quantify the 'pulse of the city'. We are not alone in believing that a new science of cities is emerging, with an understanding of how scaling laws and scientific simulations can apply to transportation systems, energy use, economic activity, health, innovation, and the full range of urban activities.[13,14,15] In this chapter, we first address the motivations for this new urban science by defining a broad set of systems of interest and describe a data taxonomy as we see them applying to our field of study. We take note where we see particular municipal interests in these data flows. We then discuss some of the technical issues where we see big data differing from traditional data analyses for urban issues, and we close with a small set of non-technical issues that we believe warrant attention if cities are to fully realize the potential of big data analytics.

Urban Science

Given the trend towards more data and increasing availability of open data, it is not a fantasy to ask "if you could know anything about a city, what do you want to know?" understanding that local governments have responsibility for education; fire; police; delivery of human services; operation of public works like streets, sewers, and solid waste and storm water management; urban planning and zoning; fostering local economic development; and development of parks and recreational opportunities to improve the quality of life. Cities deliver services to their citizens through infrastructure and through processes. We want to know how those systems operate, how they interact, and how they can be optimized.

There are three classes of urban systems about which data must be acquired.

The Infrastructure Major questions about urban infrastructure focus on its extent, condition, and performance under varying scenarios of use.

We need to know the condition of the built infrastructure: Are the bridge joints corroding? Can we find the leaky pipes? Which pavement resists excessive wear from heavy vehicles? We need to understand the operation of the infrastructure: How is traffic flowing? Is the electrical grid balanced? Is building energy efficiency performing as expected?

The Environment Major questions about the urban environment focus on the sources and fates of pollutants, the health burdens those pollutants place on vulnerable subpopulations, and the vitality of natural systems facing demands for environmental services. We need to understand whether a city's river can support recreational uses such as fishing and rowing when simultaneously allowing for nearby industrial uses. In addition to the usual meteorological and pollution variables of interest, we need to understand the full range of environmental factors, such as noise, that influence people's day-to-day experience of the city.

The People Major questions about urban populations focus on the interactions of people with each other and institutions, their interactions as organizations, as well as their interactions with the built and natural environments. Cities are built by and for people and so cannot be understood without studying the people: their movement, health status, economic activities, how they communicate, their opinions, etc.

Yet it is this ever finer temporal and spatial granularity of data about individuals and the increasing power of informatic tools to combine and mine these streams of data that stoke concerns about privacy and data access, particularly when these tools are in the hands of individuals or organizations whose interests are not perceived as being aligned with those of the data subjects. Further development of both technical tools and administrative controls that can assure privacy and security of potentially massive data flows are necessary precursors to the deeper scientific study of cities.

An Urban Data Taxonomy

From an urban science perspective, data can be thought of as falling into four broad categories according to how it is generated: transactional data, *in situ* sensor data, remote sensor data, and citizen science data. Privacy concerns can arise not only over how the data is generated, but also as a result of where the data is generated, collected, contributed and how it is correlated with other data – whether by government, private sector institutions, or individuals. We will comment on these differences, but we do not want to suggest our treatment is comprehensive.

Transactional Data The first category of urban data is the traditional transactional data – the text and numerical records – that agencies and commercial entities generate in their routine course of business. These data sources are the familiar records such as permits, tax records, public health and land use data in the public sector or sales, inventory, and customer records in the private sector that social scientists have been exploiting for decades, if not centuries. Text and numerical records can be aggregated at the city level (census, statistical bureaus), at the firm or neighborhood level (census blocks, tracts, neighborhoods), or at the individual level (retail sales records, surveys). As commerce, government, and many individual activities migrate to the digital sphere, the available volume of data is growing and the vast majority of it is 'born digital'.

For municipal governments, a major opportunity lies in extracting the full value of the traditional transactional data already in their possession. As an example, the City of New York was able to prioritize the 300 or so daily complaints about illegal housing conversions so that enforcement actions focus on those that posed the highest risk of deaths to occupants and first responders as a result of fire or structural collapse. By pulling together existing information on foreclosures, tax liens, building complaints, sanitation violations and building wall violations from multiple departments, the City increased its rate of issuing orders to vacate unsafe properties from less than 3% of onsite inspections to well over 40%.[16] Developing IT architectures and interagency agreements that allow data analysis systems to operate seamlessly across disparate agency datasets is a significant enabling technical and organizational challenge for the field.

In Situ Sensor Data Next, data collected from the local environment immediately around a sensor or scanner is the most rapidly growing category of data relevant to the interests of urban science. Enabled by progressively cheaper microprocessors and wireless communications, engineers are rapidly developing methods to instrument infrastructure and the environment or extract people's movement from commonly used personal electronic devices, such as cell phones.[17] The expanding 'internet of things' enabled by the ease of scanning barcodes or QR codes and the plummeting price of RFID tags will only accelerate the stream of data related to object identity, location, and time of last movement. Questions of ownership of such data streams are discussed by Stodden in Chapter 5 of this volume.

For municipal governments, who have a more complete toolkit for influencing the local built environment than do either federal or state governments, a major opportunity lies in understanding with increasing spatial and temporal resolution how their urban infrastructure is being used. As an example, researchers used three-week-long mobile phone billing records generated by 360,000 San Francisco Bay Area users (6.56% of the population, from one carrier) to measure transient origin–destination matrices for vehicles.[18] Their dataset, which is two orders of magnitude larger in terms of population and time of observation than the most recent surveys, allowed them to allocate conclusively the major traffic flows on congested roads to a very few geographical sources of drivers. This suggests that traffic engineers should focus their efforts to modify commuting behavior on just those few driver sources, rather than implementing measures seeking to change behavior within the full commuting region.

Beyond fixed *in situ* sensors to record light levels, temperature, loading, pollution, etc., personal sensors that record location, activity, and physiology are becoming available. Detailed personal time series data are starting to be voluntarily made public by athletes using Fitbit activity monitors or by those in the quantified-self communities. Newly emerging applications of portable, unobtrusive assistive health care technologies for monitoring those with physical or cognitive impairments raise privacy concerns, but also present new opportunities for municipalities to improve the provision of human services.[19] Social media streams, such as Facebook, Twitter, and Foursquare, may be considered a specialized subset of this data category, particularly when postings of activity or sentiment are geocoded. The privacy concerns presented by social media streams have been adequately commented upon by other authors in this volume, except to say that citizens, by virtue of their far more frequent interaction with local government compared to state or national governments and greater feeling of control, may be more willing to share information with local agencies if they see improved services in exchange.

Remote Sensor Data Cameras and other synoptic sensors are a rich new source of data relevant to urban science. There is an ongoing proliferation of video cameras at points of commerce and automatic teller machines and at portals for pedestrians and vehicles. Despite an estimated 30 million cameras in public spaces in the United States, very little of the video collected is routinely analyzed, other than as needed for post-event forensics. Traffic scene surveillance for congestion and license plate monitoring may be the major exceptions. Rapid automated analysis of camera feeds is computationally challenging, but computer vision enabled by unsupervised machine learning is beginning to open up new opportunities, with real-time labeling of objects in natural scenes possible.[20] Privacy concerns precipitated by the ease of unauthorized discovery of webcams are of regulatory interest in the United States. The Federal Trade Commission settled a case in 2013 against TRENDnet, which sold its Internet-connected SecurView cameras for purposes ranging from home security to baby monitoring, after hackers posted live feeds of nearly 700 consumer cameras on the Internet, showing activities such as babies asleep in their cribs and children playing in their homes.[21]

For municipalities, their business improvement districts (BIDs) may be among the earliest, most enthusiastic adopters, aside from police departments, of facial recognition software and other video feeds for monitoring activities in city centers. BIDs provide supplemental services, such as cleaning streets, providing security, making capital improvements, constructing pedestrian and streetscape enhancements, and marketing the area.[22] They are non-profit or quasi-governmental entities authorized by local governments to which businesses within that district's boundaries pay an additional tax in order to fund projects. For some urban science studies, BIDs may be data sources as rich as those of city agencies.

Persistent remote sensing also offers new possibilities for urban science. While transient remote sensing of urban features from satellites or aircraft is well established,[23] persistent observation from urban vantage points is an intriguing possibility. Instrumentation on a tall building in an urban center can 'see', modulo shadowing, tens of thousands of buildings within a 10 kilometer radius, without the mass, volume, power, or data rate constraints of airborne platforms. As an example, varying sampling rates in the visible spectrum allows for the study of a range of phenomena. At low sampling rates, one can watch new lighting technologies penetrate a city and correlate what is known about early adopters or lagging adopters from municipal permitting databases to tease out the behavioral and financial components of energy-efficient lighting technology diffusion. At very high sampling rates, transients observable in the lights might provide a measure of other plug loads that would only be accessible with expensive submetering. Moderate sampling rates can observe daily behavioral information. Visible, infrared, hyperspectral, and radar imagery are all phenomenologies to be explored for urban scenes, as are RADAR and LIDAR (Light Detection and Ranging).

The synoptic and persistent coverage of such modalities, together with their relatively easy and low-cost operation, may offer a useful complement to *in situ* sensing. Privacy issues can be addressed, in part, by careful design of the spatial resolution of the collected images. Clearly, unexpected monitoring from a public vantage point raises issues of data collection, which

Strandburg addresses in Chapter 1 of this volume. Information collections of which individuals are unaware have far greater potential to disturb once those collections are revealed: so particular care needs to be taken in the design, approval, and socialization of remote-sensing campaigns.

Citizen Science Data Participatory (crowd-sourced) data streams are a potentially important tool of urban science.[24] There is a long history of successful citizen science where amateur scientists have made significant contributions. One of the longest running is the Audubon Christmas Bird Count, a repository of early winter bird observations recorded since 1900 that has been used by academic researchers and federal, state, and local wildlife and land planning agencies.[25] The last decade has seen a huge expansion in the sorts of scientific endeavor where non-professionals can contribute, thanks to the extraordinary development of information technology. Activities have moved beyond donation of spare computing resources, such as SETI@home for analysis of radio telescope data, to the participatory sensing of environment phenomena, such as noise.[26] Mobile participatory platforms collecting a variety of location, photos, and text inputs will require many of the tools of big data to fully exploit their output.

For municipal governments, citizen science activities can provide data at a geospatial resolution unobtainable with tools normally available to agencies. One example is participatory urban tree inventories using mobile platforms, such as OpenTreeMap, which allows volunteers to input information (location, species ID, photos) for individual trees. Tree-level data is more useful from a forestry management perspective than the aggregate tree canopy coverage commonly available from overhead imaging techniques. While measurement errors have been studied for manually collected field data,[27] understanding the validity of data streams from these new tools is evolving. Privacy issues related to any personally identifiable information about volunteers need to be carefully considered in the design of the mobile application,[28] and campaigns that permit data collection in the United States by children under age 13 can require compliance with the Children's Online Privacy Protection Act of 1998.

How Is Big Data Different?

Aside from their origins, traditional microdata resulting from censuses, sample surveys, administrative records, and statistical modeling differ from big data in several technical aspects as noted by Capps and Wright.[29] Much of the usual microdata encompass records numbering in the hundreds of millions, while big datasets are many orders of magnitude greater. The computational challenges associated with massive data management are substantially different from those for static datasets in terms of scale and throughput. Technical advances are required to scale data infrastructure for curation, analytics, visualization, machine learning, data mining, as well as modeling and simulation to keep up with the volume and speed of data.[30]

Official statistics and datasets tend towards periodic cycles of input, analysis, and release – a corporation's quarterly earnings report or the Bureau of Labor Statistics' Employment Situation Summary on the first Friday of every month – while much of the data relevant to urban science flows continuously. Many government agencies or corporations would like to analyze that data in real time for operational reasons. Traditional microdata, including surveys, tends to be labor intensive, subject to human error, and costly in their collection, while big data are often born digital and seem relatively cheap by comparison.[31]

Surveys, which form the foundation of official statistics, "are the result of careful data collection design with clearly defined uses, while big data come with unknowns (e.g. uses are less clear, data are less understood, data are of unknown quality, and representativeness is largely unknown)."[32] Capps and Wright also note that with respect to surveys, response assumes permission to use. Big data, as with much traditional administrative data, come as byproducts of other primary activities without asking explicitly and thus without any assumed permission to use beyond uses compatible with the purpose for which the data was collected. In fact, one might argue that the exploitation of data originally acquired for another purpose is a hallmark of big data in an urban science setting given the potential scale of data generated and held by municipalities.

Somewhat counterintuitively, sheer scale is one of the few characteristics of big data that can help limit some – though clearly not all – privacy or confidentiality issues. Massive datasets at the petascale and above are challenging to transfer, since the high-capacity, wide-scale communications networks required are extremely expensive to maintain. In physics and astronomy where such datasets are common, analyses are sent to the data rather than propagating copies of data for independent use. As a result, disclosure risk measures need only be implemented for the training data used to develop estimation procedures and for the final results prior to transmission back to the analyst. Subsets of massive data collections do remain vulnerable to unauthorized and undetected copying, but that risk is improbable for the full collection. In the urban science arena, data from sensor networks and simulation data could reach such scales.

Privacy Risks The value of any large urban dataset is enhanced through its association with other data. Observations are linked through location and time, as well as through entity (person, firm, vehicle, structure). The power of such linkage in producing new information is significant and increasingly well recognized. For example, knowing an individual's ZIP code localizes that person to 1 in 30,000 (the average population of a ZIP code).[33] Linking a ZIP code with a birthdate reduces the pool to approximately one in 80, while further connecting gender and year-of-birth are sufficient, on average, to uniquely specify an individual.

However, data mining, data linking, and statistical analyses are not the only source of risk presented by big data technologies. Reliance upon distributed or cloud computing resources can create additional privacy risk including risks from unencrypted intermediate datasets resident in the cloud,[34] but other authors in this volume, particularly Landwehr in Chapter 10, will address security issues in detail. Tools for visualizing massive datasets are becoming increasingly powerful, allowing users to explore datasets with hundreds of millions of records interactively in real time as a recently developed TaxiViz tool demonstrates.[35] Data for 540 million trips can be interrogated graphically in real time. Each trip record consists of: trip id, taxi id, driver id, pickup location, dropoff location, pickup date and time, dropoff date and time, traveled distance, fare amount, tip amount, and toll amount. Taxi and driver ids were anonymized to avoid the linking of records to the actual taxi medallion and taxi driver's license. Users can draw arbitrarily small regions of interest onto the underlying map, defining a region of trip origin and a region of trip destination. The visualization tool then shows taxi rides meeting those origin–destination criteria. Applied to a passenger who regularly catches a cab in front of their house or apartment and is dropped off at their place of work, such a tool could easily allow detailed analysis of any variation in that regular pattern. Applying traditional disclosure prevention procedures, such as suppressing cells in statistical tables based upon survey data, is not straightforward in an exploratory data visualization tool. Depending upon exposure risks of the particular data involved, the software engineering required to limit allowable queries could be quite sophisticated. Techniques such as parallel coordinates are beginning to be explored as a method to allow privacy-preserving data visualizations.[36]

Realizing the Value of Urban Data

Agencies, businesses, and researchers are better able to turn the deluge that is big data into useful information and understanding when access to data is at its most open. The Open Knowledge Foundation sets out a vision that "a piece of data or content is open if anyone is free to use, reuse, and redistribute it – subject only, at most, to the requirement to attribute and/or share-alike."[37] In practice, there are multiple situations where issues of safety, security, liability, confidentiality, or proprietary concerns limit the realization of that vision. In this section, we identify a few steps we believe municipalities and the data science community can take to facilitate the use of public data.

To correlate data, it must be brought together. But organizational barriers, within and between, institutions abound. As Wilbanks notes in Chapter 11, in his discussion of frameworks for sharing data, use is frequently "governed by a hodge-podge of contractual instruments" flowing from the data collection. Where data is not open by default, the transaction costs, particularly in length of time to reach closure, arising from negotiation of data transfer agreements with multiple agencies and bureaus can be substantial. Municipal governments can facilitate the usability of their public data by developing standardized data-sharing agreements appropriate to the laws and regulations of their jurisdictions in consultation with interested individuals, civic organizations, and businesses. In addition, municipalities can design their data systems to enable sharing by building internal IT interfaces as if they were citizen-facing external interfaces.[38] Doing so makes it easy to open data to external connections, whether to the research community, the hacker community, or technology vendors, once the decision to do so is made.

Once brought together, tools exist for managing disclosure risk, which Karr and Reiter discuss in Chapter 13. Many of these have been developed by statistical agencies and in settings, such as education, medicine, and financial services, where statutory mandates act as a driver. Methods continued to be developed for estimating re-identification risks in these specialized settings.[39] A current challenge is to adapt those tools so that data scientists can implement them on their broader range of datasets without excessive computational penalties. We should also recognize that all disclosure does not have equal potential for harm nor are individuals and organizations uniform in their tolerance of disclosure. If the risk estimation tools can be developed, it would be helpful to use consistent, transparent language for communicating disclosure risks akin to what the climate community uses to make distinctions between levels of confidence in scientific understanding and the likelihoods of a specific result, e.g. 'very high confidence' means a statement has at least a 9 out of 10 chance of being correct, or 'extremely unlikely' means a less than 5% probability of the outcome.[40]

Finally, work by the information science, management information systems, and e-government research communities has documented barriers to value creation from open data platforms, among them are problems of diverse user needs and capabilities, the limitations of internally oriented data management techniques, untested assumptions about information content and accuracy, and issues associated with information quality and fitness for use.[41] Fortunately, there are signs that municipalities are building their own capacity for data science, with groups established in multiple cities. Prominent examples include the Mayor's Office of Data Analytics and the Center for Innovation through Data Intelligence in New York City and a Mayor's Office of New Urban Mechanics in both Boston and Philadelphia, but Chief Information Officers in cities as large as Chicago and as small as Asheville, NC, are taking steps to develop data science capacities. Having capable, in-house data scientists who can demonstrate to their fellow civil servants the value big data has for solving practical problems may be one of most significant steps any municipal government can take in breaking down the barriers to value creation.

Conclusions

In closing, we wholeheartedly agree with the National Research Council's Committee on the Analysis of Massive Data assessment that "it is natural to be optimistic about the prospects" for big data.[42] We believe the tools of big data combined with increasingly open data will improve our scientific understanding of the cities that will be home to 67% of the world's population by 2050[43] and could contribute as much as $3 trillion annually to world economic growth. But those benefits are not foregone conclusions. Concerns over privacy precipitated by these developing big data technologies will lead to a reassessment, and possibly a rebalancing, of access as uses evolve and benefits to society at large are weighed against costs to individuals. Even public records (property, criminal, court, birth, death, marriage, divorce records, licenses, deeds, mortgages, corporate records, business registration) open for hundreds of years may have access restricted or some of the personally identifiable information they contain suppressed given that those records, now issued in electronic format, have become accessible and transmissible in ways that were never previously possible or expected.[44]

However, we should note that not all potential threats to realizing the value of big data are privacy related. Quantification often brings unexamined power and prestige to public policy debates,[45] so caution in the interpretive power of data analyses is crucial, given the real potential for harm in some cases.[46] In the late 1960s, New York City Mayor John Lindsay hired consultants from the RAND Corporation to help modernize municipal service delivery and achieve budget savings. RAND recommended an overhaul of fire station locations and the number of engines responding to fires, based on flawed firefighter response time data. When fire broke out in the Bronx, firefighters were unable to respond in time, and fires ended up burning out of control.

For those of us who are interested improving our scientific understanding of how cities operate, our goal should be not just relevant research but impact, but we need to approach this goal with a degree of humility. And so, urban data scientists need to be aware continually of the context from which data comes, the context in which analyses are used to make decisions, and the context within which privacy concerns are balanced.

NOTES

1. "Cisco Visual Networking Index: Forecast and Methodology, 2012–2017," May 29, 2013. Available at http://www.cisco.com/en/US/solutions/collateral/ns341/ns525/ns537/ns705/ns827/white_paper_c11-481360.pdf.
2. Constance L. Hays, "What Wal-Mart Knows About Customers' Habits," *The New York Times*, November 14, 2004.
3. G. King, "Ensuring the Data-Rich Future of the Social Sciences," *Science* 331, no. 6018 (2011): 719–721, doi: 10.1126/science.1197872.
4. Robert M. Groves, "Three eras of survey research," *Public Opinion Quarterly* 75, no. 5 (2011): 861–871, doi: 10.1093/poq/nfr057.
5. For CUSP, the main open government data portals are http://www.data.gov/ for federal data, https://data.ny.gov/ for New York State data, and https://nycopendata.socrata.com/ for New York City data. Data.gov tracks countries with national open data sites that provide access to machine-readable data (http://www.data.gov/opendatasites). This is not solely a phenomenon of Western developed countries. Africa has at least two open data efforts: Open Data for Africa (http://opendataforafrica.org/) supported by the African Development Bank Group and Africa Open Data (http://africaopendata.org/) developed with support from the non-governmental sector.
6. Mick P. Couper, Eleanor Singer, Frederick G. Conrad, and Robert M. Groves, "Experimental Studies of Disclosure Risk, Disclosure Harm, Topic Sensitivity, and Survey Participation," *Journal of Official Statistics* 26, no. 2 (2010): 287–300.
7. Datakind (http://www.datakind.org/) and Code for America (http://codeforamerica.org/) are examples of non-governmental organizations that seek to engage data scientists in projects for the public and non-profit sector that will lead to better decision making and greater social impact.
8. Executive Order 13642, *Making Open and Machine Readable the New Default for Government Information*, 78 FR 28111, May 14, 2013. Office of Management and

Budget Memorandum M-13-13, *Open Data Policy – Managing Information as an Asset*, May 9, 2013. Available at http://www.whitehouse.gov/sites/default/files/omb/memoranda/2013/m-13-13.pdf.

9. David Kesmodal, "Monsanto to Buy Climate Corp. for $930 Million," *Wall Street Journal*, October 2, 2013; and Quentin Hardy, "Big Data in the Dirt (and the Cloud)," *The New York Times*, October 11, 2011.
10. Mayur Patel, Jon Sotsky, Sean Gourley, and Daniel Houghton, "The Emergence of Civic Tech: Investments in a Growing Field," The John S. and James L. Knight Foundation, December 4, 2013. Available at http://www.knightfoundation.org/media/uploads/publication_pdfs/knight-civic-tech.pdf (accessed December 27, 2013).
11. James Manyika, Michael Chui, Diana Farrell, Steve Van Kuiken, Peter Groves, and Elizabeth Almasi Doshi, *Open Data: Unlocking Innovation and Performance with Liquid Information* (McKinsey Global Institute, October 2013).
12. *The World Factbook 2013–14* (Washington, DC: Central Intelligence Agency, 2013). Calculation based upon 2012 gross world product of US$84.97 trillion (purchasing power parity), inflated by 2.9%.
13. M. Batty, K. W. Axhausen, F. Giannotti, A. Pozdnoukhov, A. Bazzani, M. Wachowicz, G. Ouzounis, and Y. Portugali, "Smart Cities of the Future," *European Physical Journal – Special Topics* 214 (2012): 481–518, doi:10.1140/epjst/e2012-01703-3.
14. Luís M. A. Bettencourt, José Lobo, Dirk Helbing, Christian Kühnert, and Geoffrey B. West, "Growth, Innovation, Scaling, and the Pace of Life in Cities," *PNAS* 104, no. 17 (2007): 7301–7306, doi:10.1073/pnas.0610172104.
15. L. Bettencourt, J. Lobo, and D. Strumsky, "Invention in the City: Increasing Returns to Patenting as a Scaling Function of Metropolitan Size," *Research Policy* 36 (2007): 107–120, doi:10.1016/j.respol.2006.09.026.
16. Office of the Mayor, "Mayor Bloomberg and Speaker Quinn Announce New Approach to Target Most Dangerous Illegally Converted Apartments" (press release), PR-193-11, The City of New York, June 7, 2011. Available at http://www.nyc.gov/cgi-bin/misc/pfprinter.cgi?action=print&sitename=OM&p=1390075778000.
17. Marta C. Gonzalez, Cesar A. Hidalgo, and Albert-Laszlo Barabasi, "Understanding Individual Human Mobility Patterns," *Nature* 453, no. 5 (2008): 779–782, doi:10.1038/nature06958.
18. P. Wang, T. Hunter, A. M. Bayen, K. Schechtner, and M. C. Gonzalez, "Understanding Road Usage Patterns in Urban Areas," *Scientific Reports* 2 (2012): article 1001, doi:10.1038/srep01001.
19. T. Giannetsos, T. Dimitriou, and N. R. Prasad, "People-centric Sensing in Assistive Healthcare: Privacy Challenges and Directions," *Security and Communication Networks* 4 (2011): 1295–1307, doi:10.1002/sec.313.
20. Clément Farabet, Camille Couprie, Laurent Najman, and Yann LeCun, "Learning Hierarchical Features for Scene Labeling," *IEEE Transactions on Pattern Analysis and Machine Intelligence* 35, no. 8 (2013): 1915–1929, doi:10.1109/TPAMI.2012.231.
21. Federal Trade Commission, "Marketer of Internet-Connected Home Security Video Cameras Settles FTC Charges It Failed to Protect Consumers' Privacy"

(press release), September 4, 2013. Available at http://www.ftc.gov/opa/2013/09/trendnet.shtm (accessed December 30, 2013).

22. Richard Briffault, "A Government for Our Time? Business Improvement Districts and Urban Governance," *Columbia Law Review* 99, no. 2 (1999): 365–477.
23. Xiaojun Yang, *Urban Remote Sensing: Monitoring, Synthesis and Modeling in the Urban Environment* (Hoboken, NJ: Wiley-Blackwell, 2011), doi:10.1002/9780470979563.
24. S. Buckingham Shum et al., "Towards a Global Participatory Platform," *European Physical Journal – Special Topics* 214 (2012): 109–152, doi:10.1140/epjst/e2012-01690-3.
25. Erica H. Dunn et al., "Enhancing the Scientific Value of the Christmas Bird Count," *The Auk* 122 (2005): 338–346.
26. Nicolas Maisonneuve, Matthias Stevens, and Bartek Ochab, "Participatory Noise Pollution Monitoring using Mobile Phones," *Information Polity* 15 (2010): 51–71, doi:10.3233/IP-2010-0200.
27. Nathalie Butt, Eleanor Slade, Jill Thompson, Yadvinder Malhi, and Terhi Riutta, "Quantifying the Sampling Error in Tree Census Measurements by Volunteers and Its Effect on Carbon Stock Estimates," *Ecological Applications* 23, no. 4 (2013): 936–943, doi:10.1890/11-2059.1.
28. Salil S. Kanhere, "Participatory Sensing: Crowdsourcing Data from Mobile Smartphones in Urban Spaces," in *Distributed Computing and Internet Technology*, 19–26 (Berlin: Springer, 2013).
29. C. Capps and T. Wright, "Toward a Vision: Official Statistics and Big Data," *Amstat News*, August 1, 2013. Available at http://magazine.amstat.org/blog/2013/08/01/official-statistics/ (accessed September 19, 2013).
30. National Research Council, *Frontiers in Massive Data Analysis* (Washington, DC: The National Academies Press, 2013). Available at http://www.nap.edu/catalog.php?record_id=18374.
31. Capps and T. Wright, "Toward a Vision: Official Statistics and Big Data."
32. Ibid.
33. Latanya Sweeney, "K-anonymity: A Model for Protecting Privacy," *International Journal of Uncertainty, Fuzziness and Knowledge-Based Systems* 10, no. 5 (2002): 557–570, doi:10.1142/S0218488502001648.
34. Xuyun Zhang, Chang Liu, Surya Nepal, Suraj Pandey, and Jinjun Chen, "A Privacy Leakage Upper-Bound Constraint Based Approach for Cost-Effective Privacy Preserving of Intermediate Datasets in Cloud," *IEEE Transactions on Parallel and Distributed Systems* 24, no. 6 (2013): 1192–1202, doi:10.1142/S0218488502001648.
35. N. Ferreira, J. Poco, H. T. Vo, J. Freire, and C. T. Silva, "Visual Exploration of Big Spatio-Temporal Urban Data: A Study of New York City Taxi Trips," *IEEE Transactions on Visualization and Computer Graphics* 19, no. 12 (2013): 2149–2158, doi:10.1109/TVCG.2013.226.
36. Aritra Dasgupta and Robert Kosara, "Privacy-Preserving Data Visualization Using Parallel Coordinates," in *Proc. Visualization and Data Analysis (VDA)*, 78680O-1–78680O-12 (International Society for Optics and Photonics, 2011), doi:10.1117/12.872635.

37. Open Knowledge Foundation, "Open Definition." Available at http://opendefinition.org/od/ (accessed on January 5, 2014).
38. Michael Chui, Diana Farrell, and Steve Van Ku, "Generating Economic Value through Open Data," in *Beyond Transparency: Open Data and the Future of Civic Innovation*, ed. Brett Goldstein and Lauren Dyson (San Francisco, CA: Code for America Press, 2013), 169.
39. Fida Kamal Dankar, Khaled El Emam, Angelica Neisa, and Tyson Roffey, "Estimating the Re-identification Risk of Clinical Data Sets," *BMC Medical Informatics & Decision Making* 12, no. 1 (2012): 66–80, doi:10.1186/1472-6947-12-66.
40. IPCC 2007, *Climate Change 2007: Synthesis Report. Contribution of Working Groups I, II and III to the Fourth Assessment Report of the Intergovernmental Panel on Climate Change*, ed. Core Writing Team, R. K. Pachauri, and A. Reisinger (Geneva: IPCC, 2007), Appendix II, p. 83.
41. S. S. Dawes and N. Helbig, "Information Strategies for Open Government: Challenges and Prospects for Deriving Public Value from Government Transparency," in *Electronic Government*, ed. M. A. Wimmer et al., Lecture Notes in Computer Science 6228 (Berlin: Springer, 2010), 50–60, doi:10.1007/978-3-642-14799-9_5.
42. National Research Council, 2.
43. *World Population Prospects: The 2010 Revision, Volume I: Comprehensive Tables,* ST/ESA/SER.A/313 (United Nations, Department of Economics and Social Affairs, Population Division, 2011).
44. D. R. Jones, "Protecting the Treasure: An Assessment of State Court Rules and Policies for Access to Online Civil Court Records," *Drake Law Review* 61 (2013): 375.
45. Theodore M. Porter, *Trust in Numbers: The Pursuit of Objectivity in Science and Public Life* (Princeton, NJ: Princeton University Press, 1996).
46. Joe Flood, *The Fires: How a Computer Formula, Big Ideas, and The Best of Intentions Burned Down New York City—and Determined the Future of Cities* (New York: Riverhead Books, 2010).

Data for the Public Good: Challenges and Barriers in the Context of Cities

Robert M. Goerge

Introduction

Comprehensive, high-quality, multidimensional data has the potential to improve the services cities provide, as it does with the best private service-providing businesses. City officials, politicians, and stakeholders require data to (1) inform decisions that demonstrate service effectiveness, (2) determine which services should be targeted in a geographic area, and (3) utilize limited resources to best serve residents and businesses.

Administrative data is now ubiquitous in government agencies concerned with health, education, social services, criminal justice, and employment. Local government has primarily used this data to count cases and support budget making within the programs for which the data is collected. Yet data linked across programs, where individuals and families can be tracked with multiple data sources either cross-sectionally or longitudinally, is rare. Both data scientists and the public sector currently have an excellent opportunity to use the big data of government to improve the quality and quantity of analyses to improve service delivery. This chapter describes an effort in one place to use the administrative data collected in the public sector to have an impact by informing city leadership.

Privacy rules and regulations and bureaucratic silos often prevent city officials from obtaining and using data to address some of their most intractable problems. This chapter thus addresses the barriers to sharing and acquiring information. How can municipalities unlock the potential value of data and harness the analytic resources that are generally in short supply, access data not in their control that would enhance their data-driven capacities, and comprehensively address the range of education, health, employment, and crime-related issues for which they are responsible? Officials must ultimately have better information to maximize their utilization of limited resources to improve the well-being of their residents and the effectiveness of the organizations (health care, schools, social service agencies and police departments) that serve them. This chapter will primarily focus on a successful Chicago effort, and also refer to others around the country that offer strategies for cities to make data-driven decisions.

All trends and forecasts suggest that cities globally will become even more important socially, economically, and politically.[1] Responding to the resulting management needs will require data. Income inequality is increasing and occurs to the greatest extent and closest proximity in cities.[2] While poverty in the United States is increasing more quickly in suburban areas, the vast majority of the poor still live in cities. Cities will need to provide a range of health and social services while those with higher incomes will pay the cost of those services through taxes. As cities grow and, as the resources available to them become scarcer, they need to increase both service efficiency and effectiveness.[3] The bankruptcy of Detroit and the dire fiscal situation of many other cities and school districts across the nation highlight the extremes of these trends.[4]

Cities must find ways to educate children, train unemployed adults, and keep residents safe and healthy, while at the same time supporting large and small businesses, the non-profit sector, and faith-based service organizations. The public sector needs to learn 'what works' locally, in specific neighborhoods, with specific populations, and what services, programs, or interventions need to be modified or discontinued. In order to make those determinations, high-quality data (which is often confidential and sensitive) combined from multiple sources is key.

Current Open Data Initiatives and Need for Confidential Data for Decision Making

In order to address the needs of cities, their residents, and non-profit and business sectors, confidential and sensitive data must be used and shared to create databases that can fuel better service and analysis. Data and tools needed to drive the development of 21st-century housing and commercial developments are increasingly available through open data portals in cities around the world (Koonin and Holland, Chapter 6 in this volume).[5] These current open data initiatives utilize information that is generally *not* confidential.

The fields of engineering and urban design are already using data-driven models to revamp and create new neighborhoods. Barcelona, San Francisco, Chicago, and other cities are demonstrating how the use of new tools and data enables the design of neighborhoods that are environmentally friendly and sustainable.[6] In Chicago, "LakeSim will connect existing urban design tools with scientific computer models to create detailed simulations relevant to large-scale development. Instead of planning separately for different pieces of the infrastructure, the framework will allow developers to simulate how various designs influence the environment, transportation, and business spheres in Chicago Lakeside under a range of possible scenarios, over hundreds of acres and decades of time."[7] In most cases, however, the data that is used to fuel these efforts is not confidential, because it is overwhelmingly about places and the information that is needed about the population is often in aggregate form.

The data that addresses what human services city residents would benefit from is difficult to acquire. To address and target specific social issues, personally identifiable information (PII) is needed. Given national and international concern about privacy, the key is to access PII without disclosing the identities of individuals. The challenge for cities – government and its private sector partners – is how to use data (primarily administrative, but also social media and any other private source such as utility data) on characteristics and needs of individuals and families to provide better services and to support analysis that leads to better evaluation, planning, and monitoring of city functions and program outcomes. Furthermore, cities must make complex decisions about (1) what data to develop and access from counties, states, the federal government, and private sources; (2) how to develop the capacity to use data; (3) how to present data and be transparent; and (4) how best to keep data secure so that individuals and organizations are protected.

The range of administrative data collected by cities is considerable and includes, but is not limited to, crime activity, school outcomes, public health care and surveillance, early care and education, workforce development, after-school activities, tax payments, and receipt of human services. Linking these data with place-based assets, such as businesses, homes, community services, transportation, and emergency services, provides the comprehensive data infrastructure that can inform and improve the city's fulfillment of its responsibilities to its residents.

Cities Must Capitalize on Their Investment in Collecting Data

Public sector staff, whether teachers, police officers, public health workers, or staff of contracted providers, spend an enormous amount of time collecting data, described above, that is entered into computerized systems. A return on this investment of time and money can be realized if what is often rounding error in a city budget is spent to create data warehouses that can be the basis for the analysis of the services provided.

Two major urban areas have taken important steps that are impossible to roll back because of the direct benefits that have been seen in serving individuals and families. These efforts have gone hand-in-hand with creating data warehouses that support decision making by city leaders. Unfortunately, few cities have the type of systems described below.

The Allegheny County Department of Human Services in Pennsylvania (Pittsburgh) operates a data warehouse that includes child welfare, behavioral health, public school, welfare program, corrections, housing, and aging services.[8] Front-line workers are able to access information on individuals or families from other agencies to inform their information gathering or service provision. Data is also used to conduct system-wide analyses to support research, strategic planning, needs assessment, and program evaluation. There are tight controls over who sees what data based on what front-line workers need to know.

New York City created Worker Connect which links data on a single individual family or household across child protective services, juvenile justice, aging, mental health, criminal justice, health and hospitals, and homeless services. Worker Connect gives front-line agency staff access to real-time data about individuals that allows them to enroll and process cases faster, provide quicker referrals, and spend more time on casework.[9] This system also facilitates studies such as the NYC/Cleveland/Chicago study of foster children entering the juvenile justice system described in the next section.

Legal, Bureaucratic, Jurisdictional, and Technical Barriers to Creating Better Data

Just as the data systems in New York City and Allegheny County can have benefits for an individual's well-being, analysis of the entirety of the data can have a positive impact on the well-being of subpopulations of vulnerable individuals and families. Data scientists and subject matter experts, whether inside or outside of government, with the right analytic tools and data, can have a significant impact on the decision making around schools, health care, employment, and public safety. Confidential data on individual/family service experiences, employment, and criminal records is the major resource needed by these experts to conduct the research and analysis that builds evidence. Without PII about people (names, birthdates, home addresses) and places (addresses), data linkage at the person or place level is impossible, which leaves decision makers with single-program datasets that are significantly less useful. The following are the barriers to combining data and sharing it with those who can generate the evidence needed to address cities' toughest problems.

Legal Barriers There are a variety and potentially increasing number of federal, state, and municipal laws designed to protect individuals from harm that data disclosure might cause (Strandburg, Chapter 1, and Ohm, Chapter 4, in this volume).[10] While these laws usually contain provisions allowing conditional data sharing, public attorneys are often reluctant to approve the use of data that might appear to violate the law or incur additional risk for their clients (mayors or agency heads). Educating public sector attorneys on what the actual law is concerning the conditions under which confidential data can be shared is seldom done, except perhaps by other attorneys whose clients need to access data.

Although little research has been done regarding harm that disclosure of protected data actually causes, there is strong public sentiment about protecting sensitive data on individuals – unless someone has been arrested for a crime.[11] There are instances in which individuals have been misidentified because of incorrect data or unauthorized use of data. This has occurred in homeland security and law enforcement at all levels of government. Clearly, cities must keep the data they collect secure, while at the same time making the data available to selected users for constructive purposes.

Bureaucratic Barriers Sometimes government leaders and staff are reluctant to share their data with other public sector agencies. This often results in 'federated' or virtual databases, where each dataset remains with the responsible agency and data is only combined in highly regulated ways where each party can control the outcome. In this case, data are not actually integrated into one database. This added layer of bureaucracy, unless managed extremely well, adds significant additional burden to any effort that requires data from multiple agencies, since it requires multiple permissions and actual bringing the data together in one place for analysis purposes. This model is truly an example of how data still exists in silos that do not facilitate its use.

In many instances, government leaders feel safer giving their data to an external party (a consultant or an academic organization) than to a fellow agency leader within the same jurisdiction. Lack of data sharing between two public sector agencies is often a symptom of lack of trust between the agencies. In this situation, the external party can link data from multiple agencies for the benefit of these agencies, without any one agency having another's data.

Political Barriers Elected leaders find it difficult to fund research when services are being cut. Perhaps over time, the benefits of data-driven decision making in improving services will become evident to the public. Then, our elected officials will be able to allocate tax dollars to research and analysis without fear of special interest group backlash. Fortunately, private philanthropy and monies from the federal government can often be used to support these important efforts.

Jurisdictional Barriers In order to have simple indicator data as well as comprehensive microdata on individuals to guide programmatic and fiscal decision making, cities often need to enter into agreements with state and/or county agencies to access what they require to create comprehensive data. City school districts, criminal courts, jails, prisons, and public housing are often not under a mayor's jurisdiction. Also, data from welfare programs (TANF, SNAP, Medicaid. LIHEAP) are not typically available to cities unless cities are also counties in county-administered states (e.g. NYC, San Francisco). So, to obtain welfare data on residents, cities must employ mechanisms and state agency agreements (sometimes established and sometimes not) to access data. Many agreements are cumbersome and need to be renewed on a yearly basis, which creates an opportunity to negotiate the terms – but also takes time as the paperwork moves slowly through the bureaucracy. As funding becomes increasingly scarce, government agencies are requiring payment for data – even between two government agencies! The need to support data efforts within agencies has led to fee-for-service in the data-sharing arena.

Technical Barriers The technical barriers to integrating data are declining over time as tools are being developed in both the proprietary and open source space at a rapid rate. Clearly there is fast turnover in the computing infrastructure, both hardware and software, applications, dashboards, and other tools used to process, analyze, and visualize data. The challenge will be for cities to decide where and how to deploy their data resources and personnel so that data is continually available, but is also continually upgraded. Furthermore, as computational power becomes less expensive and more open software is available, government has the potential to maintain state-of-the-art data infrastructures.

Given the salary discrepancies between the public and private sectors, it is often difficult to prevent the turnover of computing staff. Training staff who have both data science and policy acumen and retaining them is also important if cities are to take advantage of the data they collect. Universities must prepare the workforce of the future so that there is sufficient technical expertise available.[12] In addition, strong relationships with external organizations, such as university research centers, can provide institutional memory so that new officials and staff can maintain the practices of data-driven decision making. The next section provides an example of such a research center.

Chicago: A Case Study

This chapter continues with an example that shows how a set of stakeholding organizations took steps to build data resources that met the needs of policymakers and helped inform not only local decision makers, but key actors outside of the city, in state capitals and Washington, about what is needed to address a few of the key social problems nationwide.

Informational Needs of Policymakers

The Integrated Database on Child and Family Programs (IDB) in Illinois is one approach to addressing the data needs described above. The database is the oldest of its kind in the country, and has been continually maintained by Chapin Hall at the University of Chicago by the author and his colleagues since the mid-1980s.[13] The primary purpose of the IDB project is to inform policymakers of the circumstances of children and youth in Illinois, to provide evaluative data, and to conduct research that leads to improved policy and programs for vulnerable children and youth. In order to report on all individuals, the purpose has always been to collect data on the entire populations of individuals in public sector programs, as opposed to tracking samples. (This has resulted in the IDB holding data that the state and city no longer have available to them.)

The IDB was begun three decades ago by a group of state, foundation, and academic leaders who believed strongly that administrative data has important value in social research. In particular, the goal is to improve the decisions that public sector employees – from front-line workers through agency leadership – make about vulnerable children and adults and ultimately improve the lives of Illinois children and families.

Chapin Hall, in partnership with state, county, and city agencies, has brought together and linked a wide range of datasets at the individual, family, and case levels to achieve what was described in the previous paragraph through research and analysis. These include data from over a dozen state and city agencies and nearly all social programs, with records for over 10 million individuals. Program data includes maternal and child health data, Women, Infants and Children Nutritional Program, Supplemental Nutritional Assistance Program and Food Stamps, Temporary Assistance to Needy Families and Aid to Families with Dependent Children, Medicaid, abuse and neglect reports, child welfare services, juvenile justice, adult incarcerations, arrests, employment and earnings (Unemployment Insurance wage data), mental health services, alcohol and substance abuse treatment, child care subsidies, and special education, and data on early childhood programs (e.g. Head Start), K–12 student information, and postsecondary education in Chicago. These data have been linked at the individual level in most cases since at least 1990. Data is multigenerational, which means that children, parents, and often grandparents in the data are linked. Most of these data have addresses available and are thereby able to be geocoded and spatially analyzed.

Over time, the IDB has expanded to include adults and families in public sector programs. The city of Chicago, as both the economic engine and home of families with the most severe problems in Illinois, has been a focus of the research undertaken by most users of the IDB. For example, the K–12 data includes well over 80% of the school-age children in Chicago since 1991.

Maintenance Challenge

All the barriers described at the beginning of this chapter were experienced by Chapin Hall in building the IDB. (This author has literally been told, "There's a new sheriff in town and nobody is getting this data.") Nevertheless, the IDB continues to have the key ingredients for success including (1) strong support from executive leadership across government agencies, (2) a long and growing track record of success, (3) no data breaches, (4) lawyers and laws usually on the side of data sharing, and (5) independent funding for most of the work.

For the data to have impact, a long-term relationship between the public officials and the researchers that is mutually beneficial is paramount. The researchers benefit by understanding the problems better and being able to communicate their findings to officials in a manner that can be used in practical ways. Public officials also keep researchers 'honest' in that policy or programmatic recommendations that sound good in journal articles may be simply not practical. This reality check increases the integrity of the academic mission of building knowledge. Public officials benefit from having expertise that they cannot always buy and, through researchers, who are often around longer than the officials, can access the institutional memory around the data activities of a city. Because there was strong support from officials above the "new sheriff," data continue to flow without pause.

The next section addresses how Chapin Hall has used and uses the data. We continue to overcome and manage barriers to maintain the effort and conduct research projects and analyses for government agencies. This is necessary because the laws, rules, and context for government activity change constantly and affect the construction of the IDB. Appendix Bcontains additional examples of efforts to promote data-driven decision making in cities that face many of the same barriers but often address them in different ways.

Uses of the Integrated Data Base

Researchers at Chapin Hall, the University of Chicago, and many other institutions have used the IDB for evaluations, dissertations, and other studies leading to peer-reviewed journal articles and formal reports to policymakers. Research ranges from purely descriptive to using data in randomized control trials. In addition, the IDB has been used in multistate and multicity studies. Parts of the IDB were linked to the U.S. Census Bureau's American Community Survey data, a representative sample of the entire population, in four states so that families who did and did not participate in particular government programs could be studied.[14] The following sections describe how the IDB has been used in other studies to inform policy and practice.

Families in Multiple Systems

While states and cities collect data for ongoing program management, they often lack the capacity to transform these data into information that aids policy decisions and program development. For example, in 2008, Illinois state officials hypothesized that a minority of families accounted for the majority of service costs, but they could not identify the characteristics of these families and quantify these costs *across agencies and programs*. Illinois has little information on where to target their interventions to both alleviate problems now and prevent problems in the future because the state does not have the capacity to reconstruct family service histories.

Chapin Hall was asked by the governor's office to conduct a study that would clearly identify families in multiple systems and multiple programs with the aim of understanding the costs associated with all program utilization. The programs that we analyzed were those with the highest per diem costs, including foster care, adult and juvenile incarceration, mental health services, and substance abuse treatment. Even with the support of the governor, it was necessary to convince each of five state agency directors that the value of the project justified the effort that their staff (primarily lawyers) would have to expend in making it happen. The necessity also arose because the governor's office had no research funds and required the agencies to pay for an equal part of the research. Permission and funding was secured from five state agencies, and one additional agency declined. While the data from the declining agency (birth certificates) would have been beneficial, it would have added only demographic data and did not affect the core research questions.

The results of the analysis allow the state to target resources geographically and individually.[15] The study revealed that 23% of extended families being served in multiple systems (health, mental health, criminal and juvenile justice, and child welfare) account for 64% of the service intervention resources *and* utilize 86% of the funding resources. Spatial analysis showed Chicago has the densest geographic pockets of these families, but there are also areas in smaller urban areas around the state that have high rates. For example, there are over 10 census tracts where more than two-thirds of all children live in families participating in multiple systems. These data are crucial for targeting service delivery and focusing on the small number of multisystem families who are utilizing the majority of resources.

Chicago Public Housing Transformation

From 2004 to 2007, Chapin Hall and other researchers in Chicago met with all child- and family-serving agencies in Chicago to address the impact of public housing transformation on the children and families affected by it.[16] The transformation involved razing high-rise public housing structures and providing support for families to move to other parts of the city. These meetings occurred roughly every six weeks, and as questions arose, Chapin Hall staff analyzed demographic and geographic data in the IDB to best describe service utilization. In particular, the IDB analyses resulted directly in multiple instances of two city agencies partnering to address identified problems. The data showed that certain schools and specific communities would need additional resources to serve the particular educational and emotional needs of children newly displaced to their community.

A key factor in bringing all of the agencies to the table and gaining their participation was a pledge that all the information Chapin Hall researchers produced would remain confidential for planning purposes – and that nothing would be published without the written approval of the agencies providing data for a particular analysis. This exemplifies the concern that public officials have about opening up research that may prompt additional demands on them. Appendix A summarizes a set of analyses and actions that resulted from the discussions among agency leadership and researchers during this project.

Three-City Study of Foster Children Entering the Juvenile Justice System

Children who receive child welfare services are at risk for later delinquency and involvement with the juvenile justice system. Individuals who become involved in both systems appear to have increasingly complex needs, but may be less likely to receive comprehensive, coordinated care because of agency boundaries. Understanding the pathway from child maltreatment to participation in the juvenile justice system can help cities address one of their most vexing problems –how youth become perpetrators of crime and violent behavior.

The Three-City Study of Foster Care Children Entering the Juvenile Justice System Project consisted of separate, but highly comparable, analyses of administrative data from three large urban localities: Cook County in Illinois, Cuyahoga County in Ohio, and New York City in New York. The analyses pursued three goals: (1) determine how many children who experience out-of-home foster care placements become involved in the juvenile justice system; (2) understand individual characteristics (sex, race/ethnicity) and child welfare involvement (placement type, number of spells, age of first child welfare placement) that may distinguish children in foster care who are later involved in the juvenile justice system from those who are not; and (3) compare patterns of results across the three locations to suggest common and distinct trends. The main challenge in this project was a technical one: to make the data comparable across the three sites, so that valid comparisons could be made. The lack of comparability or knowledge of comparability is a major impediment to multisite studies.

Population and Poverty Estimates and the Inadequacy of Census Data for Cities

The IDB has also been used to create information to substitute for data that was available for many decades from the Census Bureau. For decades, cities have relied on decennial census data for good measures (albeit 10 years apart) of their population and housing characteristics at the block level. In the last decade, the Census Bureau began collecting long-form data on a continual rolling-sample basis across the country. The current American Community Survey (ACS) only provides 5-year averages at the census tract level for the data that was received by cities at the block or block-group level for each decennial census. The administrative data in the IDB and place-level ACS data made it possible to build statistical models to calculate census tract population and poverty estimates for 0–18-year-olds in Chicago. City officials require this data to make informed service resource allocation decisions, which affect school openings and closings, child care slot distribution, senior citizen programs, and many other services based on population characteristics and density.

The federal government and states are well served by the ACS and other Census Bureau data, because they make decisions using state- and county-level data, which are available annually. Much has been written and much said about this issue and some, including commercial firms, have taken to marketing their own estimates of population using available data in combination with administrative data that states and private sector firms maintain.[17] Unfortunately, it is generally true that few data sources are consistent over time as data-collecting agencies rarely take into consideration all the uses of their data by others.

Supply and Demand for Early Care and Education

Leaders in Chicago agencies who are responsible for managing early childhood programs and combining funding streams to provide high-quality care assert that the availability of one type of program in a community is dependent on the number of other programs in that community. This is the result not only of the number of eligible children, but also the complexities of federal and state regulations and the availability of funding for each program. Chapin Hall researchers combined data from the population and poverty estimates described above with data from the IDB, including child care licensing data, and Head Start, pre-K, and child care utilization, to develop an ongoing analysis of the supply and demand for early care and education programs in Chicago at the census tract level.[18] Since the attempt is made to keep this data up to date, the challenge is to get data from the data providers in a timely fashion. As resources become increasingly scarce, even if the data is collected, it is often a burden for agencies to extract it for analytic purposes or to share it with other organizations. Computer professionals are seen as non-essential personnel in some public sector agencies. Therefore, there is often significant turnover in who manages the data.

Using Census Survey Data Linked with Administrative Data

In many ways, linking population data with administrative data provides an ideal data resource for cities seeking to understand who does and does not use their services and what the outcomes are for each group. By integrating individual-level U.S. Census Bureau survey data with individual-level social program administrative data from Illinois, Maryland, Texas, and Minnesota, Chapin Hall and colleagues at three other universities and the Census Bureau developed a comprehensive model of employment support program eligibility and take-up.[19] The resulting dataset was a representative sample of all families in these states combined with administrative data on child care subsidies, income maintenance, employment, and nutritional programs at the individual and family levels. Researchers examined individual, family, household, and neighborhood characteristics that affected program take-up; the conditions under which low-income families utilize these programs; and the effect of participation in these programs on employment.

Development of Necessary Data Warehouses

Cities need to move beyond their role as data collectors and become data integrators, stewards, and users. Their decades as mere data collectors have shown no significant return. The role of data steward means delivering data to individuals, either within or outside city government, who can use it to provide policymakers with the information needed to best serve city residents.

Cities have proven they can do this when there is a compelling need. The question is whether it can be sustained. Chicago city staff combined numerous sources of data by location in real time to address the security challenge during the NATO summit in 2012.[20] Much of the data was confidential and never shared outside those agencies involved in the NATO summit security. Most of the data was place based, but some included confidential identifying information. Data on crime incidents, 911 and 311 reports, building conditions, municipal vehicle locations, transportation and traffic, Twitter streams, and other place-based data were also included in order to have information ready in case of a crisis. Similarly, other cities are making tremendous efforts in the public safety arena, with efforts like CompStat showing promise in addressing particular types of crime in specific locations.

It is possible that the new emphasis on data-driven decision making and evidence-based practice (including predictive analytics in policing and experimental efforts in education) will allow cities to build the data infrastructure and skill sets needed to use and sustain it with less effort than was needed when some current efforts first started decades ago. Executive orders and city ordinances may be necessary, but in most cases, creating better data is already allowed by the law. The following are the core ingredients in empowering cities with the data they need.

Sustainability

Sustainability is a major challenge when elected officials change in cities. New leadership needs to be educated about the importance of existing data efforts, and new relationships based on mutual trust and need must be formed. In the worst case, efforts begun during one administration are discontinued in the next. Transition to a new administration does not necessarily mean an interruption in the data flow, however. When strong relationships with middle-level staff and legal agreements are in place, data sharing and utilization can continue without pause. Often, and ideally, data delivery, with the proper legal permissions, is a hands-off process that occurs whether or not anybody is watching.

Strong Leadership

Most mayors and politicians are risk averse, and will not risk a potential class-action lawsuit or bad publicity around privacy. However, mayors and agency directors have the power to promote data sharing between and among city agencies. Mayor Bloomberg in New York and Mayor Emmanuel in Chicago both exemplify the positive impact strong leaders can have on data-driven city management and decision making.

"Where there's a will, there's a way." If an organization that has data wants to provide it to another organization for the purpose of improving services, there is usually a legal way to share the data. Although there are many potential legal and bureaucratic barriers to data sharing, if strong leaders (influential over the use of data resources) want data sharing to happen, it will most likely occur. A recent GAO report provides some examples of barriers that are perceived rather than actual.[21] For the most part, the legal and regulatory framework does not ultimately prevent data sharing. Sufficient discretion is given to public sector organizations that the attorneys who write the contracts can find a way to accommodate data use and sharing either inside or outside of government.

Building Strong Relationships

Trust between organizations can often take years to develop, through personal connections and organizational relationships based on civic standing, reputation, and experience. The process often involves many meetings, discussions, and negotiations. It may also involve refuting the fear that sharing data with external organizations may lead to data being indiscriminately disseminated.

Some collaborations across the country have had great success in linking confidential data because they have built enduring relationships. For example, participants in the University of Pennsylvania's Actionable Intelligence for Social Policy represent a group of cities and states that have combined datasets on individuals across multiple public sector programs.[22] In all cases, these efforts have depended on the sharing of personally identifiable information attached to service records to build comprehensive histories of problems, disabilities, assets, and service receipt of individuals and families. These state and local efforts have built incrementally over time both relationships and data resources to address questions important to policymakers while contributing to the knowledge base in multiple fields of social science.

APPENDIX A

The data Chapin Hall presented during the project's first two years led to a number of specific actions. These included efforts to (1) improve educational opportunities for Chicago Housing Authority (CHA) children, (2) increase Head Start enrollment among CHA children, (3) improve labor market outcomes of CHA parents, and (4) increase CHA children's enrollment in All Kids. Each of these efforts is described briefly below.

IMPROVE EDUCATIONAL OPPORTUNITIES FOR CHA CHILDREN

- Contrary to the perceptions of some school administrators, the data Chapin Hall presented indicated that relatively few schools had experienced a large influx of children whose families had relocated because of the Plan for Transformation. However, relocated children did comprise more than 5% of the student population at a small number of schools.
- CHA representatives met with the principals of some of those schools to discuss ways of improving the academic performance of the relocated children.
- Some of those schools participated in a pilot program that aimed to increase coordination between the schools and the CHA Service Connector.
- Chapin Hall also found that CHA children were half as likely to be enrolled in selective, magnet, or charter schools as their non-CHA peers, and that CHA children whose families had relocated continued to be concentrated in underperforming schools even after relocation.
- As a result, CHA and Chicago Public Schools (CPS) began to work together to increase the percentage of CHA children enrolled in selective, magnet, or charter schools, and to encourage CHA parents to consider the performance of the schools their children will be attending when deciding where to relocate.

INCREASE HEAD START ENROLLMENT AMONG CHA CHILDREN

- Chapin Hall's analysis showed that only one-third of CHA three-year-olds and 44% of CHA four-year-olds were enrolled in an early childhood program in fall 2005.
- The Chicago Department of Children and Youth Services (DCYS) responded by targeting seven community areas into which a significant number of CHA families had relocated, with the goal of enrolling all CHA three- and four-year-olds in those communities in Head Start by September 2006.

IMPROVE LABOR MARKET OUTCOMES OF CHA PARENTS

- Chapin Hall's analysis of Unemployment Insurance wage data from the Illinois Department of Employment Security (IDES) suggested that CHA household heads fall into one of three groups: approximately half have no earnings from employment in any given year; one quarter work sporadically; and one quarter are regularly employed, but generally not earning enough to escape poverty.
- CHA used that information to develop a new employment initiative that involved collaboration between case managers from the Service Connector and the Illinois Department of Human Services and included different interventions for each of the three groups.
- CHA reported high take-up rates in the two communities where the initiative was piloted.
- Service Connector and IDES agreed to discuss how precertification of CHA residents and relocatees, who are categorically eligible for employer tax credits, could increase their attractiveness to potential employers.

INCREASE CHA CHILDREN'S ENROLLMENT IN ALL KIDS

- Chapin Hall presented data indicating that a significant percentage of CHA children were not enrolled in Medicaid or Kid Care and that these children were probably not covered by insurance provided by a parent's employer.
- Department of Healthcare and Family Services responded by targeting CHA developments and community areas with the most non-enrolled children for All Kids enrollment.
- To facilitate this process, some Service Connector case managers were trained to enroll CHA children in All Kids.

APPENDIX B

UNIVERSITY OF CHICAGO CRIME LAB

- *Opportunity:* Test a promising program to reduce youth violence and improve school outcomes through randomized control trial (RCT).
- *Challenge:* (1) Getting consent to link administrative data to the control group individuals, which is challenging because it never results in high enough consent rates for sufficient power to discern differences between control and treatment groups. (2) With consent, researchers could link administrative data on arrests and school outcomes to treatment and control group individuals. Obtaining school and police data is difficult because of legal restrictions and organizational burden.
- *Solution:* (1) University IRB allowed only treatment group to provide consent for use of their administrative data. Control group did not need to provide consent. Potential benefit outweighed potential harm to control group. (2) One agency had to be taught how to do record linkage and a second agency allowed researchers to receive confidential data to link treatment and control groups.
- *Gain:* Added to the evidence base on what interventions reduced violent behavior and increased school graduation. Showed how to work with university IRB to allow RCT to be implemented in a rigorous manner.

SF YOUTH DATABASE

- *Opportunity:* Learn about at-risk youth to serve them better.
- *Challenge:* (1) Have three agency directors agree to share their data; (2) actually produce data so that linkage can be done; (3) analyze data so that learning can occur to change practice and policy.
- *Solution:* (1) Agency directors did agree to share data; (2) through cumbersome methods of physically sharing data, data was linked; (3) consultants were engaged to analyze data to facilitate learning.
- *Gain:* Learned how much multiagency youth became involved in criminal justice system and the need to intervene earlier with these youth.

NYC IMPACT OF SHELTER USE AND HOUSING PLACEMENT ON HOMELESS ADULTS

- *Opportunity:* Change how homeless adults were served.
- *Challenge:* (1) Acquire data on adults using NYC shelters; (2) match to national death registry.
- *Solution:* (1) Obtained data by working with NYC agencies; (2) matched individuals in shelters to death records despite some data challenges.
- *Gain:* Learned that prompt resolution of homelessness many contribute to reduced mortality.

(See Stephen Metraux, Nicholas Eng, Jay Bainbridge, and Dennis P. Culhane, "The Impact of Shelter Use and Housing Placement on Mortality Hazard for Unaccompanied Adults and Adults in Family Households Entering New York City Shelters: 1990–2002," *Journal of Urban Health: Bulletin of the New York Academy of Medicine* 88 (2011): 1091–1104.)

CHICAGO DATA DICTIONARY

- *Opportunity:* Learn what data is available in City of Chicago databases.
- *Challenge:* (1) List databases/systems/datafiles in City of Chicago code and sister agencies; (2) develop a tool to store and manipulate the data to drive future data use and decision making.
- *Solution:* (1) Passed a city ordinance to conduct this work, contracted with research organization with experience doing this, aggregated existing lists of databases, met with agency personnel to identify contents of databases; (2) built data dictionary application to store all of this information.
- *Gain:* For the first time, there is a process in the city for all information about the data that it collects and is responsible for that can be accessed on a public-facing website.

NOTES

1. Bruce Bruce and Jennifer Bradley, *The Metropolitan Revolution: How Cities and Metros Are Fixing Our Broken Politics and Fragile Economy* (Washington, DC: Brookings Institution, 2013).
2. Lisa Gennetian, Jens Ludwig, Thomas McDade, and Lisa Sanbonmatsu, "Why Concentrated Poverty Matters," *Pathways,* Spring 2013, 10–13, http://www.stanford.edu/group/scspi/_media/pdf/pathways/spring_2013/Pathways_Spring_2013_Gennetian_Ludwig_McDade_Sanbonmatsu.pdf.
3. "U.S. Cities Growing Faster than Suburbs," *Real Time Economics* (*Wall Street Journal* blog), May 23, 2013, http://blogs.wsj.com/economics/2013/05/23/u-s-cities-growing-faster-than-suburbs/.
4. For insight into school district bankruptcies, see Kristi L. Bowman, "Before School Districts Go Broke: A Proposal for Federal Reform," *University of Cincinnati Law Review* 79 (2011): 895.

5. See http://www.data.gov/opendatasites for a list of cities all over the world that have open data sites.
6. See http://sustainablecitiesrcn.files.wordpress.com/2013/11/urban-ecodesign-in-the-city-of-barcelona-quaderns-temes-de-disseny-elisava-2012.pdf; http://urbanccd.org/articles/computation-enabled-design-chicago-lakeside-development; http://www.sf-planning.org/index.aspx?page=3051.
7. See http://worldlandscapearchitect.com/lakesim-prototype-connects-urban-planning-with-scientific-models-for-large-scale-development/#.UqUQdGRDuGI.
8. See http://www.alleghenycounty.us/uploadedFiles/DHS/about/DataWarehouseHistory.pdf; "Human Services: Sustained and Coordinated Efforts Could Facilitate Data Sharing while Protecting Privacy," GAO-13-106 (Washington, DC: U.S. Government Accountability Office, February 8, 2013).
9. Isidore Sobkowski and Roy S. Freedman, "The Evolution of Worker Connect: A Case Study of a System of Systems," *Journal of Technology in Human Services* 31, no. 2 (2013): 129–155, doi:10.1080/15228835.2013.772010.
10. Somin Sengupta, "No U.S. Action, So States Move on Privacy Law," *The New York Times,* October 31, 2013, www.nytimes.com/2013/10/31/technology/no-us-action-so-states-move-on-privacy-law.html?_r=0; "Human Services: Sustained and Coordinated Efforts."
11. Even the identities of perpetrators of abuse or neglect against their own children are not disclosed, except in heinous cases or murder.
12. E.g. see the University of Chicago's Master of Science in Computational Analysis and Public Policy, http://capp.sites.uchicago.edu/.
13. Chapin Hall at the University of Chicago has, since its inception in 1985 as a research and policy center, focused on a mission of improving the well-being of children and youth, families, and their communities. This is done through policy research – by developing and testing new ideas, generating and analyzing information, and examining policies, programs, and practices across a wide range of service systems and organizations. Primary colleagues instrumental in the creation of the IDB include Lucy Bilaver, Bong Joo Lee, John Van Voorhis, Mairead Reidy, and Nila Barnes.
14. Bruce Meyer and R. Goerge, "Errors in Survey Reporting and Imputation and Their Effects on Estimates of Food Stamp Program Participation," Working Paper 11-14 (Washington, DC: U.S. Census Bureau, Center for Economic Studies, 2011).
15. R. Goerge, C. Smithgall, R. Seshadri, and P. Ballard, "Illinois Families and Their Use of Multiple Service Systems," Issue Brief (Chicago, IL: Chapin Hall at the University of Chicago, 2010).
16. Agencies that participated in this project were (*state agencies*) Illinois Department of Children and Family Services, Illinois Department of Corrections, Illinois Department of Employment Security, Illinois Department of Healthcare and Family Services, Illinois Department of Human Services; (*local agencies*) Chicago Department of Children and Youth Services, Chicago Department of Human Services, Chicago Department of Public Health, Chicago Housing Authority, Chicago Public Schools, Cook County Juvenile Court, Mayor's Office of Budget

and Management; (*University of Chicago*) Chapin Hall Center for Children, Consortium on Chicago School Research, Office of Community and Government Affairs, School of Social Service Administration.

17. For a perspective on using the ACS, see http://www.prb.org/Publications/Articles/2009/2010censustestimony.aspx; Geolytics is an example of a commercial firm that provides yearly estimates of population at the block level.
18. R. Goerge, J. Dilts, D. Yang, M. Wassermann, and A. Clary, "Chicago Children and Youth 1990–2010: Changing Population Trends and Their Implications" (Chicago: Chapin Hall Center for Children at the University of Chicago, 2007).
19. R. Goerge, A. Harris, L. Bilaver, K. Franzetta, M. Reidy, D. Schexnayder, D. Schroeder, J. Staveley, J. L. Kreader, S. Obenski, R. Prevost, M. Berning, and D. Resnick, "Employment Outcomes for Low-Income Families Receiving Child Care Subsidies in Illinois, Maryland, and Texas" (Chicago: Chapin Hall at the University of Chicago, 2009).
20. See http://datasmart.ash.harvard.edu/news/article/chicagos-windygrid-taking-situational-awareness-to-a-new-level-259.
21. "Highlights of a Forum: Data Analytics for Oversight and Law Enforcement," GAO-13-680SP (Washington, DC: U.S. Government Accountability Office, July 15, 2013), http://www.gao.gov/products/GAO-13-680SP.
22. See http://www.aisp.upenn.edu/.

8 A European Perspective on Research and Big Data Analysis

Peter Elias

Introduction

Chapters 6 and 7 have highlighted the research potential of large datasets, particularly those which derive from administrative systems, monitoring devices, and customer databases, outlining the legal, ethical, and practical issues involved in gaining access to and linking between such data for research with potentially wide social and economic value. While the focus in these earlier chapters has been primarily upon the development of such data as research resources within the United States, many of the legal and ethical issues outlined have wider relevance. Here the focus shifts to an examination of a number of these matters from a European perspective.

A clearly stated ambition within Europe is to create a research environment (the European Research Area)[1] in which research interests are promoted via cross-border access to microdata, free from legal constraints and other obstacles which form impediments to this ambition such as the languages used[2] and differences in the scientific culture of research. This chapter explores the legal obstacles to wider cross-border data access for researchers based in different European countries and illustrates with examples how these are being addressed. The chapter is presented in two parts. The first part gives an historical overview of the progress that has been made across Europe to develop a harmonised approach to legislation designed to provide individuals and organisations with what has become known as the 'right to privacy'. It charts the immediate postwar efforts to establish the right to a private life and traces the impact this has had in terms of the use for research of electronic records containing personal information. In so doing an attempt is made not simply to highlight the forces that have helped shape the new legislation that the European Union (EU) is about to introduce, but to address the question as to what constitutes 'misuse of personal data' in terms of privacy concerns and to gauge how important this is for European citizens.

The second part examines the impact that these legislative developments have had and are continuing to have on cross-border access to microdata for research purposes. It identifies the tension between the ambitions of the EU to create a European Research Area within which research communities gain access to and share data across national boundaries and the desire within the EU to have a harmonised legislative framework which provides protection from misuse of personal information. How do these competing aims impact upon research plans and what mechanisms are being introduced to facilitate cross-border access to personal data for research? What part has the research community played in helping to negotiate the legal labyrinths that hinder cross-border access to and sharing of personal data? To shed light on these questions two developments are described which illustrate the approaches that are being pursued. The first of these focuses upon personal microdata which form 'official statistics'[3] and shows how academics have been working closely with national statistical agencies to overcome legal obstacles and promote cross-border research access to de-identified micro records. The second describes an initiative being undertaken within one EU member state, the United Kingdom, to promote wider access to and linkage between administrative data held by government departments and agencies operating under different national jurisdictions *within* one country. This example indicates the complex nature of the processes that are required to maintain the balance between the protection of privacy on the one hand and, on the other, the need for good research access to personal information in order to address important research issues.

The chapter concludes by looking forward, identifying the further work that needs to be done at the European level to achieve better transnational access to data held by government agencies, the private sector, and the academic community whilst providing adequate safeguards through which individuals can retain protection of their right to a private life.

The Evolution of a 'Right to Privacy' across Europe (1948–1981)

At the end of the Second World War many European countries had experienced destruction on a massive scale and most were financially impoverished. Keen to avoid any repetition of the punitive postwar conditions that had given rise to the growth of militaristic fascist movements after the First World War, the U.S. Marshall Plan of 1947 provided a huge macroeconomic injection, reflecting the desire of the United States to see the formation of a more integrated Europe. This in turn meshed with ideas about the formation of a 'United States of Europe' which were growing among Western European countries – a political and economic alliance which could form a counterweight to the postwar dominance of the USSR across Eastern Europe.

The International Committee of the Movements for European Unity was established in 1946. As an umbrella organisation, this committee facilitated discussions between a number of countries (Belgium, France, the Netherlands, Luxembourg, and the United Kingdom) which led to a congress held in 1948 calling for political and economic union between the states of Europe. France and Belgium pushed for the creation of a federal Europe, whereas the United Kingdom expressed its wishes to see a consultative framework established for cooperation at the economic level. As a compromise between these two positions the Council of Europe[4] was formed in 1949, tasking its members to

> achieve a greater unity . . . for the purpose of safeguarding and realising the ideals and principles which are their common heritage and facilitating their economic and social progress.[5]

One of the most important steps taken by the Council of Europe was the development of the European Convention on Human Rights. Drawing heavily upon the Universal Declaration of Human Rights, which was adopted by the UN General Assembly in 1948, the 1950 European Convention on Human Rights (ECHR) introduced the concept of the right to privacy. Article 8 of the ECHR states that

> [e]veryone has the right to respect for his private and family life, his home and his correspondence

and

> [t]here shall be no interference by a public authority with the exercise of this right except such as in accordance with the law and is necessary in a democratic society in the interest of national security, public safety or the economic well-being of the country, for the protection of disorder or crime, for the protection of health or morals, or the protection of the rights and freedoms of others.[6]

What distinguished the ECHR from the declaration of the United Nations was the setting up of the European Court of Human Rights, giving individuals the standing to file legal claims for the restitution of their rights. This, in turn, has established a body of case law surrounding the interpretation of the 'right to respect for private and family life'. Initially these cases covered issues such as the legality of activities by the state to intercept phone conversations and the monitoring by employers of their employees' relationships with others.[7] However, the transformation in electronic communication which commenced in the late 1970s, together with the widespread adoption of electronic processing of personal information, caused some European countries to re-examine the interpretation of the 'right to a private life' and to consider the need for safeguards specific to the increasing use of electronic data. In the United Kingdom a committee was set up 'to consider whether legislation was needed to protect individuals and organisations from intrusions into their personal privacy'.[8] In its response to this committee's report, the UK government concluded that

> those who use computers to handle personal information . . . can no longer remain the sole judges of whether their own systems adequately safeguard privacy.[9]

By 1978, and following the report of a further committee,[10] moves were afoot to create a data protection authority which would have responsibility for drafting new codes, breaches of which would be criminal offences. These moves were echoed at the European level by the Council of Europe and more widely within the Organization for Economic Cooperation and Development (OECD). Early in 1978 the OECD established a group of experts, tasked to elaborate a set of principles governing the protection of personal data. Working in close collaboration with the Council of Europe, seven basic principles of data protection were defined. These were:

1. There should be limits to the collection of personal data, which should be collected by fair and lawful means and, where possible, with the consent of the data subject.
2. Personal data should be relevant to the purpose for which they are required, should be accurate, complete and up-to-date.
3. The purpose for which personal data are required should be specified not later than at the time of collection.
4. Personal data should not be disclosed or used for purposes other than that for which they were collected, except with the consent of data subjects.
5. Personal data should be protected by reasonable security safeguards against unauthorised access, loss, destruction, modification or disclosure.
6. Means should be established to facilitate the existence and nature of personal data and the identity and residence of the data controller.
7. Data subjects should have the right to gain access to their data, to challenge such data, to request erasure and to have the right to challenge any denial of these rights.[11]

By 1980 the Council of Europe had proposed a *Convention for the Protection of Individuals with regard to the Automatic Processing of Personal Data.*[12] This convention reflected the seven basic principles agreed by OECD member countries. Additionally the convention specified that data holders should take appropriate security measures against accidental or unauthorised destruction of data as well as unauthorised access, alteration, or dissemination. Importantly, the convention had a specific chapter on transborder flows of personal data, stating that

> [a] party shall not, for the sole purpose of the protection of privacy, prohibit or subject to special authorisation transborder flows of personal data going to the territory of another part.[13]

An important derogation from this provision relates to the situation where national legislation covers and protects specific types of personal data and there is no equivalent legislative protection in the country to which data might flow.

Neither the OECD nor the Council of Europe has legal authority to impose these conditions within national legislation. Signatories to the Council of Europe convention were agreeing to a provision that they would implement domestic legislation to realise its principles. However, there was no guarantee that this would happen in a consistent way across Europe. Nonetheless, the convention, which was adopted by most of the Council of Europe countries,[14] represented a landmark in establishing a framework designed to safeguard an individual's right to privacy in the electronic age. In 1981, those who drafted the convention could have had little idea of the rapid pace of technological change that lay ahead.

The Right to Privacy and the European Union (1982–2014)

Throughout the 1980s the European Commission, the body that forms the administrative heart of the European Economic Community (now the European Union), was slow to pick up on the principles enunciated by the OECD and the Council of Europe. In part this reflected the fact that the European Economic Community (EEC) was concerned more with trade and economic considerations. By 1985 the Council of Europe's Convention on Personal Data Protection came into effect, but its adoption among the EEC signatory countries was uneven. Recognising this, and in line with its widening remit to cover more than trade and other economic objectives, the EEC published a draft data protection directive in 1990.

European Union *Directives* and *Regulations*

It is important to clarify the difference between two legal instruments, *directives* and *regulations* that have been available to the EEC and now the EU. A *directive* places a legal requirement on member states to achieve a specific result. It does not specify how the result should be achieved; leaving countries with a degree of leeway with respect to the national legislation that might be required to achieve the stated aim of the directive. A *regulation* does not require member states to pass legislation to achieve its aim. It is a legal instrument that applies to all member states and which takes precedence over relevant national legislation.

The European Union, formed when the 12 member states of the EEC signed the Treaty of Maastricht in 1995, wished to ensure that the principles enunciated in the Council of Europe convention should become enshrined in the articles establishing the European Union. After a considerable period of consultation within and across member states of the EEC, the EU Data Protection Directive was passed in 1995.[15] While helpful in providing the framework for national legislation, the directive did not legally bind member states in a direct way. Under the principle known as 'transposition', each country should find ways to implement a directive such that its minimum conditions are met. Additionally, an individual does not have recourse to legal remedies if they consider that their privacy has been breached unless such remedies exist within national legislation.

The 1995 directive has been hugely influential in shaping the legislation in EU countries with regard to the implementation of the principles set out by the OECD and the Council of Europe some 15 years earlier. As new countries have been admitted to the European Union, they are required to ensure that the aims of the 1995 directive are met. The directive additionally gives citizens the right to access their personal data and to request it to be removed from processing if incomplete or inaccurate.

Despite its influence, the European Commission has concluded that, given the immense technological changes that had occurred since it was passed, the directive should be reformed and its operation strengthened by positioning it more directly within EU legislation through the creation of a new regulation.

The EU General Data Protection Regulation

A proposed new law, the General Data Protection Regulation (GDPR), was published by the European Commission in January 2012, aiming for its adoption by 2014 and implementation over the following two years. In addition to the fact that this moves data protection from a set of requirements to be achieved via national legislation to a legal requirement superseding national legislation, the other major changes from the preceding directive cover:

- the scope of the legislation (it will apply to all organisations and individuals based outside the EU if they process personal data of EU residents);
- the establishment of national data protection authorities to be coordinated by a European Data Protection Board;
- the need for consent (valid consent for data to be collected and the purposes for which it will be used must be explicit rather than implicit);
- the right to be forgotten (personal data must be removed from use if consent is withdrawn).

While much remains to be clarified before the proposed regulation is enacked it is clear that the need for consent could have a serious impact upon the collection, access to, and sharing of research data.[16] Some indication of these concerns can be gathered from documents prepared by those organisations with research interests.[17] Of particular concern here is Article 83 of the proposed regulation, which is termed a 'derogation' – an exception to the requirements of the proposed regulation regarding the need for explicit consent for data collection and to the restrictions on the processing of data which are deemed 'sensitive'.[18] For this derogation to apply, specific conditions are to be placed on the processing of personal data for 'historical, statistical and scientific research purposes':

> Within the limits of this Regulation, personal data may be processed for historical, statistical or scientific research purposes only if:
> (a) these purposes cannot be otherwise fulfilled by processing data which does not permit or not any longer permit the identification of the data subject
> (b) data enabling the attribution of information to an identified or identifiable data subject is kept separately from the other information as long as these purposes can be fulfilled in this manner.[19]

Additionally, the proposed regulation would empower the European Commission to further specify the criteria and requirements associated with the processing of personal data for historical, statistical, or scientific research purposes. The regulation provides for the imposition of sanctions upon those individuals or organisations that fail to comply with the new Europe-wide legislation. Unintentional breaches of the regulation could lead to a warning from data protection authorities and/or regular audits of data protection practices. Stiffer penalties can be imposed (of up to €100 million or 5% of turnover in the case of organisations). Penal sanctions that would be directly applicable to individuals fall outside the competence of the European Union.

Many amendments have been submitted to the European Commission seeking to remove this derogation or to limit it only to personal data concerning the health status of the data subject. If adopted, such amendments would limit research based on personal data to those data for which data subjects had given personal consent. For research on a wide range of topics in the social, economic, behavioural, and medical sciences, analysis of large linked datasets is an essential part of the research process. Even if consent had been obtained at the time data were originally collected, the reuse of such data for a new purpose would require consent to be obtained again.

Finally, it should be noted that the GDPR does not oblige data controllers to provide access to personal information for scientific research purposes. Such access would be legal, but cannot be claimed.

The Views of EU Citizens

For legislation to work effectively it must reflect an established need for a legal instrument and it must have widespread support for the outcome it seeks to achieve. In 2010, and as part of the preparatory work to inform the development of the GDPR, the European Commission organised a survey on attitudes to and awareness of personal privacy. More than 25,000 Europeans aged 15 and over were interviewed across 27 member states.[20] Among many areas which were explored, questions were asked about the following:

- the nature of information considered personal;
- perceived necessity of disclosing personal information;
- knowledge of regulations governing the processing of personal data.

Regarding the responses to categories of data deemed personal, the EU average responses showed that financial and medical information are ranked higher than other categories of data, with about three-quarters placing these two categories top of the list. There were large variations between countries, with more than three-quarters of respondents from northwest Europe stating that financial information is personal compared with just over 40% in Poland and Romania. Similar variation was found for the view that medical information is personal.

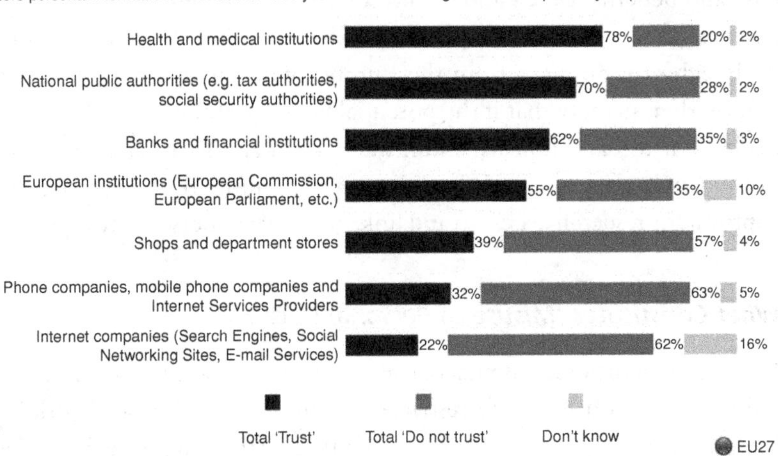

Figure 1 The views of EU citizens: trust in the ability of various types of institutions to protect personal information. *Source:* TNS Opinion and Social, *Special Eurobarometer 359: Attitudes on Data Protection and Electronic Identity in the European Union*, Brussels, 2011.

Three-quarters of Europeans interviewed agreed that disclosing personal information is an increasing part of modern life. Again, respondents in northwest Europe were more inclined to this view than those in central and southern Europe. Unsurprisingly, only one-third of Europeans are aware of the existence of a national public authority protecting their rights regarding their personal data.

Figure 1 illustrates the wide variation that exists in the public perception of the extent to which different organisations can be trusted to protect personal information.

While the survey contained no specific questions relating to the use for research of personal data, it is clear from responses to questions about who collects personal data that there is a reasonably high degree of trust in health and medical institutions and national public authorities to protect personal information. Given that these organisations are a major source of personal data with research value, it is incumbent upon them to maintain (and possibly increase further) this trust.

Some member states have also undertaken research into public attitudes towards access to and linkage between personal information. A UK report for the Wellcome Trust was based upon qualitative research designed to understand the general public's attitudes to different types of personal data and data linking.[21] The research looked at whether health data are viewed differently from other types of data, and what are the perceived risks and benefits, to self and society, of linking different kinds of data for research purposes and other purposes. The conclusions of this small-scale study echo the findings from the Europe-wide survey conducted some years earlier, namely that if the potential benefits of the research could be made clear to data subjects, and if the risks of any harm arising to them from inappropriate use of linked data were minimised, there was general approval for research access to and linkage between personal data.

What Constitutes 'Misuse of Personal Data'?

For research purposes it is vital that access to and linkage between personal information should not be restricted by the new European legislation. Researchers rarely need to know the identities of those who are the subjects of their research, working usually with what are termed 'de-identified data'. But linkage between different datasets does require that information, which is unique to individuals, is available in the data sources to be linked. Data security is key to the maintenance of public trust in the use of personal data, and security measures must be maintained in order to avoid misuse. It is important therefore to identify what constitutes a 'misuse of personal information' in the context of research. There is no legal definition of research misuse of personal data,[22] and it has been left to data controllers to establish their own rules over what constitutes misuse. Typically, these rules relate to inadvertent or deliberate disclosure of identities, unauthorised data use, and the failure to maintain data security. More specifically, they cover:

- placing within the public domain information which reveals (or has the potential to reveal) the identities of data subjects;
- using personal information from data subjects for some purpose other than research which has been planned and agreed with data controllers;
- failing to maintain personal information in a secure condition, such that others who do not have access to such data for research purposes may gain access.

It should be noted that these conditions do not make any statement about the purpose of the research nor of the 'fitness for purpose' of the data to be used for research and the quality and accuracy of the research findings stemming from the research. These are matters which should be regulated by data controllers and may be specific to local conditions and cultural sensitivities or which will be revealed by peer review of research findings.

There is, as yet, no clear framework for the elaboration of these rules in a consistent manner across Europe. As a corollary, there is no agreement on the nature of the penalties to be imposed as a result of a breach of these rules.

Where Do We Now Stand?

Since the late 1940s, European countries and Europe-wide institutions have been concerned to protect individual privacy, particularly insofar as such privacy relates to the electronic processing of personal data. Through various conventions, guidelines, directives, and now a regulation, the countries of Europe have been moving gradually towards a harmonised and consistent approach to the application of a right to privacy. In so doing, a balance has been sought between the adoption of policies, procedures, and laws which provide this right, and the value to society of research which makes use of personal data to address issues which are of public benefit. For the study of a wide range of issues including health-related matters, the distribution of social and economic disadvantage, educational progress, environmental conditions, and political developments, access to relevant personal data is vital for research. A major step forward is now under way to harmonise European legislation on data protection. This need for a balance is recognised in the proposed European regulation but the derogation which facilitates such use of personal data by freeing research use from the need for consent is under attack from a range of interest groups and lobbyists who see this as the 'thin end of a wedge' – a set of clauses which allows for the processing of personal data without the consent of data subjects and for research purposes which may be spurious. While these concerns are understandable, it is up to the research communities, their funders, and their policy beneficiaries to illustrate how these concerns will be addressed and the concomitant risks to individual privacy minimised. The next section of this chapter presents two examples that illustrate how this can be achieved.

Overcoming Obstacles to Data Access and Sharing

The preceding sections of this chapter have focussed specifically upon the development of the legal definition of a 'right to privacy' across Europe and give some indication of the restrictions that this can place upon the research use of personal data. Currently, it is national legislation and the national interpretation of the EU concept of 'data protection' that creates the biggest obstacle to cross-border data sharing. For some countries of the European Union, such as Germany with its federal structure, the ambitions of federally funded research efforts to conduct comparative research within the country and between states (Länder) is restricted by the legal power of the states relative to the federal government.[23] Similarly, within the United Kingdom, the devolution of power in some administrative areas from the UK government to the devolved nations (Scotland, Wales, and Northern Ireland) has created data-sharing issues. Given that problems can exist within countries, it will come as no surprise to find that EU institutions face difficulties in permitting transnational access for research as the next section will demonstrate.

Cross-Border Access to Microdata Designated as Official Statistics

Nowhere is this tension between the need for research access to personal data and national privacy laws more evident than at the Statistical Offices of the European Communities (Eurostat). While member states are required to provide harmonised statistical data to Eurostat, which are then made publically available in aggregate form, access to the national official microdata records supplied to Eurostat remains subject to national laws. This, in turn, led to a situation where official microdata records that may be generally available for research in one country could not be released by Eurostat without the consent of all countries.

Under considerable pressure from research communities across the European Union, resolution of this problem was high on the Eurostat agenda for many years. In 2002 a Commission Regulation came into force which permitted 'authorised bodies' (e.g. universities and recognised research institutes) to have access to anonymised microdata from a limited number of EU-wide surveys, subject to approval by all national statistical authorities.[24] While this did provide an avenue for cross-border research access, the process of gaining access was slow and expensive, often taking more than six months from initial application and with no guarantee of a successful outcome. This regulation was repealed in 2013 with a new regulation designed to improve the speed and efficiency of cross-border access to official microdata records for research purposes.[25] Under this new regulation, Eurostat has responsibility for overseeing the operation of any proposed cross-border data access request. The national statistical authorities that provided the data to Eurostat are consulted on each research proposal submitted by researchers interested in access to EU confidential personal information. If the statistical authority of one country does not accept a particular research proposal, the data for this country is then excluded from the cross-national database to be supplied for the stated research purpose. The decision of one country now has no influence on the access to other countries' data.

To take advantage of the changing EU landscape with respect to cross-border access to official microdata, **Data without Boundaries** (DwB) was set up as an Integrating Activity.[26] With a four-year budget covering the period 2011–2015, DwB is an initiative led primarily by academic research groups to overcome the legal and other barriers to cross-border research based upon access to microdata records held by national statistical institutes. There are three main strands to the work plan for DwB. First, it encourages transnational access to data held at national statistical institutes by funding the travel and subsistence costs for researchers and paying any usage charges they may face. Second, it is promoting the development of a network of remote access centres, through which access can be given to authorised researchers without their having to travel to the place where national official microdata records are stored. Remote access is provided by various means, including the use of thin-client technology, submission of analytical code by e-mail, and the use of encryption devices. Third, it has a range of additional activities (standards development, training workshops, etc.) all of which are designed to promote wider use of European official microdata for research.

While these developments are to be welcomed, it is unlikely that an access facility operated by Eurostat, providing remote access to official EU microdata by European researchers will become a realistic proposition in the short run. Nevertheless, the hope is that, as some countries develop remote access to data held in their national statistical institutes (e.g. as is now the case in the Netherlands and the United Kingdom), other countries will perceive the benefits that ensue in terms of the research it promotes.

Overcoming National Obstacles to Research Data Access and Sharing

As the experience of DwB has revealed, the major obstacles to trans-national data access often lie at the national level. However, it is not simply a question of the need for revised national legislation, or for a new European regulation to be implemented. National issues can be tricky and complex as will be illustrated by recent developments within the United Kingdom to improve research access to and linkage between public sector administrative datasets.

Administrative data in the United Kingdom, as in most other European countries and the United States, are derived from a range of activities which constitute the everyday functioning of public agencies. Raising tax revenues from individuals and businesses and from the sale of goods, paying social security benefits and pensions, recording educational progress, monitoring judicial systems all provide an 'electronic trail' that has significant research potential given that the data already exist, are usually population wide, and are continuous through time. While individual government departments and agencies may use such data for performance monitoring or service delivery, the data are often underutilised as research resources. Attempts to gain access to such data for research purposes have sometimes been denied, especially where requests have been made to link data held by different departments.

In the area of medical statistics, personal data which are the byproduct of an administrative system are now being linked to generate new and powerful research resources. While progress has been slow to digitise the UK's National Health Service, as new national datasets started to become available the health research community acted quickly to establish the legal and ethical framework for access to and linkage between pseudonymised health records. Based on this experience, and recognising the vast research potential of other categories of administrative data, a group of research funding agencies convened a workshop in May 2011 to review progress made across the biomedical sciences in respect of the research use of data derived from administrative systems in health services (e.g. hospital admissions, GP practice visit, patient prescribing data, etc.) and to consider the much slower rate of progress in areas of wider socioeconomic interest (incomes, social security, education, housing, environmental conditions, etc.). Given the lack of progress in the latter area the workshop advocated that a taskforce should be set up to detail the problems and to propose solutions.

In its report, the Administrative Data Taskforce[27] identified the major obstacles surrounding access to, linking between, and sharing of personal information for research purposes. These were

- the legal status of those bodies holding data;
- the lack of agreed and common standards covering data security and the authentication of potential research users;
- the need for public support for the use for research of de-identified personal information;
- the need for a coordinated governance structure for all activities associated with accessing, linking, and sharing personal information.

The report of the taskforce was welcomed by the UK government, which allocated sufficient funds to allow for a Research Data Centre to be established in each of the four countries of the United Kingdom, coordinated and supported by a new national service (the Administrative Data Research Service) and with the formation of a UK-wide governance structure currently under construction. Importantly, a legal team was set up to draft new legislation which would permit data sharing and linkage by those bodies which currently faced legal barriers to such activities.

This example demonstrates how complex obstacles to data sharing can be overcome if there is sufficient impetus from the scientific community. An important element of this impetus comes from demonstration of the research value of new forms of data and the public benefit that derives from the research they underpin.

EU Privacy Legislation and Research – Looking Ahead

For the past two decades European countries have made concerted efforts to prevent the misuse of personal information held in electronic formats. However, the approach that was adopted has led to a situation where there are varying degrees of protection in different countries. Coupled with the technological changes that have taken place over this same period, such as cloud storage and processing of data, the development of high-speed data networks, and the flow of personal data through social media, the European Commission has taken the bold step of proposing the introduction of a new legal instrument that would supersede all national legislation, thereby providing a common legal framework for the protection of personal information. The proposed new law, the General Data Protection Regulation, is making steady progress through the European institutions and is likely to be passed by the European Parliament and adopted by the European Council in 2014. Fundamental to the new law is the notion that individuals should give consent for their personal data to be processed. Recognising that this could have serious consequences for research, a derogation has been included in the legislation which will permit the processing of personal information for research purposes if consent is impractical or infeasible.

In the widescale public consultation associated with the introduction of the new legislation, concerns have surfaced about the way this derogation will operate. What is classed as research? What if researchers fail to protect the information they are processing even if it has been de-identified? In other words, will this derogation provide a loophole, allowing the processing of personal information in ways which, although legal, could cause reputational or physical harm to individuals? The fact that amendments to the new law have been put forward to strike out this derogation indicates that some unease exists among sections of the European citizenry. A survey of the adult population across 27 countries indicates that although there is a reasonable degree of trust among the population in the ability of different institutions to hold their personal data in a secure manner, for data collected by some institutions, notably private sector organisations, more than half the population does not trust them to hold data securely. In this situation, research use of such data may give rise to public concern about the nature of the derogation for research in the new law.

Looking ahead, the next few years are likely to be difficult for researchers who wish to build significant new research resources from the growing volumes of 'big data' that can be linked via personal identifying information (e.g. official statistical sources, private sector databases, charitable bodies) especially where such data may cross national boundaries. Those countries that traditionally have had a cautious approach to research access to administrative data (e.g. Germany) are unlikely to modify their practices following adoption of the new law. Conversely, those countries that have a more open approach (e.g. the Netherlands) may now have to reconsider how they work as the European Commission develops the protocols that will sit alongside the new legislation – protocols that will define with more clarity how the derogation for research will be operated.

The uncertainties that will prevail at the European level should not stop individual member states from moving ahead with ambitious plans to link large datasets held by the public and private sectors. Foremost in this respect is the work currently being undertaken within the United Kingdom to facilitate linkage at the individual level between comprehensive health records, tax and social security data, education records, and criminal justice records. The goal within the United Kingdom is to establish mechanisms that allow researchers to build large-scale longitudinal databases on individuals and organisations using detailed information on individuals that already exists within different public and private sector bodies. This work is driven by the research community, working closely with data controllers to ensure that the steps taken to move towards this goal are legal, ethical, well governed, and capable of producing benefits for the well-being of the citizenry.

Translating the actions of any single member state into Europe-wide procedures which allow cross-border access to large linked datasets is going to prove slow and cumbersome, even though the new General Data Protection Regulation is designed to improve efficiency in this respect. However, the time is right for the development of ethical guidelines for governance of the research use of data, particularly where such data arise from administrative systems, customer databases, monitoring devices, or communications and where transnational access for research is proposed. While the research community must take a clear lead in this area,[28] the OECD, European Commission, and Council of Europe have shown previously that they can coordinate the resources and expertise needed to develop such guidelines and that they have the collective voice required for their implementation.

NOTES

1. The European Research Area is described by the European Commission as 'a unified research area open to the world, based on the internal market, in which researchers, scientific knowledge and technology circulate freely'; European Commission, *Proposal for a Regulation of the European Parliament and of the Council on the protection of individuals with regard to the processing of personal data and on the free movement of such data* (General Data Protection Regulation), COM (2012) 11, Brussels, January 25, 2012.
2. The European Union has 24 official and working languages.
3. Data that are collected via national statistical agencies, often via legal instruments that require individuals and organisations to provide the information that is requested.
4. Not to be confused with the *Council of the European Union*, also informally known as the *EU Council*, which is where national ministers from each EU country meet to adopt laws and coordinate policies, and the *European Council*, another EU institution, where EU leaders meet around four times a year to discuss the European Union's political priorities. The **Council of Europe** is not an EU institution.
5. *Statute of the Council of Europe*, London, May 5, 1949, Article 1, http://conventions.coe.int/Treaty/en/Treaties/Html/001.htm.
6. *European Convention on Human Rights* (ECHR), Rome, November 4, 1950, Article 8.1; ECRH, Article 8.2.
7. See e.g. Niemietz v. Germany, 251 Eur. Ct. H.R. (ser.A) (1992) and Copland v. UK, 253 Eur. Ct. H.R. (sec.IV) (2007).
8. Known as the report of the Younger Committee: UK House of Lords, Committee on Privacy, *Report of the Committee on Privacy*, Kenneth Younger, chair, Cmnd 5012, July 1972.
9. UK Government, White Paper, Cmnd 5353, 1975. Some indication of the scale of change that was under way comes from a survey conducted as part of the work of the Younger Committee. In April 1971 it was estimated that the total number of computers in the United Kingdom in use or on order for all purposes was 6,075. By 1995 it was estimated that at least half a million data users were obliged to register under the 1984 Data Protection Act (see n.14).
10. UK Home Department, *Report of the Data Protection Committee* (Lindop Report), Cmnd 7341, 1978.

11. *OECD Guidelines on the Protection of Privacy and Transborder Flows of Personal Data*, Organisation for Economic Co-operation and Development, September 1980.
12. 'Automatic processing' is defined in the convention as 'the following operations if carried out in whole or in part by automated means: storage of data, carrying out of logical and/or arithmetic operations on those data, their alteration, erasure, retrieval or dissemination' (European Convention on Personal Data Protection, Strasbourg, January 1981, Article 2).
13. European Convention on Personal Data Protection, Article 12.
14. In the United Kingdom, it led to passage of the 1984 Data Protection Act.
15. Officially known as Directive 95/56/EC 'on the protection of individuals with regard to the processing of personal data and on the free movement of such data'.
16. For an overview of the more than 3,000 amendments which have been submitted in the committees involved with the regulation, plus analysis of the weight of opinion towards strengthening or weakening the regulation, see http://lobbyplag.eu/lp.
17. Jeremy Farrar et al., "Data Protection" (letter to the editor), *The Times* (London), January 29, 2014; see also http://www.wellcome.ac.uk/News/Media-office/Press-releases/2014/WTP055581.htm.
18. Data that relate to issues such as sexual orientation and religious and political beliefs.
19. General Data Protection Regulation, Article 83(1).
20. A regionally stratified sample of household addresses was used for sampling purposes. The individual selected for interview was on the 'nearest birthday' rule. TNS Opinion and Social, *Special Eurobarometer 359: Attitudes on Data Protection and Electronic Identity in the European Union*, Brussels, 2011.
21. Wellcome Trust, *Summary Report of Qualitative Research into Public Attitudes to Personal Data and Linking Personal Data*, London, 2013. Available at http://www.wellcome.ac.uk/stellent/groups/corporatesite/@msh_grants/documents/web_document/wtp053205.pdf.
22. As opposed to a 'breach of personal data processing', which is defined in European law and relates to the requirement of data controllers to prevent a breach of security leading to accidental or unlawful destruction, loss, unauthorised disclosure of access to personal information (E-privacy Directive 2002/58/EC, Article 2(i)).
23. As an example, data collected within the National Educational Panel Study, a multicohort longitudinal study of educational progress, may not be used for cross-state comparative research that identifies specific states.
24. Commission Regulation (EC) No 831/2002 on Community Statistics, concerning access to confidential data for scientific purposes.
25. Commission Regulation (EU) No 557/2013 as regards access to confidential data for scientific purposes.
26. This is the formal name for projects that help to integrate research resources and researchers across the European Union.
27. The UK Economic and Social Research Council, the UK Medical Research Council and the Wellcome Trust, *The UK Administrative Data Research Network: Improving Access for Research and Policy*, Report from the Administrative Data

Taskforce, December 2012, http://www.esrc.ac.uk/_images/ADT-Improving-Access-for-Research-and-Policy_tcm8-24462.pdf.

28. See e.g. P. Elias and B. Entwisle, *New Data for Understanding the Human Condition: International Perspectives* (Paris: OECD, 2013), http://www.oecd.org/sti/sci-tech/new-data-for-understanding-the-human-condition.htm.

9 The New Deal on Data: A Framework for Institutional Controls

Daniel Greenwood, Arkadiusz Stopczynski, Brian Sweatt, Thomas Hardjono, and Alex Pentland

Introduction

In order to realize the promise of a Big Data society and to reduce the potential risk to individuals, institutions are updating the operational frameworks which govern the business, legal, and technical dimensions of their internal organizations. In this chapter we outline ways to support the emergence of such a society within the framework of the *New Deal on Data*, and describe future directions for research and development.

In our view, the traditional control points relied on as part of corporate governance, management oversight, legal compliance, and enterprise architecture must evolve and expand to match operational frameworks for big data. These controls must support and reflect greater user control over personal data, as well as large-scale interoperability for data sharing between and among institutions. The core capabilities of these controls should include responsive rule-based systems governance and fine-grained authorizations for distributed rights management.

The New Realities of Living in a Big Data Society

Building an infrastructure that sustains a healthy, safe, and efficient society is, in part, a scientific and engineering challenge which dates back to the 1800s when the Industrial Revolution spurred rapid urban growth. That growth created new social and environmental problems. The remedy then was to build centralized networks that delivered clean water and safe food, enabled commerce, removed waste, provided energy, facilitated transportation, and offered access to centralized health care, police, and educational services. These networks formed the backbone of society as we know it today.

These century-old solutions are, however, becoming increasingly obsolete and inefficient. We now face the challenges of global warming, uncertain energy, water, and food supplies, and a rising population and urbanization that will add 350 million people to the urban population by 2025 in China alone.[1] The new challenge is how to build an infrastructure that enables cities to be energy efficient, have secure food and water supplies, be protected from pandemics, and to have better governance. Big data can enable us to achieve such goals. Rather than static systems separated by function – water, food, waste, transport, education, energy – we can instead regard the systems as dynamic, data-driven networks. Instead of focusing only on access and distribution, we need networked and self-regulating systems, driven by the needs and preferences of citizens – a 'nervous system' that maintains the stability of government, energy, and public health systems around the globe. A *control* framework should be established which enables data to be captured about different situations, those observations to be combined with models of demand and dynamic reaction, and the resulting predictions to be used to tune the nervous system to match those needs and preferences.

The engine driving this nervous system is big data: the newly ubiquitous digital data now available about so many aspects of human life. We can analyze patterns of human activity within the digital breadcrumbs we all leave behind as we move through the world: call records, credit card transactions, GPS location fixes, among others.[2] These data, which record actual activity, may be very different from what we put on Facebook or Twitter; our postings there are what we choose to tell people, edited according to the standards of the day and filtered to match the persona we are building. Although mining social networks can give great insight into human nature,[3] the value is limited for operational purposes.[4]

The process of analyzing the patterns within these digital breadcrumbs is called 'reality mining.'[5] The Human Dynamics research group at MIT found that these patterns can be used to tell us if we are likely to get diabetes,[6] or whether we are the sort of person who will pay back loans.[7] By analyzing them across many people, we are discovering that we can begin to explain many things – crashes, revolutions, bubbles – that previously appeared unpredictable.[8] For this reason, the magazine *MIT Technology Review* named our development of reality mining one of the 10 technologies that will change the world.[9]

The New Deal on Data

The digital breadcrumbs we leave behind are clues to who we are, what we do, and what we want. This makes personal data – data about individuals – immensely valuable, both for public good and for private companies. As the European Consumer Commissioner, Meglena Kuneva, said recently, "Personal data is the new oil of the Internet and the new currency of the digital world."[10] The ability to see details of so many interactions is also immensely powerful and can be used for good or for ill. Therefore, protecting personal privacy and freedom is critical to our future success as a society. We need to enable more data sharing for the public good; at the same time, we need to do a much better job of protecting the privacy of individuals.

A successful data-driven society must be able to guarantee that our data will not be abused – perhaps especially that government will not abuse the power conferred by access to such fine-grained data. There are many ways in which abuses might be directly targeted – from imposing higher insurance rates based on individual shopping history,[11] to creating problems for the entire society, by limiting user choices and enclosing users in information bubbles.[12] To achieve the potential for a new society, we require the *New Deal on Data*, which describes workable guarantees that the data needed for public good are readily available while at the same time protecting the citizenry.[13]

The key insight behind the New Deal on Data is that our data are worth more when shared. Aggregate data – averaged, combined across population, and often distilled to high-level features – can be used to inform improvements in systems such as public health, transportation, and government. For instance, we have demonstrated that data about the way we behave and where we go can be used to minimize the spread of infectious disease.[14] Our research has also shown how digital breadcrumbs can be used to track the spread of influenza from person to person on an individual level. And the public good can be served as a result: if we can see it, we can also stop it. Similarly, if we are worried about global warming, shared, aggregated data can reveal how patterns of mobility relate to productivity.[15] This, in turn, equips us to design cities that are more productive and, at the same time, more energy efficient. However, to obtain these results and make a greener world, we must be able to see people moving around; this depends on having many people willing to contribute their data, if only anonymously and in aggregate. In addition, the Big Data transformation can help society find efficient means of governance by providing tools to analyze and understand what needs to be done, and to reach consensus on how to do it. This goes beyond simply creating more communication platforms; the assumption that more interaction between users will produce better decisions may be very misleading. Although in recent years we have seen impressive uses of social networks for better organization in society, for example during political protests,[16] we are far from even starting to reach consensus about the big problems: epidemics, climate change, pollution – big data can help us achieve such goals.

However, to enable the sharing of personal data and experiences, we need secure technology and regulation that allows individuals to safely and conveniently share personal information with each other, with corporations, and with government. Consequently, the heart of the New Deal on Data must be to provide both regulatory standards and financial incentives enticing owners to share data, while at the same time serving the interests of individuals and society at large. We must promote greater idea flow among individuals, not just within corporations or government departments.

Unfortunately, today most personal data are siloed in private companies and therefore largely unavailable. Private organizations collect the vast majority of personal data in the form of mobility patterns, financial transactions, and phone and Internet communications. These data must not remain the exclusive domain of private companies, because they are then less likely to contribute to the common good; private organizations must be key players in the New Deal on Data. Likewise, these data should not become the exclusive domain of the government. The entities who should be empowered to share and make decisions about their data are the people themselves: users, participants, citizens. We can involve both experts and use the wisdom of crowds – users themselves interested in improving society.

Personal Data: Emergence of a New Asset Class

One of the first steps to promoting liquidity in land and commodity markets is to guarantee ownership rights so that people can safely buy and sell. Similarly, a first step toward creating more ideas and greater flow of ideas – idea liquidity – is to define ownership rights. The only politically viable course is to give individual citizens key rights over data that are about them, the type of rights that have undergirded the European Union's Privacy Directive since 1995.[17] We need to recognize personal data as a valuable asset of the individual, which can be given to companies and government in return for services.

We can draw the definition of ownership from English common law on ownership rights of possession, use, and disposal:

- *You have the right to possess data about yourself.* Regardless of what entity collects the data, the data belong to you, and you can access your data at any time. Data collectors thus play a role akin to a bank, managing data on behalf of their 'customers'.

- *You have the right to full control over the use of your data.* The terms of use must be opt in and clearly explained in plain language. If you are not happy with the way a company uses your data, you can remove the data, just as you would close your account with a bank that is not providing satisfactory service.
- *You have the right to dispose of or distribute your data.* You have the option to have data about you destroyed or redeployed elsewhere.

Individual rights to personal data must be balanced with the need of corporations and governments to use certain data-account activity, billing information, and the like to run their day-to-day operations. The New Deal on Data therefore gives individuals the right to possess, control, and dispose of copies of these required operational data, along with copies of the incidental data collected about the individual, such as location and similar context. These ownership rights are not exactly the same as literal ownership under modern law; the practical effect is that disputes are resolved in a different, simpler manner than would be the case for land ownership disputes, for example.

In 2007, one author (AP) first proposed the New Deal on Data to the World Economic Forum.[18] Since then, this idea has run through various discussions and eventually helped to shape the 2012 Consumer Data Bill of Rights in the United States, along with a matching declaration on Personal Data Rights in the European Union.

The World Economic Forum (WEF) echoed the European Consumer Commissioner Meglena Kuneva in dubbing personal data the 'new oil' or new resource of the 21st century.[19] The 'personal data sector' of the economy today is in its infancy, its state akin to the oil industry during the late 1890s. Productive collaboration between government (building the state-owned freeways), the private sector (mining and refining oil, building automobiles), and the citizens (the user-base of these services) allowed developed nations to expand their economies by creating new markets adjacent to the automobile and oil industries.

If personal data, as the new oil, is to reach its global economic potential, productive collaboration is needed between all stakeholders in the establishment of a *personal data ecosystem.* A number of fundamental uncertainties exist, however, about privacy, property, global governance, human rights – essentially about who should benefit from the products and services built on personal data.[20] The rapid rate of technological change and commercialization in the use of personal data is undermining end-user confidence and trust.

The current personal data ecosystem is feudal, fragmented, and inefficient. Too much leverage is currently accorded to service providers that enroll and register end-users. Their siloed repositories of personal data exemplify the fragmentation of the ecosystem, containing data of varying qualities; some are attributes of persons that are unverified, while others represent higher quality data that have been cross-correlated with other data points of the end-user. For many individuals, the risks and liabilities of the current ecosystem exceed the economic returns. Besides not having the infrastructure and tools to manage personal data, many end-users simply do not see the benefit of fully participating. Personal privacy concerns are thus addressed inadequately at best, or simply overlooked in the majority of cases. Current technologies and laws fall short of providing the legal and technical infrastructure needed to support a well-functioning digital economy.

Recently, we have seen the challenges, but also the feasibility of opening private big data. In the Data for Development (D4D) Challenge (http://www.d4d.orange.com), the telecommunication operator Orange opened access to a large dataset of call detail records from the Ivory Coast. Working with the data as part of a challenge, teams of researchers came up with life-changing insights for the country. For example, one team developed a model for how disease spreads in the country and demonstrated that information campaigns based on one-to-one phone conversations among members of social groups can be an effective countermeasure.[21] Data release must be carefully done, however; as we have seen in several cases, such as the Netflix Prize privacy disaster[22] and other similar privacy breaches,[23] true anonymization is extremely hard – recent research by de Montjoye et al. and others[24,25] has shown that even though human beings are highly predictable, we are also unique. Having access to one dataset may be enough to uniquely fingerprint someone based on just a few data points, and this fingerprint can be used to discover their true identity. In releasing and analyzing the D4D data, the privacy of the people who generated the data was protected not only by technical means, such as removal of personally identifiable information (PII), but also by legal means, with the researchers signing an agreement that they would not use the data for re-identification or other nefarious purposes. Opening data from the silos by publishing static datasets – collected at some point and unchanging – is important, but it is only the first step. We can do even more when data is available in real time and can become part of a society`s nervous system.

Epidemics can be monitored and prevented in real time,[26] underperforming students can be helped, and people with health risks can be treated before they get sick.[27]

The report of the World Economic Forum[28] suggests a way forward by identifying useful areas on which to focus efforts:

- *Alignment of key stakeholders* Citizens, the private sector, and the public sector need to work in support of one another. Efforts such as NSTIC[29] in the United States – albeit still in its infancy – represent a promising direction for global collaboration.
- *Viewing 'data as money'* There needs to be a new mindset, in which an individual's personal data items are viewed and treated in the same way as their money. These personal data items would reside in an 'account' (like a bank account) where they would be controlled, managed, exchanged, and accounted for just as personal banking services operate today.
- *End-user centricity* All entities in the ecosystem need to recognize end-users as vital and independent stakeholders in the co-creation and exchange of services and experiences. Efforts such as the User Managed Access (UMA) initiative[30] provide examples of system design that are user-centric and managed by the user.

Enforcing the New Deal on Data

How can we enforce this New Deal? The threat of legal action is important, but not sufficient; if you cannot see abuses, you cannot prosecute them. Enforcement can be addressed significantly without prosecution or public statute or regulation. In many fields, companies and governments rely on rules governing common business, legal, and technical (BLT) practices to create effective self-organization and enforcement. This approach holds promise as a method by which institutional controls can form a reliable operational framework for big data, privacy, and access.

One current best practice is a system of data sharing called a 'trust network', a combination of networked computers and legal rules defining and governing expectations regarding data. For personal data, these networks of technical and legal rules keep track of user permissions for each piece of data and act as a legal contract, specifying what happens in case of a violation. For example, in a trust network all personal data can have attached labels specifying where the data come from and what they can and cannot be used for. These labels are exactly matched by the terms in the legal contracts between all of the participants, stating penalties for not obeying them. The rules can – and often do – reference or require audits of relevant systems and data use, demonstrating how traditional internal controls can be leveraged as part of the transition to more novel trust models. A well-designed trust network, elegantly integrating computer and legal rules, allows automatic auditing of data use and allows individuals to change their permissions and withdraw data.

The mechanism for establishing and operating a trust network is to create system rules for the applications, service providers, data, and the users themselves. System rules are sometimes called 'operating regulations' in the credit card context, 'trust frameworks' in the identity federation context, or 'trading partner agreements' in a supply value chain context. Several multiparty shared architectural and contractual rules create binding obligations and enforceable expectations on all participants in scalable networks. Furthermore, the design of the system rules allows participants to be widely distributed across heterogeneous business ownership boundaries, legal governance structures, and technical security domains. However, the parties need not conform in all or even most aspects of their basic roles, relationships, and activities in order to connect to a trust network. Cross-domain trusted systems must – by their nature – focus enforceable rules narrowly on commonly agreed items in order for that network to achieve its purpose.

For example, institutions participating in credit card and automated clearing house networks are subject to profoundly different sets of regulations, business practices, economic conditions, and social expectations. The network rules focus on the topmost agreed items affecting interoperability, reciprocity, risk, and revenue allocation. The knowledge that fundamental rules are subject to enforcement action is one of the foundations of trust and a motivation to prevent or address violations before they trigger penalties. A clear example of this approach can be found in the Visa Operating Rules, which cover a vast global real-time network of parties agreeing to rules governing their roles in the system as merchants, banks, transaction processors, individual or business card holders, and other key system roles.

Such rules have made the interbank money transfer system among the safest systems in the world and the backbone for daily exchanges of trillions of dollars, but until recently those were only for the 'big guys'.[31] To give individuals a similarly safe method of managing personal data, the Human Dynamics group at MIT, in partnership with the Institute for Data Driven Design (co-founded by John Clippinger and one author (AP)) have helped to build an open Personal Data Store (openPDS).[32] The openPDS is a consumer version of a personal cloud trust network now being tested with a variety of industry and government partners. The aim is to make sharing personal data as safe and secure as transferring money between banks.

When dealing with data intended to be accessible over networks – whether big, personal, or otherwise – the traditional container of an institution makes less and less sense. Institutional controls apply, by definition, to some type of institutional entity such as a business, governmental, or religious organization. A synopsis of all the BLT facts and circumstances surrounding big data is necessary in order to know what access, confidentiality, and other expectations exist; the relevant contextual aspects of big data at one institution are often profoundly different from those at another. As more and more organizations use and rely on big data, a single formula for institutional controls will not work for increasingly heterogeneous BLT environments.

The capacity to apply appropriate methods of enforcement for a trust network depends on clear understanding and agreement among the parties about the purpose of the system and the respective roles or expectations of those connecting as participants. Therefore, some contextual anchor is needed to have a clear basis for establishing an operational framework and institutional controls appropriate for big data.

Transitioning End-User Assent Practices

The way users grant authorization to share their data is not a trivial matter. The flow of personal information such as location data, purchases, and health records can be very complex. Every tweet, geotagged picture, phone call, or purchase with credit card provides the user's location not only to the primary service, but also to all the applications and services that have been authorized to access and reuse these data. The authorization may come from the end-user or be granted by the collecting service, based on umbrella terms of service that cover reuse of the data. Implementation of such flows was a crucial part of the Web 2.0 revolution, realized with RESTful APIs, mash-ups, and authorization-based access. The way personal data travels between services has arguably become too complex for a user to handle and manage.

Increasing the range of data controlled by the user and the granularity of this control is meaningless if it cannot be exercised in an informed way. For many years, a poor model has been provided by End User License Agreements (EULAs), long incomprehensible texts that are accepted blindly by users trusting they have not agreed to anything that could harm them. The process of granting meaningful authorization cannot be too complex, as it would prevent a user from understanding her decisions. At the same time, it cannot be too simplistic, as it may not sufficiently convey the weight of the privacy-related decisions it captures. It is a challenge in itself to build end-user assent systems that allow users to understand and adjust their privacy settings.

This gap between the interface – single click – and the effect can render data ownership meaningless; one click may wrench people and their data into systems and rules that are antithetical to fair information practices, as is prevalent with today's end-user licenses in cloud services or applications. Managing the long-term tensions fueled by 'old deal' systems operating simultaneously with the New Deal is an important design and migration challenge during the transition to a Big Data economy. During this transition and after the New Deal on Data is no longer new, personal data must continue to flow in order to be useful. Protecting the data of people outside of directly user-controlled domains is very hard without a combination of cost-effective and useful business practices, legal rules, and technical solutions.

We envision 'living informed consent', where the user is entitled to know what data is being collected about her by which entities, empowered to understand the implications of data sharing, and finally put in charge of the sharing authorizations. We suggest that readers ask themselves a question: *Which services know which city I am in today?* Google? Apple? Twitter? Amazon? Facebook? Flickr? Some app I authorized a few years ago to access my Facebook check-ins and have since forgotten about? This is an example of a fundamental question related to user privacy and assent, and yet finding an accurate answer can be surprisingly difficult in today's ecosystem. We can hope that most services treat data responsibly and according to user authorizations. In the complex network of data flows, however, it is relatively easy for data to leak to careless or malicious services.[33] We need to build solutions that help users to make well-informed decisions about data sharing in this environment.

Big Data and Personal Data Institutional Controls

The concept of 'institutional controls' refers to safeguards and protections implemented through legal, policy, governance, and other measures that are not solely technical, engineering, or mechanical. Institutional controls in the context of big data can perhaps best be understood by examining how such controls have been applied to other domains, most prevalently in the field of environmental regulation. A good example of how this concept supports and reflects the goals and objectives of environmental regulation can be found in the policy documents of the Environmental Protection Agency (EPA), which gives the following definition in its Institutional Controls Glossary:

> Institutional Controls – Non-engineering measures intended to affect human activities in such a way as to prevent or reduce exposure to hazardous substances. They are almost always used in conjunction with, or as a supplement to, other measures such as waste treatment or containment. There are four categories of institutional controls: governmental controls; proprietary controls; enforcement tools; and informational devices.[34]

The concept of an 'institutional control boundary' is especially clarifying and powerful when applied to the networked and digital boundaries of an institution. In the context of Florida's environmental regulation, the phrase is applied when a property owner's risk management and clean-up responsibilities extend beyond the area defined by the physical property boundary. For example, a recent University of Florida report on clean-up target levels (CTLs) states, "in some rare situations, the institutional control boundary at which default CTLs must be met can extend beyond the site property boundary."[35]

When institutional controls apply to "separately owned neighboring properties" a number of possibilities arise that are very relevant to management of personal data across legal, business, and other systemic boundaries. Requiring the party responsible for site clean-up to use "best efforts" to attain agreement from the neighboring owners to institute the relevant institutional controls is perhaps the most direct and least prescriptive approach. When direct negotiated agreement is unsuccessful, then use of third-party neutrals to resolve disagreements regarding institutional controls can be required. If necessary, environmental regulation can force the acquisition of neighboring land by compelling the party responsible to purchase the other property or by purchase of the property directly by the EPA.[36]

In the context of big data, institutional controls are seldom, if ever, imposed through government regulatory frameworks such as are seen in environmental waste management oversight by the EPA.[37] Rather, institutions applying measures constituting institutional controls in the big data and related information technology and enterprise architecture contexts will typically employ governance safeguards, business practices, legal contracts, technical security, reporting, and audit programs and various risk management measures.

Inevitably, institutional controls for big data will have to operate effectively across institutional boundaries, just as environmental waste management must sometimes be applied across real property boundaries and may subject multiple different owners to enforcement actions corresponding to the applicable controls. Short of government regulation, the use of system rules as a general model is one widely understood, accepted, and efficient method for defining, agreeing, and enforcing institutional and other controls across BLT domains of ownership, governance, and operation.

Following on from the World Economic Forum's recommendation to treat personal data stores in the manner of bank accounts,[38] a number of infrastructure improvements need to be realized if the personal data ecosystem is to flourish and deliver new economic opportunities:

- *New global data provenance network* In order for personal data stores to be treated like bank accounts, origin information regarding data items coming into the data store must be maintained.[39] In other words, the provenance of all data items must be accounted for by the IT infrastructure on which the personal data store operates. The databases must then be interconnected in order to provide a resilient, scalable platform for audit and accounting systems to track and reconcile the movement of personal data from different data stores.
- *Trust network for computational law* For trust to be established between parties who wish to exchange personal data, some degree of 'computational law' technology may have to be integrated into the design of personal data systems. This technology should not only verify terms of contracts (e.g. terms of data use) against user-defined policies but also have mechanisms built in to ensure non-repudiation of entities who have accepted these digital contracts. Efforts such as the UMA initiative are beginning to bring better evidentiary proof and enforceability of contracts into technical protocol flows.[40]
- *Development of institutional controls for digital institutions* Currently, a number of proposals for the creation of virtual currencies (e.g. BitCoin,[41] Ven[42]) have underlying systems with the potential to evolve into self-governing 'digital institutions'.[43] Such systems and the institutions that operate on them will necessitate the development of a new paradigm to understand aspects of institutional control within their context.

Scenarios of Use in Context

Developing frameworks for big data that effectively balance economic, legal, security, and other interests requires an understanding of the relevant context and applicable scenarios within which the data exists.

A sound starting point from which to establish the applicable scenarios of use is to enumerate the institutions involved with a given set of big data, and develop a description of how or why they hold, access, or otherwise intermediate the data. Although big data straddles multiple BLT boundaries, one or more institutions are typically able to, or in some situations required to, manage and control the data. The public good referred to in the title of this book can be articulated as design requirements or even as certification criteria applicable to those institutions that operate the systems through which the big data is computed or flows.

It may be also be necessary to narrowly define certain aspects of the scenario in which the data exist in order to establish the basic ownership, control, and other expectations of the key parties. For example, describing a transaction as a financial exchange may not provide enough relevant detail to reveal the rights, obligations, or other outcomes reasonably expected by the individuals and organizations involved. The sale of used cars via an app, the conduct of a counseling session via Google Hangout, and the earning of a master's degree via an online university all represent scenarios in which the use case of a financial exchange takes place. However, each of these scenarios occurs in a context that is easily identifiable: the sale of goods and deeper access to financial information if the car is financed; the practice of therapy by a licensed professional accessing and creating confidential mental health data; or e-learning services and protected educational records and possibly deeper financial information if the program is funded by scholarship or loans. The scenarios can also identify the key elements necessary to establish existing consumer rights – the people (a consumer and a used car dealer), the transaction (purchase of a used car), the data (sales and title data, finance information, etc.), and the systems (the third-party app and its relevant services or functions, state DMV services, credit card and bank services, etc.). The rights established by relevant state lemon laws, the Uniform Commercial Code, and other applicable rules will determine when duties arise or are terminated, what must be promised, what can be repudiated, by whom data must be kept secure, and other requirements or constraints on the use of personal data and big data. These and other factors differ when a transaction that seems identical operates within a different scenario, and even scenarios will differ depending on which contexts apply.

The following four elements are critical for defining high-level goals and objectives:

1. Who are the *people* in the scenario (e.g. who are the parties involved and what are their respective roles and relationships)?
2. What are the relevant *interactions* (e.g. what transactions or other actions are conducted by or with the people involved)?
3. What are the relevant *data* and datasets (e.g. what types of data are created, stored, computed, transmitted, modified, or deleted)?
4. What are the relevant *systems* (e.g. what services or other software are used by the people, for the transactions, or with the data)?

Inspired by common law, the New Deal on Data sets out general principles of ownership that both guide and inform basic relationships and expectations. However, the dynamic bundle of recombinant rights and responsibilities constituting 'ownership' interests in personal data and expectations pertaining to big data vary significantly from context to context, and even from one scenario to another within a given general context. Institutional controls and other system safeguards are important methods to ensure that there are context-appropriate outcomes that are consistent with clearly applicable system scenarios as well as the contours and foundations for a greater public good. The New Deal on Data can be achieved in part by sets of institutional controls involving governance, business, legal, and technical aspects of big data and interoperating systems. Reference scenarios can be used to reveal signature features of the New Deal on Data in various contexts and can serve as anchors in evaluating what institutional controls are well aligned to achieve a balance of economic, privacy, and other interests.

The types of requirements and rules governing participation by individuals and organizations in trust networks vary depending on the facts and circumstances of the transactions, data types, relevant roles of people, and other factors. Antecedent but relevant networks such as credit card systems, trading partner systems, and exchange networks are instructive not only for their many common elements but also as important examples of how vastly different they are from one another in their contexts, scenarios, legal obligations, business models, technical processes, and other signature patterns. Trust networks that are formed to help manage big data in ways that appropriately respect personal data rights and other broader interests will similarly succeed to the extent they can tolerate or promote a wide degree of heterogeneity among participants for BLT matters that need not be uniform or directly harmonized. In some situations, new business models and contexts will emerge that require fresh thinking and novel combinations of roles or types of relationships among transacting parties. In these cases, understanding the actual context and scenarios is critical in customizing acceptable and sustainable BLT rules and systems. Example scenarios can describe deeper fact-based situations and circumstances in the context of social science research involving personal data and big data.[44] The roles of people, their interactions, the use of data, and the design of the corresponding systems reflect and support the New Deal on Data in ways that deliberately provide greater immediate value to stakeholders than is typically expected.

The New Deal on Data is designed to provide good value to anyone creating, using, or benefiting from personal data, but the vision need not be adopted in its entirety before its value becomes apparent. Its principles can be adopted on a large scale in increments – an economic sector, transaction type, or data type at a time. Adopting the New Deal on Data in successive phases helps to address typical objections to change based on cost, disruption, or overregulation. Policy incentives can further address these objections, for example by allowing safe harbor protections for organizations operating under the rules of a trust network.

Predesigned use cases can provide benchmarks for determining whether given uses of personal data are consistent with measurable criteria. Such criteria can be used to establish compliance with the rules of a trust network and for certification by government for the right to safe harbor or other protections. Because the New Deal on Data is rooted in common law and the social compact, the appropriate set of rights and expectations covering privacy and other personal data interests can be enumerated, debated, and agreed upon in ways that fit the given use cases.

Conclusions

Society today faces unprecedented challenges and meeting them will require access to personal data, so we can understand how society works, how we move around, what makes us productive, and how everything from ideas to diseases spread. The insights must be actionable and available in real time, thus engaging the population, creating the nervous system of the society. In this chapter we have reviewed how big data collected in institutional contexts can be used for the public good. In many cases, although the data needed to create a better society has already been collected, it sits in the closed silos of companies and governments. We have described how the silos can be opened using well-designed and carefully implemented sets of institutional controls, covering business, legal, and technical dimensions. The framework for doing this – the New Deal on Data – postulates that the primary driver of change must be recognizing that ownership of personal data rests with the people that data is about. This ownership – the right to use, transfer, and remove the data – ensures that the data is available for the public good, while at the same time protecting the privacy of citizens.

The New Deal on Data is still new. We have described here our efforts to understand the technical means of its implementation, the legal framework around it, its business ramifications, and the direct value of the greater access to data that it enables. It is clear that companies must play the major role in implementing the New Deal, incentivized by business opportunities, guided by legislation, and pressured by demands from users. Only with such orchestration will it be possible to modernize the current system of data ownership and put immense quantities and capabilities of collected personal data to good use.

NOTES

1. Jonathan Woetzel et al., "Preparing for China's Urban Billion" (McKinsey Global Institute, March 2009), http://www.mckinsey.com/insights/urbanization/preparing_for_urban_billion_in_china.
2. David Lazer, Alex Sandy Pentland, Lada Adamic, Sinan Aral, Albert Laszlo Barabasi, Devon Brewer, Nicholas Christakis, Noshir Contractor, James Fowler, and Myron Gutmann, "Life in the Network: The Coming Age of Computational Social Science," *Science* 323 (2009): 721–723.
3. Sinan Aral and Dylan Walker, "Identifying Influential And Susceptible Members Of Social Networks," *Science* 337 (2012): 337–341; Alan Mislove, Sune Lehmann, Yong-Yeol Ahn, Jukka-Pekka Onnela, and J. Niels Rosenquist, *Pulse of the Nation: U.S. Mood throughout the Day Inferred from Twitter* (website), http://www.ccs.neu.edu/home/amislove/twittermood/ (accessed November 22, 2013); Jessica Vitak, Paul Zube, Andrew Smock, Caleb T. Carr, Nicole Ellison, and Cliff Lampe, "It's Complicated: Facebook Users' Political Participation in the 2008 Election," *Cyberpsychology, Behavior, and Social Networking* 14 (2011): 107–114.
4. Alexis Madrigal, "Dark Social: We Have the Whole History of the Web Wrong," *The Atlantic*, October 12, 2013, http://www.theatlantic.com/technology/archive/2012/10/dark-social-we-have-the-whole-history-of-the-web-wrong/263523/.
5. Nathan Eagle and Alex Pentland, "Reality Mining: Sensing Complex Social Systems," *Personal and Ubiquitous Computing* 10 (2006): 255–268; Alex Pentland, "Reality Mining of Mobile Communications: Toward a New Deal on Data," *The Global Information Technology Report 2008–2009* (Geneva: World Economic Forum, 2009), 75–80.

6. Alex Pentland, David Lazer, Devon Brewer, and Tracy Heibeck, "Using Reality Mining to Improve Public Health and Medicine," *Studies in Health Technology and Informatics* 149 (2009): 93–102.
7. Vivek K. Singh, Laura Freeman, Bruno Lepri, and Alex Sandy Pentland, "Classifying Spending Behavior using Socio-Mobile Data," *HUMAN* 2 (2013): 99–111.
8. Wei Pan, Yaniv Altshuler, and Alex Sandy Pentland, "Decoding Social Influence and the Wisdom of the Crowd in Financial Trading Network," in *2012 International Conference on Privacy, Security, Risk and Trust (PASSAT), and 2012 International Conference on Social Computing (SocialCom)*, 203–209.
9. Kate Greene, "Reality Mining," *MIT Technology Review*, March/April 2008, http://pubs.media.mit.edu/pubs/papers/tr10pdfdownload.pdf.
10. Meglena Kuneva, European Consumer Commissioner, "Keynote Speech," in *Roundtable on Online Data Collection, Targeting and Profiling*, March 31, 2009, http://europa.eu/rapid/press-release_SPEECH-09-156_en.htm.
11. Kim Gittleson, "How Big Data Is Changing The Cost Of Insurance," *BBC News*, November 14, 2013, http://www.bbc.co.uk/news/business-24941415.
12. Aniko Hannak, Piotr Sapiezynski, Kakhki Arash Molavi, Balachander Krishnamurthy, David Lazer, Alan Mislove, and Christo Wilson, "Measuring Personalization of Web Search," in *Proc. 22nd International Conference on World Wide Web (WWW 2013)*, 527–538.
13. Pentland, "Reality Mining of Mobile Communications."
14. Anmol Madan, Manuel Cebrian, David Lazer, and Alex Pentland, "Social Sensing for Epidemiological Behavior Change," in *Proc. 12th ACM International Conference on Ubiquitous Computing (Ubicomp 2010)*, 291–300; Pentland et al. "Using Reality Mining to Improve Public Health and Medicine."
15. Wei Pan, Gourab Ghoshal, Coco Krumme, Manuel Cebrian, and Alex Pentland, "Urban Characteristics Attributable to Density-Driven Tie Formation," *Nature Communications* 4 (2013): article 1961.
16. Lev Grossman, "Iran Protests: Twitter, the Medium of the Movement," *Time Magazine*, June 17, 2009; Ellen Barry, "Protests in Moldova Explode, with Help of Twitter," *The New York Times*, April 8, 2009.
17. "Directive 95/46/EC of the European Parliament and of the Council of 24 October 1995 on the Protection of Individuals with Regard to the Processing of Personal Data and on the Free Movement of Such Data," *Official Journal* L281 (November 23, 1995): 31–50.
18. World Economic Forum, "Personal Data: The Emergence of a New Asset Class," January 2011, http://www.weforum.org/reports/personal-data-emergence-new-asset-class.
19. Ibid.
20. Ibid.
21. Antonio Lima, Manlio De Domenico, Veljko Pejovic, and Mirco Musolesi, "Exploiting Cellular Data for Disease Containment and Information Campaign Strategies in Country-Wide Epidemics," School of Computer Science Technical Report CSR-13-01, University of Birmingham, May 2013.
22. Arvind Narayanan and Vitaly Shmatikov, "Robust De-Anonymization of Large Sparse Datasets," in *Proc. 2008 IEEE Symposium on Security and Privacy (SP)*, 111–125.

23. Latanya Sweeney, "Simple Demographics Often Identify People Uniquely," Data Privacy Working Paper 3, Carnegie Mellon University, Pittsburgh, 2000.
24. de Montjoye, Yves-Alexandre, Samuel S. Wang, Alex Pentland, "On the Trusted Use of Large-Scale Personal Data," *IEEE Data Engineering Bulletin* 35, no. 4 (2012): 5–8.
25. Chaoming Song, Zehui Qu, Nicholas Blumm, and Albert-Laszlo Barabasi, "Limits of Predictability in Human Mobility," *Science* 327 (2010): 1018–1021.
26. Pentland et al., "Using Reality Mining to Improve Public Health and Medicine."
27. David Tacconi, Oscar Mayora, Paul Lukowicz, Bert Arnrich, Cornelia Setz, Gerhard Troster, and Christian Haring, "Activity and Emotion Recognition to Support Early Diagnosis of Psychiatric Diseases," in *Proc. 2nd International ICST Conference on Pervasive Computing Technologies for Healthcare,* 100–102.
28. World Economic Forum, "Personal Data."
29. The White House, "National Strategy for Trusted Identities in Cyberspace: Enhancing Online Choice, Efficiency, Security, and Privacy," Washington, DC, April 2011, http://www.whitehouse.gov/sites/default/files/rss_viewer/NSTICstrategy_041511.pdf.
30. Thomas Hardjono, "User-Managed Access UMA Profile of OAuth2.0," Internet draft, 2013, http://docs.kantarainitiative.org/uma/draft-uma-core.html.
31. A Creative Commons licensed example set of integrated business and technical system rules for the institutional use of personal data stores is available at https://github.com/HumanDynamics/SystemRules.
32. See http://openPDS.media.mit.edu for project information and https://github.com/HumanDynamics/openPDS for the open source code.
33. Nick Bilton, "Girls around Me: An App Takes Creepy to a New Level," *The New York Times, Bits* (blog), March 30, 2012, http://bits.blogs.nytimes.com/2012/03/30/girls-around-me-ios-app-takes-creepy-to-a-new-level.
34. U.S. Environmental Protection Agency, RCRA Corrective Action Program, "Institutional Controls Glossary," Washington, DC, 2007, http://www.epa.gov/epawaste/hazard/correctiveaction/resources/guidance/ics/glossary1.pdf.
35. University of Florida, Center for Environmental & Human Toxicology, "Development of Cleanup Target Levels (CTLs) for Chapter 62-777, F.A.C.," Technical report, Florida Department of Environmental Protection, Division of Waste Management, February 2005, http://www.dep.state.fl.us/waste/quick_topics/publications/wc/FinalGuidanceDocumentsFlowCharts_April2005/TechnicalReport2FinalFeb2005(Final3-28-05).pdf.
36. U.S. Environmental Protection Agency, "Institutional Controls: A Guide to Planning, Implementing, Maintaining, and Enforcing Institutional Controls at Contaminated Sites," OSWER 9355.0-89, Washington, DC, December 2012, http://www.epa.gov/superfund/policy/ic/guide/Final%20PIME%20Guidance%20December%202012.pdf.
37. Ralph A. DeMeo and Sarah Meyer Doar, "Restrictive Covenants as Institutional Controls for Remediated Sites: Worth the Effort?" *The Florida Bar Journal* 85, no. 2 (February 2011); Florida Department of Environmental Protection, Division of Waste Management, "Institutional Controls Procedures Guidance," Tallahassee, June 2012, http://www.dep.state.fl.us/waste/quick_topics/publications/wc/csf/icpg.pdf; University of Florida, "Development of Cleanup Target Levels."

38. World Economic Forum, "Personal Data."
39. Thomas Hardjono, Daniel Greenwood, and Alex Pentland, "Towards a Trustworthy Digital Infrastructure for Core Identities and Personal Data Stores," in *Proc. ID360 Conference on Identity,* 2013.
40. Hardjono, "User-Managed Access UMA Profile of OAuth2.0"; Eve Maler and Thomas Hardjono, "Binding Obligations on User-Managed Access (UMA) Participants," Internet draft, 2013, http://docs.kantarainitiative.org/uma/draft-uma-trust.html.
41. Simon Barber, Xavier Boyen, Elaine Shi, and Ersin Uzun, "Bitter to Better – How to Make Bitcoin a Better Currency," in *Proc. Financial Cryptography and Data Security Conference (2012),* LNCS 7397, 399–414.
42. Stan Stalnaker, "About [Ven Currency]," http://www.ven.vc (accessed January 16, 2014).
43. Thomas Hardjono, Patrick Deegan, and John Clippinger, "On the Design of Trustworthy Compute Frameworks for Self-Organizing Digital Institutions," in *Proc. 16th International Conference on Human-Computer Interaction (2014),* forthcoming; Lazer et al., "Life in the Network."
44. See e.g. the study SensibleDTU (https://www.sensible.dtu.dk/?lang=en). This study of 1,000 freshman students at the Technical University of Denmark gives students mobile phones in order to study their networks and social behavior during an important change in their lives. It uses not only data collected from the mobile phones (such as location, Bluetooth-based proximity, and call and sms logs), but also from social networks and questionnaires filled out by participants.

10 Engineered Controls for Dealing with Big Data

Carl Landwehr

Introduction

It is one thing for a patient to trust a physician with a handwritten record that is expected to stay in the doctor's office. It's quite another for the patient to consent to place their comprehensive electronic health record in a repository that may be open to researchers anywhere on the planet. The potentially great payoffs from (for example) being able to find a set of similar patients who have suffered from the same condition as oneself and to review their treatment choices and outcomes will likely be unavailable unless people can be persuaded that their individual data will be handled properly in such a system. Agreeing on an effective set of institutional controls (see Chapter 9) is an essential prerequisite, but equally important is the question of whether the agreed upon policies can be enforced by controls engineered into the system. Without sound technical enforcement, incidents of abuse, misuse, theft of data, and even invalid scientific conclusions based on undetectably altered data can be expected. While technical controls can limit the occurrence of such incidents substantially, some will inevitably occur. When they do, the ability of the system to support accountability will be crucial, so that abusers can be properly identified and penalized and systems can be appropriately reinforced or amended.

Questions to ask about the engineered controls include:

- How are legitimate system users identified and authenticated?
- What mechanisms are employed to distinguish classes of users and to limit their actions to those authorized by the relevant policies?
- What mechanisms limit the authorities of system administrators?
- How is the system software installed, configured, and maintained? How are user and administrator actions logged?
- Can the logs be effectively monitored for policy violations?
- When policy violations are detected, what mechanisms can be used to identify violators and hold them to account?

Finally, the usability of the engineered controls – their impact on system users – must be considered. Time and again, users have demonstrated that they will find ways around controls that they see as needlessly complex or strict and that impede them from their primary goals in using the system.

Threats to Big Data: Accidental, Malicious

Threats to 'big data' sets can come from several directions. Not least of these are the threats of accidental damage or loss, for example from device failure, physical loss of a mobile device (laptop, tablet, memory stick), physical damage to a device from fire, flood, or simply dropping it, if it's small enough to carry.

The general motives for intentional compromise of datasets or systems are relatively few, but they can be manifested in many ways. Financial gain is likely to be the strongest motive for compromise; desire for fame has motivated some groups and individuals from the early days of the Internet; desire for revenge (against an employer or a lover, for example, but also against a government organization or a research colleague) can be a strong motivator; and finally simple curiosity can lead to attempts to defeat security and confidentiality controls.

The nature of the data, and what motivated individuals might glean from them, is significant in assessing threat. For example, a large set of medical records from which obviously identifying information has been removed might not seem to be of much interest to those beyond the research community. Yet if the records concerned a sexually transmitted disease or drug addiction, and if a celebrity suspected of having that condition was treated at a facility that contributed to the dataset, someone might exert considerable effort to re-identify the data.

The custodian needs to consider carefully both the value of the raw data and what kinds of inferences might be drawn from them by users or abusers. This exercise will help identify what might motivate particular individuals or groups to attack the system. Chapters 1–5 in this volume provide considerable guidance about legal and policy issues; Chapters 9 and 11 identify mechanisms that may facilitate research access to the data; Chapters 13 and 14 identify ways to limit unwanted disclosures. The custodian needs to pay even closer attention to those with administrative access to datasets, because they will likely have access that is unrestricted, or much less restricted, than the researchers to whom data may be released under particular controls. If they have the ability to disable or corrupt logs, they can also damage accountability.

Vulnerabilities: Accidental, Engineered

Nearly all of today's cyberattacks take advantage of latent vulnerabilities in software that has been legitimately installed on the user's system. Although it is possible to write rigorous specifications[1] for software and even to prove that an implementation conforms to those specifications, this is a costly process and one that is infeasible to apply to large and complex software systems. Commercially produced software, even software that has undergone standard testing regimes and is considered of good quality, will typically include a few bugs per thousand lines of code, at minimum. Only a fraction of these will be exploitable security vulnerabilities, but with the total size of codebases for both operating systems and applications in the range of hundreds of kilobytes to thousands of megabytes, it's unrealistic to think that there will not be exploitable vulnerabilities in the platforms used to store and process large datasets. A typical exploit will take advantage of an accidental flaw in the software to gain a foothold and then download a larger payload that provides the infrastructure attackers use to achieve whatever objectives they may have – for example, altering sensitive data, sending sensitive data back to a place where the attacker can retrieve it, or denying access to the data to legitimate users.

It is also possible for vulnerabilities to be engineered into operating systems, compilers, network support, and applications by those responsible for developing or maintaining them. For example, a developer may decide to leave a 'back door' into a software component that can later be used to perform remote testing or maintenance. Such an engineered-in 'feature' can be exploited by others if they learn of its existence.

A more subtle form of vulnerability that may enable the illicit transmission of information is referred to as a 'covert channel', or more commonly today as a 'side channel'. When, as a side effect of a computation, some shared resource (such as a file lock) is manipulated, and that manipulation is visible outside of the immediate context, that manipulation may be used to transmit information. For example, the power consumed by a chip as it computes a cryptographic algorithm, if monitored, may expose the key being used in the computation (Kocher et al. 1999). While these channels have rarely been used to compromise large datasets, it is nearly impossible to build a system entirely free of them.

Today's software is also largely built from components. An application programmer will naturally look for libraries and components that will reduce the new code that must be developed. Whatever code is incorporated in this way brings with it whatever vulnerabilities have been accidentally left in it or purposely engineered into it. Further, flaws may arise in the interactions of components that individually behave as intended.

The question of whether 'open source' or 'closed source' software is more or less likely to contain exploitable vulnerabilities has been hotly debated over the years. Having the source code available for anyone to inspect and analyze seems to be an advantage, but only if people with the appropriate expertise and access to the appropriate tools actually do the inspection and analysis. Evidence to date is that qualified people rarely carry out such tasks without being paid.

When the source is closed, typically some commercial entity has a financial interest in it. The owner of the software has incentives to assure its quality (including freedom from bugs and vulnerabilities) because if others find those bugs and vulnerabilities after the software is released, the owner may still have to fix them and the product's reputation will suffer. Nevertheless, experience shows that owners, even when informed of security issues in their products, do not necessarily act promptly to address them. Moreover, software owners *could* hide all sorts of things in delivered software should they wish to. Of course, if malicious software were found in a commercial product that seemed to have been placed there purposely by the developer, the company might very well lose its customers in short order, so there would be a strong incentive to make any such hidden features appear to be accidentally introduced.

Strategies

Given that there will be vulnerabilities and there may be threats against them, what is the right strategy for dealing with them? We will discuss some mechanisms for implementing strategies in subsequent sections, but what is the right overall approach?

The Canadian Conservation Institute, advising on controlling biological infestations in a museum environment, identifies an approach with five stages: Avoid, Block, Detect, Respond, Recover/Treat (Strang 1996). Approaches developed for dealing with cyber 'infestations' have much in common:

- Treasury Board of Canada (TBC 2006): Prevention, Detection, Response, and Recovery
- Microsoft (Microsoft 2010): Protect, Detect, Respond, Recover
- NIST (NIST 2013): Know, Prevent, Detect, Respond, Recover

The addition of 'know' in the NIST framework emphasizes that one must first have an inventory of what is to be protected. This requirement seems particularly germane for a big data environment. Depending on the policies to be enforced, what is to be protected may include not only the datasets themselves but also the use made of them – the sets of queries posed and the responses provided, for example. The identity of the querier may also need to be either publicly recorded or protected from view, according to policy.

If policy violations could be prevented with certainty through technical controls, the need for detection, response, and recovery would be greatly reduced, but unfortunately today they can't be. Detection of intentional breaches of computer system security is quite challenging. Usually the perpetrator is not interested in publicizing the breach, and data can be copied and removed while leaving the original completely unaltered – the money seems still to be in the vault, so to speak. Verizon reported that 70% of disclosed data breach incidents reported in 2012 were discovered by third parties, not by the custodians of the data (Verizon 2013, 53). Detection often happens only when the stolen material is used in some public way, to purchase goods or to generate publicity, for example, which may be considerably after the actual theft.

Response often means conducting a post mortem to discover exactly how the data were stolen (or in the case of an integrity attack, how they were modified) and addressing the problems exposed in order to avoid a recurrence. A key aspect of response is to establish accountability for the incident. This may lead to prosecution of the perpetrator but also may affect personnel charged with maintaining the confidentiality of the data sets. Mechanisms for establishing accountability are crucial to deterring both negligence and malice.

Recovery involves restoring the integrity of the system following an incident. The most basic technical support for recovery is maintenance of backup copies of both data and software used to process it. The recovery process must assure the integrity of the restored backup copy. Cryptographic hashes, digital signatures, and public reference copies may be used for this purpose. The complexity of the recovery process depends on the subtlety and sophistication of the attack. If data are stolen, they may simply have been copied and there is no need to recover the data *per se* at all. But if they were stolen through a piece of software maliciously inserted into the system (malware), that malware needs to be located and removed. If the malware is present in backup copies of the software, restoring the system to a previous state may not in fact eliminate the source of the breach, so care must be taken in this process.

Technical Controls

In this section, we review the kinds of technical controls available to implement policies controlling access to data and for providing accountability when access is granted. We consider the role of encryption in these controls as well as currently available mechanisms to assure software integrity.

Identification and Authentication

As we have already discussed, maintaining accountability for the use of a dataset is one of the strongest tools we have to assure the subjects of the data that their information will not be abused. To maintain accountability, we need to be able to link human beings with activities that take place in the computer, and that linkage starts with having users identify themselves and authenticating that the claimed identity is correct. The simplest and most commonly used form of this technology, and one of the most easily abused, is the user ID (for identification) and password (for authentication). Some recent research documents in some detail how people really use passwords and introduces a metric for the difficulty of guessing passwords (Bonneau 2012). Despite the desires of many, it seems unlikely that passwords will disappear soon, so it's important for security administrators to realize the implications of imposing constraints on users' choices of passwords. In general, it appears better to allow (and encourage) users to choose relatively long passwords and *not* to put constraints on the content (e.g., at least one digit, at least on special character, no words in the dictionary) instead of trying to increase the entropy of relatively short passwords (Weir et al. 2010).

More promising is the increasing use of two-factor authentication schemes, in which the user enters not only a password (something the user *knows*) but also something to indicate that he possesses some physical token (something the user *has*) as well. For many years, the prevalent way of achieving two-factor authentication was to employ a commercially provided token that would generate a new short pseudo-random challenge number every minute or so. The user's system would track which token was assigned to a particular user and could compute what challenge would appear on that token during a specific time interval. The cost of such a scheme included both paying for the commercial token and associated server-side software plus the cost to the user (in time, space, and weight) of carrying and using the token. More recently, several schemes have been developed that take advantage of the fact that most users today carry mobile phones. In this case, the system merely needs to know the user's mobile phone number and it can transmit (via audio or text message) a challenge number over that medium. The user then keys that number into the system, proving that whoever has typed in the initial user ID and password also has access to the user's mobile phone.

Biometrics can provide another form of two-factor authentication. A biometric (fingerprint, iris pattern, or facial image) is generally hard (though not necessarily impossible) to spoof, but of course a camera or other device must be available to read the biometric and the reader needs a trustworthy communication path to the biometric authentication database. Recording the observed biometric can also improve accountability.

Once the user is authenticated, there is also the question of whether another user may take over for her – the professor logs in and the graduate student takes over for her, for example. There are schemes for *continuous authentication* designed to deal with this sort of problem, but in general they are either irritating (requiring the user to re-authenticate periodically) or they require some additional mechanism such as an active token. Biometrics that a user provides as a side effect of using the system (typing characteristics, facial images, gait) are the subject of research; success could provide more usable continuous authentication.

Also relevant to this discussion are the continuing efforts to develop federated identity systems. The idea is to allow users to establish strong authentication with one of several identity providers and then have that identity forwarded to others as needed, avoiding re-authentication and providing a 'single sign-on' capability. Efforts include the earlier Liberty Alliance, and current Shibboleth and OpenID (Birrell and Schneider 2013). It seems likely that some form of federated identity management will be widespread within a few years. Risks of federation include the possibility that a weak or sabotaged provider may forward incorrect authentication data or it may be unavailable when needed.

Access Control and Information Flow Policies

Without a security or confidentiality policy, there can be no violations, only surprises. The chapters in Part I of this volume address policy considerations from many vantage points. This section is concerned with the tools available to state and enforce policies mechanically within computer systems.

Access control policies are generally stated in terms of *subjects*, *objects*, and *access modes*. In the physical world, subjects are people and objects may be the records kept in an office environment, and there may be rules about which individuals, or which job functions, are allowed to read and update various documents or fields within documents. In a computer system, a subject is typically a process, which is a program in execution in the computer, and the subject is identified with a user who has been authenticated. Objects are typically files or other containers of information that a process may need to access in order to complete its computations, and the access mode may be read, write, or execute, so a specific subject may be authorized to read, write, or execute a particular object. The set of all allowed triples (subject, object, access mode) defines a security policy for the system. Although no one ever writes a security policy that way, because it would be far too detailed and time consuming, people do write policies that determine what kinds of access different classes of users require (or that the programs invoked by users require). The *least privilege principle* states that a user (or a process) should only be granted access to those resources (files, programs) that it needs in order to do its job and no others. In practice, this principle is loosely observed, because managing very fine-grained access control is cumbersome. The general notion in an access control policy is that higher level security and confidentiality policies are enforced on the basis of which subjects are allowed to read, write, or execute which objects (Landwehr 1981).

An alternative, and perhaps more intuitive, way to express policies is in terms of *information flow*. An information flow policy specifies the allowed and prohibited flows among subjects and objects in a computer system. In an access control policy, if one wishes to assert that subject A should not be able to read the information in object B, it is insufficient simply to ban A from ever having permission to read B, since the data contained in B might be copied (by some other subject) into object C, which A might be able to read. Stating the constraint in terms of information flow restrictions is straightforward: information in object B is not allowed to flow to subject A, through any path.

Role-based Access Control Many applications have distinct sets of privileges associated with job functions. For example, a bank teller may require the ability to update information in bank accounts, while a bank vice president might not require that ability but would require the ability to generate summary information across many accounts. To simplify management of sets of privileges, the notion of 'role-based access controls' was developed (Sandhu 1996; Landwehr et al. 1984).

Usage Control With the recognition that policies often limit the use of data to particular purposes, the notion of 'usage controls' is being explored (Park and Sandhu 2004). Restrictions on the use of collected data are frequently phrased in terms of the 'purpose' of the collection; the data may be used for some purpose or purposes, but not others. Deducing the purpose behind a particular access request is difficult, but some current research aims at the automated enforcement of such policies or the automated detection of potential violations of such policies (Tschantz et al. 2012).

Prevention: Policy Enforcement via Reference Monitors

One of the foundational papers in computer security developed the notion of a 'reference monitor' to enforce security policies (Anderson 1972). The reference monitor (RM) is a component that checks every reference made by a subject to an object to determine whether it is permitted under the current security policy. To assure policy enforcement, the RM must be correct, tamperproof, and non-bypassable; that is, the RM must be small and simple enough that the correctness of its enforcement is clear ('correct'), it must not be possible to alter the RM ('tamperproof'), and it must not be possible for subjects to make references that don't go through the RM ('non-bypassable'). The cost of placing a check on every reference that a computer makes to memory was well understood to be infeasible; rather the idea was that when, for example, a program processed a request to read or write a new file, the RM would be invoked to check that the request was consistent with the current set of permissions. If so, the file would be opened and the program would be free to read (or write, or execute, or some combination of these) without further interference. Even so, commercial systems were not generally built around this concept, and attempts to build RM-based systems generally did run into performance and other technical issues. Nevertheless the concept of a reference monitor as an idealization of enforcement mechanisms remains useful. In practice, for reasons of efficiency and modular extensibility, security checks tend to be decentralized, resulting in a distributed RM. Although distribution of checks makes it more difficult to assure that the RM has the three properties required, it has been shown that the same classes of policies enforceable by centralized RMs can be enforced by RMs that make checks 'in-line', that is, distributed throughout the system (Schneider 2000).

Enforcement of information flow policies has taken more time to mature; it turns out that RMs can't necessarily enforce such policies directly, because whether information flows from a variable x to a variable y can depend not only on the current execution path but on the set of alternative execution paths *not* taken, which are not visible to the RM (for a detailed exposition, see McLean 1994; Schneider 2000). Compilers have been developed that can accept information flow policies as input and assure that only permitted flows will occur in the code they generate. To assure an information flow policy is maintained system-wide is a more substantial challenge since typically systems are composed of many components and applications, and assuring that the flow policies are properly enforced may require either clever architectures or substantial reconstruction of components. Research continues in this area.

RMs are designed to prevent violations of policy before they occur, and so they address the 'prevent' portion of an overall strategy for securing a computer system. Other measures are needed to deal with detection, response, and recovery.

Cryptography and Its Applications

Today's cryptography can be used to protect both information in transit (communications) and information at rest (stored in files or databases, for example). Publicly available cryptographic algorithms can prevent even the strongest attacker from decrypting enciphered data unless the encryption key is compromised. Thus cryptography transforms the problem of protecting a large amount of data into the problem of protecting a much smaller amount of data – the encryption key.

Two fundamental kinds of cryptographic algorithms are in use today: *symmetric* algorithms, in which the same key is used for encryption and decryption, and *asymmetric* algorithms (also known as public key algorithms) in which one key is used for encryption and a different key is used for decryption. The development and gradual deployment of public key cryptography has been a major advance in the past 30 years, facilitating secure communications and electronic commerce on the Internet. Secure hash algorithms, which can generate a short encrypted digest of a long (or short) string of bits, provide a mechanism that enables a program to verify that a particular piece of data has not been altered. This mechanism is the basis for many useful functions such as digital signatures and assuring integrity and provenance of data.

As already noted, cryptography in itself offers a way to transform one kind of problem into another. While this can be very helpful, the problems remain of generating and managing the keys essential to the process and of implementing the cryptographic algorithms correctly. As the strength of publicly available cryptography has increased, the focus of attacks has moved to exploiting errors in keying, in the cryptographic protocols used to generate and communicate keys among separated parties, and in implementations of these protocols and algorithms. In fact, going back to World War II and earlier, errors in the use of the cryptographic algorithms, rather than weaknesses in the algorithms themselves, have been a common cause of significant compromises (Kahn 1996). This fact highlights again the more general issue of *usability* of security controls in all sorts of contexts. Often the users of systems themselves will find the security measures designed into or added onto a system sufficiently inconvenient that they will find a way around them, whether it be writing down and sharing passwords or reusing the same cryptographic key for different messages. System designers ignore this lesson at the peril of the overall security of their systems.

Media Encryption Today, cryptography can help users of large datasets in a few specific ways. Sadly, it is not uncommon to see reports of lost laptops or other devices that, despite their small physical size, may hold very large volumes of data. If sensitive data are held on any such portable device, they should be encrypted in such a manner that they are not readable by whoever finds (or steals) the device. Many hard drives and even flash drives today provide built-in encryption processors so that the entire drive is automatically protected from simple theft. The user will be responsible for providing a key to the device in order to unlock it. Of course, in order to process the data, they must normally be decrypted, and even though the permanent files on the external hard drive or flash drive may be updated and re-encrypted, in most situations caches and temporary files may be created and remain on the host machine unless precautions are taken that those are encrypted as well.

Encryption for Fine-Grained Sharing Media encryption and file encryption provide for only relatively coarse-grained sharing of information. The development of attribute-based encryption (Goyal et al. 2006; Piretti et al. 2006) and more recently functional encryption (Boneh et al. 2011) enable much finer grained sharing policies to be enforced. In functional encryption, the data are encrypted once. A user with certain attributes will have an individual key that will enable her to learn a specific function of the encrypted data (which might be specific fields of a record, for example), but nothing else. Research is still advancing in this area, and attempts are underway to explore practical applications (Akinyele et al. 2011).

Computing on Encrypted Data What if you could encrypt data, perform any algorithm you wish on the encrypted data, and then decrypt the data in such a way that the decrypted data provided the results of the algorithm? In 2009, Gentry proved that this sequence is theoretically possible, though the scheme he used to demonstrate it would not be nearly efficient enough to be practical (Gentry 2009). If this scheme (called fully homomorphic encryption, FHE) were to become practical, it would enable many interesting ways of distributing trust – for example, the processor on which the results were computed would be unable to compromise any of the data used to compute the results, or the result itself. Cryptographers have been hard at work to find practical schemes to realize the promise of FHE. Functional encryption can be seen as one such example: it limits the function computed in the encrypted domain in specific ways. DARPA's PROCEED (Programming Computation on Encrypted Data) program is pushing in precisely this direction (DARPA 2010) and IARPA's SPAR (Security and Privacy Assurance Research; IARPA 2011) program has been pursuing related goals in the context of database access.

Software Integrity

The analysis of big datasets will involve programs of many sorts. Nearly all of these will have been created by people other than the person doing the analysis. How can the analyst be sure that the program being executed is the one intended? Put another way, how can one be sure of the software configuration of one's computer as a whole?

One control engineered for this purpose creates what is known as a 'root of trust' in the machine. The idea is that, starting with the point at which the machine is powered up and the initial firmware begins to load the operating system, the integrity of each new layer of software loaded into the system should have its integrity checked before it starts to run. Checking integrity in this case amounts to checking a secure hash of the bit string corresponding to the software load of the particular program. This does nothing to assure that the software is correct or lacks exploitable vulnerabilities, but it does assure that the software hasn't been altered since the secure hash was computed (presumably by the developer). Once the software begins operation, exploitable vulnerabilities can still be triggered and cause undesired effects until the next time the checksums are recomputed and checked.

To facilitate this sort of checking, a computing industry consortium (the Trusted Computing Group) developed specifications for hardware, called a Trusted Platform Module (TPM) starting in the early 2000s, and the microprocessors at the heart of many computers now incorporate TPM functions. Although the initial versions of the technology would have required the entire software load to be checked, later versions allow a 'dynamic root of trust' to be established in a way that simplifies initiation of new application environments. This technology also provides a way for one platform to attest to other, remote platforms that it is operating with a specific, known software configuration. If one is using a remote computing resource (as in 'cloud computing') and desires to verify that the cloud environment is the one expected, this technology can help. Although widely deployed, this hardware is largely unused. Microsoft's BitLocker software is an exception.

Data Provenance

Knowing where the data in a dataset came from and how they have been transformed is critically important for scientific research. Engineered controls for assuring data provenance have gained increasing attention over the past decade (Buneman et al. 2001; Muniswamy-Reddy et al. 2006), but what is in use seems primarily to be manual methods, overlaid on database management systems in some cases.

Provenance information has been characterized formally as an acyclic directed graph in which the leaves are data items and the nodes are other data and processes on which the present value of the data depends. For data that are updated frequently from a number of sources, the provenance graph could clearly become quite large.[2] On the other hand, this is exactly the information needed to resolve some scientific disputes.

This definition of provenance is closely related to what we discussed earlier as 'information flow'. Information flow is a more comprehensive idea because it aims to record implicit flows of information (flow that occurs when a variable is left unchanged because of the state of a condition that might have caused it to change, but didn't). Provenance deals only with explicit flows.

One might also observe that the TPM technology described above actually aims to establish (or perhaps to assure) the provenance of the software running on a particular platform at a particular time. Since the software is simply a bit string from the TPM's perspective, it could just as well be a file or database, and so the same technology could be applied to assure that the program processes a specific, intended dataset.[3] However, it would not help in tracing and recording changes made to the dataset in execution; it could at best record the final state by computing an encrypted checksum for the new version.

For some data, the provenance information itself might be sensitive – consider the provenance of a health record involving a sexually transmitted disease, for example. In this case, access controls will need to be applied to the provenance data. Conversely, some researchers have proposed using the provenance data as a security control (Ni et al. 2009; Martin et al. 2012).

The past few years have seen increasing efforts to investigate how provenance might be handled in a cloud computing context (Muniswamy-Reddy et al. 2010; Abbadi and Lyle 2012).

Detection and Recovery

Detecting when data have been incorrectly modified using an authorized mechanism, either accidentally or maliciously, requires logging; otherwise there is no way to detect the change. Further, the log must be available for review and it must *be* reviewed, and the log itself must be protected against accidental or malicious modification. This may sound like a recursive problem – if we can't protect the data, how can we protect the log? But mechanisms can be applied to the log that would not work for data in general. A solution does require care in organizing the logging system and in administering the privileges of system users.

Key questions to be addressed in organizing a log or audit trail for a user of large datasets include the following.

- What events must be logged? For example, one might simply log the event that a researcher checks out a dataset and checks it back in later, or one might log every access to an online dataset. The latter approach may permit faster detection of problems (assuming the log is monitored) but will result in a great deal of low-level data to be archived, and using low-level events to reconstruct higher level actions that may reflect the behavior of an attacker is a significant effort in itself.
- How can we assure all relevant events are logged? In effect, a reference monitor for the data is needed; the log file records its actions.
- How is the initiator of the logged event identified? Accountability for actions is a key control in any system. In the heart of a computer system, a process (a program in execution) is typically identified as the initiator of a specific action. It must be possible to trace back from the identity of a process to a responsible human or accountability is lost.
- How is the log itself protected from corruption? Write-once optical media are still used in some specialized applications today, but it is much

more common to send these records to one or more remote archive providers. The archive facilities evolve with technology for providing highly reliable long-term storage. An attacker, having penetrated a target system and wishing to cover her tracks, must either modify or disable local logging facilities, inhibit communications, or penetrate the archive sites as well.

- How is the log protected from compromise? The log itself is likely to contain sensitive data. For example, in a multiuser system it will incorporate actions from many different users, and no individual user may have the right to read log entries generated by others.
- Where will the log be kept? To avoid having a physical failure corrupt both the data and the log, they can be separated either on different devices in a system or the log data might be streamed to an offsite location.

Some researchers have proposed using query logs as a means for deriving user intent in accessing data as a way of enforcing purpose-based controls (Tschantz et al. 2012; Gunter et al. 2011). Access patterns present in the logs may be analyzed (using machine learning techniques, perhaps) to see whether they correspond to proper or improper usage patterns.

For static datasets, recovery from damage will simply be restoration of the original set. An offline backup will suffice in such simple cases. Where the dataset is dynamic, along with the logging processes just considered, and possibly integrated with them, there must be processes for creating backup files that can be restored. Changes made since the most recent backup may be recovered from the log files.

Very similar considerations apply to the backup files: they need to be kept in a protected location, access to them needs to be controlled, and they need to be protected from corruption and compromise. Frequently the most cost-effective way to compromise system data has been to steal a backup tape. Good practice dictates that all the backup tapes be encrypted and that the owner of the data, rather than the administrator of the backup facility, manage the keys.

Disposal

How long should data be kept? Some scientific data, such as experimental results obtained in the laboratory or observations of the natural world such as climate records, should be accessible for the indefinite future, and provenance is a key concern. Some filtered and abstracted datasets, such as released census records, fall in the same category. On the other hand, some confidential raw data have lifetimes specified by corporate or governmental policies. If data are collected or authorized for a specific use, it may be appropriate to delete them once that use has been made. What measures are available to be sure data are destroyed at the end of their intended life?

For data stored on computers, the main concern is more often losing data rather than failing to destroy it. Consequently, a great deal of attention may be paid to providing redundant backup copies, as discussed above. When one wishes to destroy the data, not only the primary copy, but also all of the backup copies, need to be dealt with. Further, as everyone should know by now, simply invoking the 'delete' function for a file doesn't typically delete anything at all; it most likely simply links the file into a 'deleted file' data structure, from which it may be removed again if the user invokes an 'undelete' function. Even when the file is 'permanently' deleted, the storage space it occupied is unlikely to be overwritten until some other file is allocated that space and actually stores something in it. Further, hardware devices often have mechanisms to deal with hardware faults so that if a fault occurs in a particular region of storage, the data are copied to a new block and the faulty block is removed from regular use. But the data in the faulty block may still be readable through special interfaces, and if they could contain important information, such as a cryptographic key, someone may exert the effort needed to recover it.

In earlier days, most archives and backups were kept on magnetic tape and could be erased directly by exposing the storage media to sufficiently strong and varying magnetic fields. Traces of data left behind after some effort to erase it are known as *remanence*, and guidelines have been produced to prescribe, for example, how many times a particular medium must be randomly overwritten before all traces of remanence are removed. Today there is a much wider range of storage media available and many (optical storage on CD-ROMs, for example) are not magnetically erasable and indeed not easily erased by any means. Even if they become unreadable for normal use, data may well be recovered from them if additional technical means are brought to bear. Computer hardware that turns up for resale has frequently turned out to include significant and readily available traces of the past owner's activities (Garfinkel 2013) and indeed computer forensic investigation frequently depends on such traces. Physical destruction of the media remains a very real approach to assuring data destruction.

The cleanest approach seems to be to encrypt all of the copies under strong cryptographic algorithms and to keep the keys separately from the data. In this case, erasing the keys used in the encryption is equivalent to erasing the records, at least until the cryptographic scheme is broken.

Many enterprises reduce the risk of losing data stored on lost or stolen smartphones in this way. All data on the smartphone is stored encrypted. A master encryption key is stored on the phone as well, but the phone will respond to a properly encoded 'remote wipe' command by zeroing the key, thereby effectively erasing all the data on the phone without requiring that the data be rewritten at all, or that the authorized user physically possess the phone. Disk drives and flash drives are available that support bulk encryption of all data on the drive; some of these provide a means to destroy or remove the key when the device is recycled. But this approach comes with the (real but decreasing) computational cost of encryption and decryption and the complexity of managing the keys.

NIST has produced useful guidelines on media sanitization that go into much more detail than is possible here; according to this guidance, overwriting a record stored on a magnetic hard drive one time is usually sufficient to render it unreadable without extraordinary effort, but this of course does not deal with the failed blocks problem. A draft revision (NIST 2012) incorporates discussions of 'cryptographic erase' and new types of media.

Future challenges for big data users in this area will probably come from their use of a multiplicity of devices, including mobile devices, to access data that they may later wish to expunge, and through the use of cloud resources to store and process datasets.

Big Data and the Cloud

The notion of a computing utility – first described by the creators of Multics in the mid-1960s (Corbato and Vyssotsky 1965) and in particular its goals of convenient remote access, continuous operation, and flexible expansion or contraction of resources – is being realized today in the 'cloud'[4] computing that Amazon, Microsoft, Apple, Google, Rackspace, and others now provide. The technology of virtualization of resources underlies cloud computing services. Each cloud user can be provided an environment that seems to be a dedicated, real computer, remotely accessed. In fact, this environment is simulated by a virtual machine; many virtual machines share very large pools of processors and storage that are managed by the cloud provider. Thus what appears a dedicated and modularly expandable resource to the cloud user is really a highly shared resource.

Researchers of big data might use cloud resources in several ways: for example, as archival storage for data and analytic results, as a source of computing horsepower to conduct analyses with software developed by the researcher, or as a source of application software services to analyze datasets stored in the cloud or elsewhere. Some threats require additional attention in public cloud environments: users are forced to rely on networked communications, they must depend on the competence of remote personnel acting for different management and possibly other alternative legal constraints, and they share resources with other tenants of unknown trustworthiness or intent. A more comprehensive list of cloud threat classes can be found in the Defense Science Board's report on the digital cloud (DSB 2013).

The ease of use that makes cloud computing attractive can mask some concerns that researchers need to consider. Issues to consider include the following.

For computing, how strong are the walls between different clients? In general, cloud providers provide 'multitenancy' – serving a large number of clients on exactly the same pieces of hardware – rather than 'sole occupancy'. The separation of different tenants is provided by the software (typically known as a hypervisor) that multiplexes the hardware. Though some attacks have been demonstrated that enable a user to break the hypervisor's separation (King et al. 2006) and others have shown that it's possible for data to be leaked from one user to another through relatively sophisticated signaling schemes using side channels (Wu et al. 2012), the risks a typical researcher takes in adopting a cloud service rather than, say, a shared, large-scale computing system at a research university do not seem to be great. In fact, universities are frequently turning to cloud providers to operate their e-mail and sometimes other computing resources.

For archival storage: where geographically do the data reside? Cloud providers typically have facilities at a variety of locations, including in different countries, both for reliability and availability, so that if one location has a problem, other locations can take over, and for economic reasons, if power is less expensive in one place than another. Policy may forbid that some data from crossing international borders or traversing certain networks. Some cloud providers may have service agreements that recognize such limitations, but *caveat emptor.*[5]

If a client uses the cloud just for storage, it may encrypt all the data before sending them to the provider and then the provider will be unable to compromise or undetectably alter them. However, the customer will also be unable to use the computing resources of the cloud provider to analyze the data unless they are first decrypted (unless computing with encrypted data, mentioned earlier, becomes a reality, in which case the cloud computing resource could be exploited without increasing the trust required in the provider).

Cloud vendors also provide services for 'private clouds' to address clients' concerns about multitenancy, among other things. For example, Rackspace markets a private cloud service in which hardware infrastructure is dedicated to a client. Amazon's Virtual Private Cloud (VPC) provides a 'logically isolated' portion of Amazon Web Services (AWS) and can be augmented with a Hardware Security Module (HSM) on which the client can store sensitive keys to be used to protect data stored on his AWS VPC. The HSM communicates over a Virtual Private Network (VPN) to the VPC; Amazon has no access to the client's keys. These augmented services frequently add cost.

One of the major cloud applications for analyzing big data is Hadoop, an open source implementation based on Google's MapReduce software (Vavilapalli et al. 2013). Hadoop uses HBase, modeled on Google's BigTable storage system. These systems enable massive computing resources to be brought to bear on problems involving very large datasets, but they generally lack access control on the datasets. The Accumulo system, originally developed at the U.S. National Security Agency on top of Hadoop and subsequently open-sourced, provides a distributed key/value store at large scale but also incorporates cell-level access controls, lacking in HBase and BigTable, so that a computation can be limited to data for which the process has the correct authorizations (Accumulo 2013).

Cloud service providers are likely to see commercial benefit in providing flexible and reasonably strong mechanisms for protecting their clients' data from other clients who may be curious about it and for enforcing policies on geolocation of data. Their infrastructures will be subject to the same kinds of vulnerabilities found in virtually all software systems today, but they may have more expert operators available to configure and manage their software systems than are available from smaller providers. For further discussion of cloud security and privacy issues, see Samarati and De Capitani di Vimercati (2010) and De Capitani di Vimercati et al. (2012).

Conclusions

Fundamental computing concepts for engineered controls on access to data and on information flows are reasonably well developed. They are perhaps not so widely deployed as they might be. The result is that for the next several years at least, an individual responsible for assuring that big datasets are accessed according to prescribed policies will need to rely on mechanisms that may be built into applications or may be lacking entirely. If they are built into applications, those applications will be layered over programming language libraries, operating systems, and perhaps hypervisors that will probably be reasonably reliable but will undoubtedly be penetrable. If they are not built into the applications, the individual will have to trust researchers who have access to the datasets to use them properly. In either case, as long as the data to be protected are not seen as of high value to an attacker, the kinds of problems likely to occur will mostly be accidental. Nevertheless, accidental disclosures of (for example) masses of location data about citizens' movements in a city could create a considerable stir. Establishing accountability for the event will be crucial; hence the custodians of the dataset should pay particular attention to assuring they have reasonable logs and audit trails in place and that those mechanisms are monitored on a regular basis.

Areas of research that could change the picture in the future include advances in practical cryptographic solutions to computing on encrypted data, which could reduce the need to trust hardware and system software. Advances in methods for building systems in which information flow, rather than access control, is the basis for policy enforcement could also open the door for better enforcement of comprehensible policies.

NOTES

1. Assuring that the specifications correctly capture the system requirements is itself a difficult and inherently manual task.
2. Note that *complete* provenance for a particular data item would include not only the programs that read and write the data, but also the provenance data for those programs themselves – version numbers, etc. – and also data about the compilers and other software used in creating them.
3. Note that this approach places trust in the correct functioning of the TPM mechanisms.
4. The cloud metaphor seems to have arisen from graphics in which a computing network was represented by an abstract cloud, to avoid the complexity of drawing all the lines. It is not a very helpful metaphor except perhaps in the sense that the cloud interface obscures the details behind it.
5. The global reach and accessibility of Internet-connected hosts makes geographic location irrelevant in relation to the probability of remote cyberattacks.

REFERENCES

Abbadi, Imad M., and John Lyle. 2011. Challenges for Provenance in Cloud Computing. In *Proc. TaPP '11, 3rd USENIX Workshop on the Theory and Practice of Provenance*, June. Available at http://www.usenix.org/events/tapp11/tech/final_files/Abbadi.pdf (accessed December 28, 2013).

Accumulo. 2013. Apache Accumulo User Manual 1.5. Available at http://accumulo.apache.org/1.5/accumulo_user_manual.html (accessed December 28, 2013).

Akinyele, J. A., M. W. Pagano, M. D. Green, C. U. Lehmann, Z. N. J. Peterson, and A. D. Rubin. 2011. Securing Electronic Medical Records Using Attribute-Based Encryption on Mobile Devices. In *ACM CCS Workshop on Security and Privacy in Smartphones and Mobile Devices*, Chicago, October. Available at http://sharps.org/wp-content/uploads/AKINYELE-CCS.pdf (accessed December 28, 2013).

Anderson, James P. 1972. *Computer Security Technology Planning Study*. ESD-TR-73-51, vol II, ESD/AFSC, Hanscom AFB, Bedford, MA, October. Available at http://seclab.cs.ucdavis.edu/projects/history/papers/ande72.pdf (accessed December 18, 2013).

Birrell, Eleanor, and Fred B. Schneider. 2013. Federated Identity Management Systems: A Privacy-based Characterization. *IEEE Security & Privacy Magazine* 11, no. 5 (September–October): 36–48.

Boneh, Dan, Amit Sahai, and Brent Waters. 2011. Functional Encryption: Definitions and Challenges. In *Proc. IACR 8th Theory of Cryptography Conference 2011*, LNCS 6597, 253–257. Heidelberg: Springer.

Bonneau, Joseph. 2012. The Science of Guessing: Analyzing an Anonymized Corpus of 70 Million Passwords. In *Proc. 2012 IEEE Symposium on Security and Privacy*, 538–552.

Buneman, Peter, Sanjeev Khanna, and Wang-Chiew Tan. 2001. Why and Where: A Characterization of Data Provenance. In *Proc. International Conference on Database Theory (ICOT) 2001*, 316–330. Heidelberg: Springer.

Corbato, Fernando J., and Victor A. Vyssotsky. 1965. Introduction and Overview of the MULTICS System. In *Proc. AFIPS Fall Joint Computer Conference 1965*, 185–197. Available at http://www.multicians.org/fjcc1.html (accessed December 28, 2013).

DARPA (Defense Advanced Research Projects Agency). 2010. Broad Agency Announcement (BAA) Programming Computation on Encrypted Data (PROCEED). July. Available at https://www.fbo.gov/utils/view?id=11be1516746ea13def0e82984d39f59b (accessed December 28, 2013).

De Capitani di Vimercati, Sabrina, Sara Foresti, and Pierangela Samarati. 2012. Managing and Accessing Data in the Cloud: Privacy Risks and Approaches. In *Proc. 7th International Conference on Risk and Security of Internet and Systems (CRiSIS)*, 1–9.

DSB (U.S. Department of Defense, Defense Science Board). 2013. *Task Force Report: Cyber Security and Reliability in a Digital Cloud*. OUSD AT&L, January. Available at http://www.acq.osd.mil/dsb/reports/CyberCloud.pdf (accessed December 28, 2013).

Garfinkel, Simson. 2013. Digital Forensics. *American Scientist* 101, no. 5 (September–October): 370ff.

Gentry, Craig. 2009. Fully Homomorphic Encryption Using Ideal Lattices. In *Proc. ACM Symposium on Theory of Computing (STOC) 2009*, 169–178.

Goyal, V., A. Sahai, O. Pandey, and B. Waters. 2006. Attribute-based Encryption for Fine-Grained Access Control of Encrypted Data. In *ACM Conference on Computer and Communications Security 2006*, 89–98.

Gunter, Carl A., David M. Leibovitz, and Bradley Malin. 2011. Experience-based Access Management: A Life-Cycle Framework for Identity and Access Management Systems. *IEEE Security & Privacy Magazine* 9, no. 5 (September–October): 48–55.

IARPA (Intelligence Advanced Research Projects Activity). 2011. Security and Privacy Assurance Research (SPAR) Program. IARPA-BAA-11-01. Available at https://www.fbo.gov/utils/view?id=920029a5107a9974c2e379324a1dcc4e (accessed December 28, 2013).

Kahn, David. 1996. *The Codebreakers: The Comprehensive History of Secret Communication from Ancient Times to the Internet*, revised edition. New York: Scribner.

King, S. T., P. M. Chen, Y.-M. Wang, C. Verbowski, H. J. Wang, and J. R. Lorch. 2006. SubVirt: Implementing Malware with Virtual Machines. In *Proc. 2006 IEEE Symposium on Security and Privacy*, 314–327.

Kocher, Paul, Joshua Jaffe, and Benjamin Jun. 1999. Differential power analysis. In *Advances in Cryptology—CRYPTO '99*, 388–397. Heidelberg: Springer.

Landwehr, Carl E. 1981. Formal Models for Computer Security. *ACM Computing Surveys* 13, no. 3 (September): 247–278.

Landwehr, Carl E., Constance L. Heitmeyer, and John D. McLean. 1984. A Security Model for Military Message Systems. *ACM Transactions on Computer Systems* 2, no. 3 (August): 198–222.

Martin, Andrew, John Lyle, and Cornelius Namiluko. 2012. Provenance as a Security Control. In *TaPP '12, Proc. 4th UNIX Workshop on Theory and Practice of Provenance*, June. Available at https://www.usenix.org/system/files/conference/tapp12/tapp12-final17.pdf (accessed December 28, 2013).

McLean, John D. 1994. A General Theory of Composition for Trace Sets Closed under Selective Interleaving Functions. In *Proc. 1994 IEEE Symposium on Security and Privacy*, 79–93.

Microsoft.. 2010. Cybersecurity for Open Government, June. Available at http://download.microsoft.com/download/1/1/F/11F98312-8E4C-4277-AF3F-B276FD6978DA/CyberSecurityWhitePaper.pdf (accessed December 27, 2013).

Muniswamy-Reddy, K., D. Holland, U. Braun, and M. Seltzer. 2006. Provenance-Aware Storage Systems. In *Proc. 2006 USENIX Annual Technical Conference*, June. Available at http://www.eecs.harvard.edu/margo/papers/usenix06-pass/paper.pdf (accessed December 28, 2013).

Muniswamy-Reddy, K.-K., P. Macko, and M. Seltzer. 2010. Provenance for the Cloud. In *Proc. FAST '10, 8th USENIX Conference on File and Storage Technologies*. Available at www.usenix.org/events/fast10/tech/full_papers/muniswamy-reddy.pdf (accessed December 28, 2013).

NIST (U.S. National Institutes of Standards and Technology). 2012. *Guidelines for Media Sanitization*, by Richard Kissel, Matthew Scholl, Steven Skolochenko, and Xing Li. Draft NIST Special Publication 800-88, Revision 1. Available at http://csrc.nist.gov/publications/drafts/800-88-rev1/sp800_88_r1_draft.pdf (accessed December 28, 2013).

NIST (U.S. National Institutes of Standards and Technology). 2013. *Improving Critical Infrastructure Cybersecurity, Executive Order 13636: Preliminary Cybersecurity*

Framework, October. Available at http://www.nist.gov/itl/upload/preliminary-cybersecurity-framework.pdf (accessed December 27, 2013).

Ni, Qun, Shouhuai Xu, Elisa Bertino, Ravi Sandhu, and Weili Han. 2009. An Access Control Language for a General Provenance Model. In *Proc. Secure Data Management (SDM) 2009,* LNCS 5779, 68–88. Heidelberg: Springer.

Park, Jaehong, and Ravi Sandhu. 2004. The $UCON_{ABC}$ Usage Control Model. *ACM Transactions on Information and System Security* 7, no. 1 (February): 128–174.

Piretti, Matthew, Patrick Traynor, Patrick McDaniel, and Brent Waters. 2006. Secure Attribute-Based Systems. In *ACM Conference on Computer and Communications Security 2006,* 99–112. Available at http://www.patrickmcdaniel.org/pubs/ccs06b.pdf (accessed December 28, 2013).

Samarati, Pierangela, and Sabrina De Capitani di Vimercati. 2010. Data Protection in Outsourcing Scenrios: Issues and Directions. In *Proc. ACM AsiaCCS 2010.* Available at spdp.di.unimi.it/papers/sd-asiaccs10.pdf (accessed December 28, 2013).

Sandhu, Ravi. 1996. Role-based Access Control Models. *IEEE Computer* 29, no. 2 (February): 38–47.

Schneider, Fred B. 2000. Enforceable Security Policies. *ACM Transactions on Information and System Security* 3, no. 1 (February): 30–50.

Strang, Thomas J. K. 1996. Preventing Infestations: Control Strategies and Detection Methods. Canadian Conservation Institute. *CCI Notes 3/1.* Available at http://www.cci-icc.gc.ca/publications/notes/3-1_e.pdf (accessed December 17, 2013).

TBC (Treasury Board of Canada). 2006. *Operational Security Standard, Management of Information Technology Security (MITS),* Sec. 16–18. Available at http://www.tbs-sct.gc.ca/pol/doc-eng.aspx?id=12328§ion=text (accessed December 17, 2013).

Tschantz, Michael C., Anupam Datta, and Jeannette M. Wing. 2012. Formalizing and Enforcing Purpose Restrictions in Privacy Policies. In *Proc. 2012 IEEE Symposium on Security and Privacy,* May.

Vavilapalli, V. K., A. C. Murthy, C. Douglas, S. Agarwal, M. Konar, R. Evans, T. Graves, J. Lowe, H. Shah, S. Seth, B. Saha, C. Curino, O. O'Malley, S. Radia, B. Reed, and E. Baldeschwieler. 2013. Apache Hadoop YARN: Yet Another Resource Negotiator. In *Proc. ACM Symposium on Cloud Computing,* October. Available at http://www.socc2013.org/home/program/a5-vavilapalli.pdf?attredirects=0 (accessed December 28, 2013).

Verizon, Inc. 2013. *2013 Data Breach Investigations Report,* April. Available at http://www.verizonenterprise.com/resources/reports/rp_data-breach-investigations-report-2013_en_xg.pdf (accessed December 29, 2013).

Weir, Matt, Sudhir Aggarwal, Michael Collins, and Henry Stern. 2010. Testing Metrics for Password Creation Policies Using Large Sets of Revealed Passwords. In *Proc. ACM Conference on Computer and Communications Security.*

Wu, Zhenyu, Xu Zhang, and Haining Wang. 2012. Whispers in the Hyper-Space: High-Speed Covert Channel Attacks in the Cloud. In *Proc. USENIX Security Conference 2012.* Available at https://www.usenix.org/system/files/conference/usenixsecurity12/sec12-final97.pdf (accessed December 28, 2013).

11 Portable Approaches to Informed Consent and Open Data

John Wilbanks

Introduction

What frameworks are available to permit data reuse? How can legal and technical systems be structured to allow people to donate their data to science? What are appropriate methods for repurposing traditional consent forms so that user-donated data can be gathered, de-identified, and syndicated for use in computational research environments?

This chapter will examine how traditional frameworks for permitting data reuse have been left behind by the mix of advanced techniques for re-identification and cheap technologies for the creation of data about individuals. Existing systems typically depend on the idea that de-identification is robust and stable, despite significant[1] evidence[2] that re-identification is regularly possible on at least some portion of a de-identified cohort. The promise that privacy can always be protected, that data can always be de-identified or made anonymous, is at odds with many of the emerging realities of our world.[3]

At issue here is a real risk to scientific progress. If privacy concerns block the redistribution of data on which scientific and policy conclusions are based, then those conclusions will be difficult to justify to the public who must understand them. We must find a balance between our ability to make and produce identifiable data, the known failure rates of de-identification systems, and our need for policy and technology supported by 'good' data. If we cannot find this balance we risk a tragedy of the data commons in which the justifications for social, scientific, and political actions are available only to a select few.[4]

Approaches and frameworks that are emerging to deal with this reality tend to fall along two contours. One uses technological and organizational systems to 'create' privacy where it has been eroded while allowing data reuse. This approach draws on encryption and boundary organizations to manage privacy on behalf of individuals. The second applies an approach of 'radical honesty' towards data contribution by acknowledging up front the tension between anonymization and utility, and the difficulty of true de-identification. It draws on the traditions of beneficence and utility as well as autonomy in informed consent to create reusable and redistributable open data, and leverages cloud-based systems to facilitate storage, collaborative reuse, and analysis of data.

Traditional Approaches Falling Behind

Most data collected commercially in the United States today lacks direct protection under the law. It is instead governed by a hodge-podge of contractual instruments such as terms of use, implemented by long and rarely read 'click-through' agreements on phones, tablets, and laptops. The parties gathering the data typically attempt to minimize the ability of the person about whom data is being gathered to comprehend the scope of data, and its usage, through a mixture of sharp design and obscure legal jargon.[5]

However, several kinds of data do receive direct legislative and/or executive protection for commercial and private use at the national level: primarily educational, financial, and health records. This chapter will focus on the last of the three.

Health records and health data – the two are often thought of separately, although they are rapidly aligning and in many cases merging – receive distinct privacy protections under the Health Insurance Portability and Accountability Act (HIPAA) passed in 1996. HIPAA contains a specific Privacy Rule regulating the use and disclosure of 'protected health information' (PHI) held by 'covered entities' involved in the administration and payment of health care. PHI is any information held about an individual by a covered entity about health status, payment, treatment, and other related information.

HIPAA lays out a specific set of kinds of data that are explicitly PHI: names, geographic indicators, dates, phone, fax, e-mail, Social Security numbers, medical record numbers, health insurance beneficiary numbers, account numbers, license numbers, vehicle identifiers (i.e. VIN or license plate), device identifiers, URLs, IP addresses, biometric identifiers, full face images, and "any other" unique number, characteristic, or code except the code used by the investigator to code the data. It is possible to distribute data about an individual, such as clinical information or treatment information, if these 18 identifiers are removed (this creates a 'safe harbor' in which the data is considered by law to be de-identified) or if the overall data, including

PHI, has been certified by an expert to be technically de-identified to an extent where the possibility of re-identification is very small.

These regulations, however, ignore a fact of data that for years was well known to experts in the field but poorly known outside: if there is either enough direct data (such as clinical information) or indirect data (such as metadata or data emerging from 'non-health' devices) that is collected and shared, even the absence of the 18 identifiers is unlikely to protect against re-identification approaches by experts. The tension is not unique to health but perhaps is most acute: we need to know a lot about an individual to properly make use of that individual's data in a scientific research context, but precisely by knowing a lot about the individual, we degrade the ability to guarantee that individual's anonymity.

As an example, suppose we were attempting to understand why it is that a cancer drug fails to achieve its outcome in 75% of those who take it. This is unfortunately common, and being able to know in advance which 25% will actually benefit from dosage, versus which will experience only toxic side effects, would be beneficial both to the patient and to the insurance system that pays for failed dosages just as it pays for successful ones.

But it requires a deep dive into those individuals' data: what are their unique genetic variations, which might provide clues as to how individual genomes affect drug response? What are their environments and lifestyles, which might provide clues as to how diet and exercise affect drug response? What kind of data is coming out of their classic clinical sampling, which might provide early signals if the drug is or isn't working? Taken together these data points provide precisely the kind of longitudinal health data that can power strong analytic models and pinpoint clusters of predictive information – but at the same time, can allow someone else to vector in on the identities of the individuals in the study.

High quantities of varied data tied to individuals will make this failure worse. As we begin to connect mobile devices to health through applications, connected hardware, fitness trackers, and even wearable electronics, it is not hard to imagine a world in which health insurance companies and mobile providers form partnerships to sense when a person is in line at McDonald's and send them a text message suggesting a day at the gym – perhaps even offering a coupon. When our grocery store loyalty cards are connected in turn to our fitness trackers, it will be very hard to hide.

But these at least are decisions that we have the power to change, or at least obscure from view. Open operating systems on mobile devices have the potential to re-create some forms of privacy in this space. But as our genomes and our medical records converge, as they inevitably will, there will be permanent records of the diseases and syndromes to which we are more or less susceptible.[6] And those records are indeed almost perfectly identifiable – our genomes are far more accurate ways to precisely identify us than even our fingerprints or our credit card numbers, which are in turn each far more unique to us than our Social Security card numbers.

The law is most concerned with whether or not John Wilbanks has been uniquely identified by a health provider. But this is irrelevant when an entire network of providers outside the traditional health system knows that my credit card has been used at a grocery to buy pizza, that my phone spends five times as many minutes in fast food restaurants as it does in athletic environments, or that my online shopping habits are yielding larger and larger pants sizes. There is only one person for whom all of those can be true, and he's probably at risk for diabetes and metabolic syndrome.

It is precisely this network-aggregated, metadata-centric approach that has become so notorious through the revelations of domestic data capture at the National Security Agency in 2013 and before. This approach falls well outside the boundaries of health legislation or regulation, but can just as easily be used to infer health information as the data that is traditionally known to be about health. Indeed, even social media postings and photographs contain data that can be converted to health data.[7] Simply put, our legislative and regulatory systems that attempt to guarantee health privacy have been overwhelmed by the technocratic infrastructure that suffuses our daily lives, and are unlikely to catch up in the near future.

The deepest irony is that the protections are strongest and the regulations most effective at the institutions where health and life sciences research take place: the covered entities. Start-up companies, telecommunications providers, and others are almost entirely unaffected as they gather metadata and actual data from which health, and identity, can be inferred. Research data that is often federally funded or involves data from federally funded research becomes subject to the laws and regulations for government data management – which is significantly more complex, but is not necessarily able to prevent re-identification attacks.

The Database of Genotypes and Phenotypes (dbGaP) was founded in 2006 and is maintained by the U.S. National Institutes of Health. It was developed to archive and distribute the results of studies that have investigated the interaction of individual genetic variations (the genotype) with the observable health and traits of those individuals (the phenotype). These studies hold the promise of decoding which minute differences in genetic makeup affect whether or not a cancer drug will work before it is given, or what kinds of mutations mean a rapid acceleration of a disease such as Parkinson's or Alzheimer's.[8]

Recipients of U.S. NIH funding related to genome-wide association studies and whole genome sequencing studies are expected to deposit the results in dbGaP, based on policies first promulgated in 2007 and updated in 2009:

> Consistent with the NIH mission to improve public health through research and the longstanding NIH policy to make data publicly available from the research activities that it funds, the NIH has concluded that the full value of sequence-based genomic data can best be realized by making the sequence, as well as other genomic and phenotype datasets derived from large-scale studies, available as broadly as possible to a wide range of scientific investigators.[9]

The idea behind dbGaP was to support two levels of data access: 'pooled' or 'aggregated' data that is not granular to the individual level (analogous to ZIP-code-level statistics in census data) in a public layer and individual study information in a more controlled layer. The controlled-layer data would still be de-identified according to HIPAA, vetted by data access committees, and released only to authorized investigators who agreed to the terms and conditions of use. These data are available behind a firewall and can be accessed after application for use by investigators.

The first two years of the database saw nearly 500 complete downloads of the total open layer of data.[10] But in late 2008, a paper was published that demonstrated a feasible re-identification method:[11] how to plausibly identify at least some individuals inside large sets of aggregate genomic information that had been de-identified according to the laws and regulations. The NIH quickly moved to shift aggregate genomic data from the open layer of dbGaP to the controlled layer. Follow-on studies revealed that re-identification vectors were possible.[12]

The reason this is a problem is that the studies involved did not contemplate the risks of re-identification on a platform like dbGaP. The data came online under a dizzying variety of informed consent terms and procedures, which typically focused on the study itself rather than data sharing and were written by investigators whose field is health – not re-identification. In particular, the forms failed to notify participants of the risk of participation in a study destined for online archives.

This is a real problem. Our ability to understand how minute individual variations in genetics affect health will depend on having the right to perform research on individual variations. But our data governance systems are behind the technology curves of genetic data generation, data distribution over the internet, and re-identification.

We must acknowledge that participation in genetic studies carries a risk of identification, one that increases with the volume of data generated about an individual and as time improves the tools for re-identification – often in fields far removed from the genome. Thus, we face a real test of our capacity to design and implement new approaches to individual-level data governance in health, ones that both facilitate the acceleration of knowledge creation but also provide honest, transparent guidance to study participants.

Technological Control Frameworks

Given the complexity of HIPAA and other federal approaches to privacy, as well as the varying degrees of health data protection across the 50 states of the United States, many attempts to increase personal data privacy while facilitating sharing do not attempt to change legislation, harmonize policy, or otherwise engage in government. They instead attempt to encode a more contemporary approach to data gathering and sharing inside technological, structural, organizational, legal, and normative ***frameworks*** that work inside existing laws and attempt to leverage those laws to reach the goal of reconciling research with privacy. These frameworks have in common the idea of harm prevention – of preventing improper, harmful, or bad uses of data – through ***implementation of controls***.

The concept of control as the guiding principle of these frameworks is essential to their design and implementation. Control can be implemented via software, contract, terms of use, intellectual property, liability, economics, or other means. Each attempts to regulate the allowable behaviors of data users to protect the privacy of individuals whose data is being used.

These frameworks are discussed in more detail elsewhere in this volume, but it is worthwhile to quickly examine some of the most relevant systems in the context of health and to look at their advantages and disadvantages depending on the kind of data being controlled.

One of the most popular control frameworks is 'differential privacy', which provides access to statistical databases for queries while using algorithmic methods to reduce the odds of re-identification of records in the database itself. The goal is to respond to queries to the database as accurately as possible without compromising identity. The advantages of this approach are several: the approaches are well known and tested, the addition or subtraction of a single individual's data is unlikely to affect the accuracy of a query response (allowing for withdrawal of an individual from a clinical study, for example, without significantly compromising the overall queries already run), and there is good research on how to protect against excessive distortion of the data.[13]

The disadvantages, however, are real. The connection between accuracy and identifiability creates a documented tension between the statistical addition of 'noise' to data to increase the difficulty of identification and the actual utility of the database. When the data concern movie reviews, this noise is innocuous. When the data are genetic variations – the individual mutations that make each of us unique – then the addition of noise to the data creates real problems. Trust that the underlying data accurately represent the genomic profile of the individuals is essential to the analysis, and thus adding noise via 'fake' mutations to the system to make identifiability harder in turn reduces the research utility of the data. Indeed, the entire point of a genetic study looking for health outcomes is in many ways to use accurate genetic data tied to outcomes in order to find patterns of individual mutations correlated to outcomes, and statistical approaches to differential privacy very likely would confuse the issue. As one author notes, standard anonymization techniques are not applicable to genomic information as it is the "ultimate identity code."[14]

A technological framework often proposed to solve the same problem is homomorphic encryption (HE).[15] HE enables predetermined queries to run on encrypted data that obtain, in turn, an encrypted result. That result can then be decrypted in a way that exactly matches the results that the data user would have seen if running the query on non-encrypted data. This would allow a user to run genome-wide queries on many genomes, find the individual variations, but not know from which genomes the variation patterns came.

A key advantage to HE is that it allows for one person to run the query but not be able to decrypt the answer (allowing for an expert, or an expert system, to execute encrypted queries on behalf of a less skilled operator, who would then be the only one able to read the result). This is a real advantage in health and life sciences research, as the researchers working on these kinds of data are very unlikely to have sufficient training to manage complex encryption systems on their own. Another advantage of HE is its innate compatibility with cloud architectures to which large-scale data processing is rapidly moving.

A key disadvantage is speed. HE is inherently slow and gets slower the more information that is encoded in the ciphertext. Placing a single whole-sequence genome into HE would create serious performance problems, much less placing tens of thousands of genomes (as required for performing significantly powerful analysis). HE is also comparatively novel as a technology and lacks widespread support from vendors or expertise in implementation.

A third technical framework depends on distributed storage of the data. Private, sensitive information is stored across multiple databases held by multiple entities – for example, part of an individual's genome would be held in Boston, while another part of the same individual's genome would be held in New York. Each of the two parties would have incomplete, semi-random data. But when a technical framework integrates the data, 'sensitive' values may be extracted. This is known as 'secret sharing'[16] and may be analogized to the peer-to-peer approaches for sharing content on the web such as Bit Torrent.

A key advantage of secret sharing is that predefined queries can run on the shared data without transmission of the data itself, without complex or slow encryption, and without adding statistical noise to the information. Another advantage is the successful deployment of the secret sharing approach in auctions[17] and financial markets,[18] which increases the availability and support of tools for its implementation.

Organizational Structures for Control

Technical frameworks for privacy, on their own, are typically not sufficient to create a controlled environment for sharing. Organizations must operate the frameworks to ensure they are running correctly, watch for violations, punish transgressions, and perform other functions. The various technological frameworks map to some organizational structures better than others. Of the technologies noted here, only secret sharing, with its inherent distributed nature, aligns well with a non-organizational control structure.

Differential privacy maps well to a marketplace paradigm, in which data-sharing arrangements are structured as markets that can either 'hedge' an individual against harm by taking a stake in the market or even directly receive financial benefits in the event of data usage harms. This creates an interesting model for organizations, both for-profit and not-for-profit, to operate data marketplaces where individuals choose to store their data and have it be bought and sold for use inside the framework run by the organization.

Another structural paradigm that is implied is one we know well: banking. If data is an economic asset (as the market idea recognizes, and as it is often treated in the daily world of commerce and social media) then it is an asset in which an individual might wish to invest in hope of a return. Health record banking[19] is a well-developed concept, and related models such as land trusts, conservation trusts, and development trusts are each being explored for technology-mediated, data-centric health collaboration.

There are also some established business models for access to sensitive data. One is the liability regime, in which a data holder provides access to data that may (or may not) be identifiable, but the data user must agree to punitive terms and conditions that govern the kinds of queries that may be run, restrict data that may be downloaded, and outline penalties for failure to comply. PatientsLikeMe is a good example of this, though the model is a longstanding and established one in data provision companies. Another is the services model, in which security and privacy are maintained by the data aggregator holding all data internally, and performing fee-for-service research as a consulting service rather than allowing access to the data directly.

In all of these structural-organizational frameworks, the organization would install terms of use around any entry into the market, presumably through contracts. Users would be verified and subject to some financial liability if they breach the contract by running queries they are not allowed to run, or attempting re-identification attacks, or by somehow rebuilding elements of the raw data and republishing them elsewhere.

The key weakness in any of these models is the inability to lock down the data should the data somehow 'leak' in violation of contractual restrictions or technical restrictions. The organization in charge of data aggregation would clearly have the right to sue the transgressor, but the copy that is then 'in the wild' would be nearly impossible to track down and protect.

This reflects a structural element of the intellectual property regime around data, which is inherently less able to lock down reuse copyrights and patents. Data is considered by most jurisdictions to fall under the regime of a fact, rather than a creative work or an invention, and thus sits in the public domain by default. This is a powerful and important default status, as it prevents facts from being owned, laws of nature from being enclosed, and ensconces them in the commons of the mind. But it also makes it hard to license *conditionally* – which is the root of open source software and Creative Commons licensing.

Conditional licensing is the use of a contract to note that reuse is allowed, but only if certain conditions are met, such as attribution, or non-commercial use, or even that downstream works must be relicensed under the same terms. The key is that in copyright, such terms are easily enforceable. The conditions follow the song, or the software. But these types of conditions for data don't travel with the data as it is propagated across a network.[20] The public domain intellectual property status of data thus has a sharp tang in the privacy sphere: data are unlikely to receive the kinds of protections that can be used to take down a dataset that was private but has been made public either by accident or by malice.

Commons-based Frameworks: Open Consent

The number of foundations, endowments, and other non-profit groups investing in health IT has exploded over the past ten years. Small, nimble groups formed by patients and large foundations created by technology entrepreneurs have joined traditional funders such as state and federal governments, creating a significant overall increase in funding to basic life sciences and translational health research. At the same time, a dizzying variety of data began to flood the market. We can use mobile devices to track our heart rate, blood pressure, weight, physical activity, gym visits, sleep, and more. Physicians are under pressure to complete the transition to electronic health records. Companies are aggressively moving to provide services ranging from genotyping to self-tracking to community disease management.

In the midst of this, expectations remain that some form of privacy or anonymity should be the goal, even if it is next to impossible to guarantee.[21] The frameworks explored in the previous section attempt to address the imbalance between that goal and the reality of identification. But it is also possible that studies can create different ideas, or locate participants who are less worried about privacy than about advancing health and science.

To the extent these expectations do exist in health studies, they have emerged from the methods used to enroll participants: the documents and processes used to obtain informed consent. The NIH's data-sharing policy explicitly calls out protecting privacy and confidentiality as critical, which creates in turn a direct implication of a control-based framework. And thus as the ability to capture information explodes, and the cost of capture drops, the consent structures have remained resolutely dogged in assuring people their privacy will be protected – even though that assurance most definitely cannot be kept for all.

It is important that scientists recognize this, and find a solution. Given the wealth of clinical and genetic information now collected in a clinical trial it is becoming apparent that there are a variety of secondary uses of clinical trial data that could greatly enhance the rate of scientific progress in a variety of ways not foreseen by the original study developers. The same is true of epidemiological and/or observational studies.

This is particularly true of genetic data: the American Society of Human Genetics stated they are "acutely aware that the most accurate individual identifier is the DNA sequence itself or its surrogate here, genotypes across the genome. It is clear that these available genotypes alone, available on tens to hundreds of thousands of individuals in the repository, are more accurate identifiers than demographic variables alone; the combination is an accurate and unique identifier."[22]

But a response has emerged to this problem, one that embeds the reuse of the information as a higher goal than the guarantee of privacy or the prevention of re-identification. These are often called 'open' consents, but the consent is simply part of a large ***commons-based framework*** intended to share data, rather than to control its use.

Commons-based frameworks for reuse attempt to recruit individuals who have not only benefited from the explosions in investment and technology, but who also understand the risks and uncertainties of making their data available for reuse. The key complexity comes from the uncertainty – from the 'unknown unknowns' that may emerge as risks downstream, years after data is made available for reuse. This is consistent with the complexity of the data itself, which is beyond most of our attempts to comprehend as individuals, as well as with the unknown benefits of reusable health data. We simply do not yet understand the system that is emerging well enough to precisely assess either its true benefits or true risks over the long term. Commons-based frameworks must live with this fundamental uncertainty, and participants must understand this uncertainty enough to provide informed consent for data redistribution and reuse.

'Open' consent was first developed by the founders of the Personal Genome Project (PGP) at Harvard Medical School. The PGP study asks participants to post health information (both as records and as interviews), performs whole genome sequencing, and creates an immortalized cell line that is available under liberal terms and at a low cost from a cell culture repository. PGP has the potential to create a data resource where a user might identify a promising set of variations tied to outcomes computationally, then order the cells and test the hypothesis in vitro within days.

The study's model of consent starts not from the premise of preventing harm, or controlling use, but instead from the idea that participants in genotype–phenotype studies must understand that the data collected not only can be, but is intended to be accessed, shared, and linked to other sets of information. PGP participants must complete a thorough and, for many, difficult consent protocol proving they understand these intentions as well as the risks involved in participation, the bulk of which remain unknown or unknowable at the moment of consent.

Privacy is not guaranteed in the PGP, and identifiability is called out as a possibility. Participants can withdraw, and their data will be taken down from the study's servers, but no guarantees are made that their data is gone from the web: if it has been downloaded and redistributed, the study can no longer control its presence or its use.

Commons-based Frameworks: Portable Consent

Inspired by the PGP's groundbreaking work, but desiring a simpler process to create informed consent, Sage Bionetworks began work on a 'portable' approach to informed consent, called Portable Legal Consent or PLC.

Portable Legal Consent was developed as a tool to allow patients to tell the doctors, researchers, and companies experimenting on them that they, the patients, have rights with respect to the data generated from their bodies. PLC states that what the patient desires is for the data to be shared broadly in the public domain, to serve scientific progress as a whole, regardless of the particular individual or institution that makes the breakthrough.

PLC emerged from the 2011 Sage Bionetworks Commons Congress, where a working group focused on the need for standardized approaches to privacy and patient empowerment. It became clear that two approaches were needed: one to populate computational platforms with individual-level data that can be used to perform computational research with as few barriers as possible, and one to empower patients to take control of their own data. The first approach became PLC, while the second approach informs the PEER platform hosted by the Genetic Alliance.

The legal inspirations for PLC were the informed consent process developed by the Personal Genome Project (from which it draws both ideas, and even some text in its FAQ). PLC also drew on the idea of 'human readable' interfaces to complex legal documents that Creative Commons pioneered. PLC embedded the idea that study investigators should disclose risks about research inside the idea that data should be something that can be remixed, to allow unexpected discoveries to emerge from the combination of earlier studies by later scientists.

If a person completed the PLC process, she had an informed consent that traveled with her from one upload of data into an environment that allowed many studies – that is portable from a research perspective, and that she controls. PLC meant that the data she chose to upload into a common genomics repository would be able to support a broad range of genomic and health research without the unintended fragmentation of data created by traditional consent systems.

Participants in the study enrolled via a web-based 'wizard' that extracted the key elements of the consent – the obligations placed on researchers (and the significant limitations of those obligations), the freedoms granted by participants to data users, the risks involved – into a clear, layperson-friendly structure. Participants completed the wizard and watched a short video before indicating consent and being presented with the consent form to sign. All text, software, multimedia, user interface designs, and related systems were provided as public domain documents under the CC0 public domain waiver so that study designers outside Sage Bionetworks may take up and remix the elements of PLC as they wish.

The wizard that served as a consent tutorial was, by the numbers, effective at communicating the risks and benefits of data sharing based on the survey data, though it is very likely that the population who enrolled was self-selected to bias in favor of pre-existing comprehension and risk tolerance. Concepts that were essential to the study, such as 're-identification', received plain-language definitions submitted for ethical review and approval that appeared on mouse-over by the user, a method shown to increase comprehension in online environments.[23]

Participants were required to tick checkboxes next to key statements (such as agreeing to allow redistribution of their data) to indicate assent to individual elements of the study rather than a single digital signature on a single legal form. The statements were grouped into pages indicating key freedoms granted to researchers, and key risks understood by participants, and included a page where participants who felt uncomfortable could exit the tutorial before being presented with the opportunity to sign any binding consent documents or upload files. Only after passing through all the pages and specifically indicating desire to sign the form was a participant presented with a digital version of a traditional informed consent document. Every word on every screen, including the script of the video, was approved by an independent Institutional Review Board (IRB).

After signing the form, participants proceeded to upload data files including electronic health records, genomes and genotypes, lifestyle data from mobile devices, and anything else they found relevant to their health. The data were to be de-identified and syndicated to Sage's Synapse platform, a technology designed to facilitate computational reuse and collaboration on complex health and biological data. But the heterogeneity of the formats of data, particularly the use of large image files (including photographs) of medical records, made de-identification virtually impossible without rekeying the data by hand, and thus data were not syndicated by default to researchers.

PLC was completely voluntary – one need not enroll, and if enrolled, one need not upload any data. Several hundred enrolled and uploaded at least one data file. A survey was run concurrently of users and indicated strong comprehension of the risks and conditions of the study – preliminary data indicate that more than 90% of those completing both the consent process and the survey understood key issues around data upload, study withdrawal, and permissions granted to data users.

The PLC study assigned a unique identifier to every enrolled participant (the author's is Individual 1418165/syn1418165), and PLC was IRB-approved to send emails to the unique identifiers to ask follow-up questions, recruit participants for follow-on studies, and more. This constituted a 're-mail' process that resolves between the ID and the e-mail address held apart from the syndicated data as well as some interface designs that protect against recipients accidentally revealing personal information in replies.

However, PLC's bioethical promise to engage people in research was greater than the scientific power it enabled during the study lifetime. As a variety of federal and international projects move to make health data more standardized and 'liquid', as well as to create positive rights to access health data, the situation will change – but for now, very few people have meaningful health data about themselves, and only a small percentage of those have that data in a computationally useable format. PDFs of health records, or scanned images, constituted the vast majority of donated data, but these are of little use in a computational collaboration.

Thus it is important to connect the lessons of PLC and other privacy-compatible research engagement systems to emerging policy and advocacy movements that increase personal access to health data. As standardized clinical trial data sharing moves into practice and standard electronic medical record formats become the norm, the scientific power of data donation will begin to match the ethical power of engagement in consent processes like PLC.

Portable Legal Consent 2.0

Our goal with future versions of PLC is to leverage the desire to participate in new forms of clinical study to generate not just patient engagement, but the creation of commonly pooled resources that are scientifically useful as well. The capital markets see data primarily as an economic asset, not as a research one, and may well be able to find value even in the kinds of heterogenous data donated through something like PLC 1.0 – indeed, treating it as metadata for clinical trial recruitment or even drugs marketing.

But the social value we hope to create comes from the data as a vector to connect groups of patients with groups of experts who can analyze the data and are committed to returning results to patients. Thus, future versions focus on making PLC compatible with many different sorts of clinical studies, whether initiated by patients themselves or by more traditional investigators, on Sage Bionetworks' BRIDGE patient engagement platform.

First, modularity is an essential element of forthcoming versions of PLC. There are multiple approaches to making consent, or privacy separate from consent, modular – most focus on differential approaches that indicate one use is acceptable while another is not. In PLC, the overwhelming response has been not a desire for differential consent that allows only academic use, or consent only for a certain field of use; instead participants are far more concerned about the ***kind of data*** they have and how they would like to donate it. Thus, modularization of PLC will proceed as a function of data classes rather than a complex set of differential permissions.

This has multiple benefits. The most important is that it allows a far simpler consent experience for many users. The most complex aspects of the existing consent process revolve around the donation of genomic sequence information, which is highly identifiable no matter how much de-identification we perform around the metadata for the file. This is also the kind of data that is most ambiguous in terms of possible future harms, while simultaneously the kind of data that the fewest people have.

By breaking sequence data apart from other data classes the consent experience can be radically shortened and simplified for most participants. Only if a participant has sequence data and wishes to donate it would she move through the module related to sequences. Other benefits include the ability to rapidly repurpose user interface and legal code for other studies that are gathering data whose classes are represented in PLC 2.0, and the ability to begin applying automatic de-identification processes to data.

Feedback indicates at least five desired data classes:

- Genetic sequence
- Clinical information
- Medical record
- Patient-reported outcomes
- Personal sensor data (including but not limited to mobile device)

It is important to note that many of these data classes have fuzzy borders. Medical records contain clinical information and are moving to expand to the capture of personal sensor data and genetic sequence. Patient-reported outcomes also may contain clinical information and personal sensor data.

The choice of these labels for the modules does not ignore this reality but reflects instead a choice to focus on the file formats associated with each kind of data. It is easy to know by file format if one is dealing with a sequence, or an electronic medical record, or the output from a mobile application or fitness device, and that makes it easier to algorithmically assign consent modules and/or de-identification processes to that file.

Second, we are moving to tie the consent forms themselves more tightly to the system that qualifies users of the data under consent. In PLC 1.0, users of the data need only have a Google account or otherwise verify their identity in a lightweight form. In 2014, users inside Sage Bionetworks' Synapse system will also have to pass a comprehension test to access certain features, including the ability to access data available under liberal terms and use programmatic tools such as the API. This shift recognizes that the risk–benefit relationship in the consent process is not solely the responsibility of the data donor, but also of the data user.

Conclusion

Emerging methods for data generation increasingly fall outside traditional legal frameworks for health data protections. More and more, citizens are able to generate data about themselves directly, whether by purchasing it as a consumer service, installing applications on their phones and computers, wearing devices, or more. This data has enormous scientific value but currently sits well outside the legal and regulatory frameworks typically associated with science. Whatever systems emerge for data reuse must be extensible and flexible enough to integrate with the data that is to come, not just the data we have today. And the likely outcome is not a single monoculture, but a diverse ecosystem that features technical, organizational, structural, and commons-based approaches as design choices available to study investigators, data users, and citizens.

In the short term, the most likely benefit of that ecosystem will be a greater ability to understand how specific changes from one genotype to the next affect health outcomes – so that clinicians understand why one person responds well to a drug, but another does not. In the long term, anyone who can analyze the data will have the capacity to become a genomic explorer whether they have lots of money/funding or not. The biggest impacts of that change are by definition hard to describe right now, just as it was hard at the dawn of the Internet to imagine the modern Web, or at the dawn of the Web to imagine the smartphone revolution.

NOTES

1. K. El Emam, E. Jonker, L. Arbuckle, and B. Malin, "A Systematic Review of Re-Identification Attacks on Health Data," *PLoS ONE* 6 (12): e28071.
2. C. Christine Porter, "De-Identified Data and Third Party Data Mining: The Risk of Re-Identification of Personal Information," *Shidler Journal of Law, Commerce and Technology* 5 (September 23, 2008): 3.
3. Paul Ohm, "Broken Promises of Privacy: Responding to the Surprising Failure of Anonymization" (August 13, 2009); *UCLA Law Review* 57 (2010): 1701; University of Colorado Law Legal Studies Research Paper No. 9–12.
4. Jane R. Bambauer, "Tragedy of the Data Commons," *Harvard Journal of Law and Technology* 25 (2011): 1–67.
5. Avi Charkham, "5 Design Tricks Facebook Uses to Affect Your Privacy Decisions," *TechCrunch*, August 25, 2012, http://techcrunch.com/2012/08/25/5-design-tricks-facebook-uses-to-affect-your-privacy-decisions/.
6. Among many others, see R. Sachidanandam, D. Weissman, S. C. Schmidt, J. M. Kakol, L. D. Stein, G. Marth, S. Sherry, J. C. Mullikin, B. J. Mortimore, D. L. Willey, S. E. Hunt, C. G. Cole, P. C. Coggill, C. M. Rice, Z. Ning, J. Rogers, D. R. Bentley, P. Y. Kwok, E. R. Mardis, R. T. Yeh, B. Schultz, L. Cook, R. Davenport, M. Dante, L. Fulton, L. Hillier, R. H. Waterston, J. D. McPherson, B. Gilman, S. Schaffner, W. J. Van Etten, D. Reich, J. Higgins, M. J. Daly, B. Blumenstiel, J. Baldwin, N. Stange-Thomann, M. C. Zody, L. Linton, E. S. Lander, D. Altshuler, and International SNP Map Working Group, "A Map of Human Genome Sequence Variation Containing 1.42 Million Single Nucleotide Polymorphisms," *Nature* 409, no. 6822 (February 15, 2001): 928–933.
7. See e.g. "We Eat Less Healthy than We Think" at http://data.massivehealth.com/#infographic/perception – data derived from a social application that takes pictures of foods and ratings by more than 500,000 users.
8. This section owes an enormous debt to Dan Vorhaus's online writing, available at the *Genomics Law Report* at http://www.genomicslawreport.com/index.php/author/dvorhaus/.

9. See U.S. National Institutes of Health, "Notice on Development of Data Sharing Policy for Sequence and Related Genomic Data," Notice No. NOT-HG-10-006, release date October 19, 2009. At the time of writing, the policy is under review yet again, in part because of privacy issues emerging since the last update. See "Input on the Draft NIH Genomic Data Sharing Policy," NIH Notice No. NOT-OD-14-018, release date September 27, 2013.
10. Catherine Clabby, "DNA Research Commons Scaled Back," *American Scientist* 97, no. 2 (March–April 2009): 113.
11. Nils Homer, Szabolcs Szelinger, Margot Redman, David Duggan, Waibhav Tembe, Jill Muehling, John V. Pearson, Dietrich A. Stephan, Stanley F. Nelson, and David W. Craig, "Resolving Individuals Contributing Trace Amounts of DNA to Highly Complex Mixtures Using High-Density SNP Genotyping Microarrays," *PLOS Genetics* 4, no. 8 (August 29, 2008): e1000167.
12. Kevin B. Jacobs, Meredith Yeager, Sholom Wacholder, David Craig, Peter Kraft, David J. Hunter, Justin Paschal, Teri A. Manolio, Margaret Tucker, Robert N. Hoover, Gilles D. Thomas, Stephen J. Chanock, and Nilanjan Chatterjee, "A New Statistic and Its Power to Infer Membership in a Genome-Wide Association Study using Genotype Frequencies," *Nature Genetics* 41 (2009): 1253–1257.
13. Khaled El Emam and Fida Kamal Dankar, "Protecting Privacy Using k-Anonymity," *Journal of the American Medical Informatics Association* 15, no. 5 (2008): 627–637.
14. Liina Kamm, Dan Bogdanov, Sven Laur, and Jaak Vilo, "A New Way to Protect Privacy in Large-Scale Genome-Wide Association Studies," *Bioinformatics* 29, no. 7 (April 1, 2013): 886–893.
15. Daniele Micciancio, "Technical Perspective: A First Glimpse of Cryptography's Holy Grail," *Communications of the ACM* 53, no. 3 (2010): 96.
16. Ibid., 14.
17. P. Bogetoft et al., "Secure Multiparty Computation Goes Live," in *Proc. Financial Cryptography 2009*, ed. R. Dingledine and P. Golle, Lecture Notes in Computer Science 5628 (Heidelberg: Springer, 2009), 325–343.
18. D. Bogdanov et al., "Deploying Secure Multi-party Computation for Financial Data Analysis," in *Proc. Financial Cryptography 2012*, Lecture Notes in Computer Science 7397 (Heidelberg: Springer, 2012), 57–64.
19. Health Record Banking Alliance, "National Infrastructure for HIE Using Personally Controlled Records," White paper, January 2013. Available at http://www.healthbanking.org/docs/HRBA%20Architecture%20White%20Paper%20Jan%202013.pdf.
20. Thinh Nguyen, Science Commons, "Freedom to Research: Keeping Scientific Data Open, Accessible, and Interoperable." Available at http://sciencecommons.org/wp-content/uploads/freedom-to-research.pdf.
21. Lee Rainie, Sara Kiesler, Ruogu Kang, and Mary Madden, "Anonymity, Privacy, and Security Online. A Report of the Pew Internet and American Life Project," September 5, 2013, http://www.pewinternet.org/Reports/2013/Anonymity-online/Summary-of-Findings.aspx.

22. American Society of Human Genetics, "ASHG Response to NIH on Genome-Wide Association Studies," Policy Statement, November 30, 2006, http://www.ashg.org/pdf/policy/ASHG_PS_November2006.pdf.
23. P. Antonenko and D. Niederhauser, "The Influence of Leads on Cognitive Load and Learning in a Hypertext Environment," *Computers in Human Behavior* 26, no. 2 (March 2010): 140–150.

Part III Statistical Framework

If big data are to be used for the public good, the inference that is drawn from them must be valid for different, targeted populations. For that to occur, statisticians have to access the data so that they may understand the data-generating process, know whether the assumptions of their statistical model are met, and see what relevant information is included or excluded. It is clear from earlier chapters in this book that the utility of big data lies in being able to study small groups in real time, using new data analytic techniques, such as machine learning or data mining. These demands pose real challenges for anonymization and statistical analysis. The essays in this part of the book identify the issues, spell out the statistical framework for both analysis and data release, and outline key directions for future research.

A major theme of the essays is that neither the data-generating process nor the data collection process is well understood for big data. As Kreuter and Peng argue, almost all statistical experience with human subjects is based on survey data, and over time statisticians have parsed the sources of error neatly into a total survey error framework. But the data-generating process of many data streams – such as administrative data or big data – is less transparent and is not under the control of the researcher; therefore, access to the data itself is critical to building the necessary understanding. Continuous effort will be needed to develop standards of transparency in the collection of big data. Transparency is also needed on the 'back end' – any linkage, data preparation and processing, analysis, and reporting – to ensure reproducibility. Kreuter and Peng point out that much more research is needed on linkage and matching, because the resulting knowledge will not only enrich possible analysis, but also help to evaluate the quality of the linked sources.

Another major theme is that the long history of providing access to survey data, with many different strategies for reducing the risk, gives very little leverage on the problem of providing access to big data. Big data carry greater disclosure risks than the typical survey sample. The core of the problem is that traditional risk assessment is based on assumptions about what a possible intruder knows about the released data (response knowledge), how easily someone can be identified (uniqueness), and whether that intruder is malicious. In the case of big data, an undefined – maybe large – number of people know identities in the data (response knowledge), and it is unknown how many of them might be malicious. Big data also captures information about many more individuals and can include many more variables, so that nearly everyone in the data is unique in the population (uniqueness).

The chapters here begin to parse out the key elements that need to be studied in this context. The first is the content: Karr and Reiter distinguish what must be protected from what might be protected. The second is the people: Karr and Reiter treat analysts, who use the data for statistical analysis, differently from intruders, who try to re-identify individuals. The third is definitions: Karr and Reiter think that big data may change attitudes about privacy, and change what information is considered sensitive (in the past, salary; now, medical records and DNA sequences). Dwork goes a step further with this argument, because "'big data' mandates a mathematically rigorous theory of privacy, a theory amenable to measuring – and minimizing – cumulative privacy loss, as data are analyzed, re-analyzed, shared, and linked". She points out that we need a definition of privacy, equipped with a measure of privacy loss. Data usage and research with microdata should be accompanied by publication of the amount of privacy loss, that is, its privacy 'price'. Any realistic definition of privacy must recognize that researchers want to learn useful facts from the data they analyze. In her view, it does not matter whether a particular person is in the data, because a generalized result may affect that person all the same: "that's the heart of differential privacy."

Randomization of the original data is one key element in protecting privacy. As Dwork points out "the property of being differentially private depends *only* on the data protection algorithm – something that the data curator, or 'good guy', controls." The preferred solution for anonymizing data is to generate redacted (synthetic) data. Although some examples are presented, all authors agree, that methods for generating massive synthetic databases merit further research. It is impossible to simultaneously preserve privacy and to release a synthetic dataset that answers too many research questions with much accuracy.

Nonstatistical approaches could draw on existing experience with the dissemination of administrative or complex, combined data (such as linked employer-employee data). The authors outline indirect ways in which these kinds of data are made available to researchers. The main mechanisms are remote access solutions or safe centers, where restricted and controlled access is possible. For many big datasets, the confidentiality risks of dissemination may be so high that it is nearly impossible to share unrestricted-use microdata. The ideal scenario is that big data are held by trusted or trustworthy institutions. As Karr and Reiter note, the "data access model of the future will be to take the analysis to the data rather than the data to the analyst or the analyst to the data. . . . Truly big data are too big to take to the users."

To overcome the dilemma of preventing privacy loss while enabling detailed empirical research, Karr and Reiter propose an integrated system for big data access as an integrated system including

1. unrestricted access to *highly redacted data* followed with
2. means for approved researchers to access the *confidential data via remote access* solutions, glued together by
3. *verification servers* that allows users to assess the quality of their inferences with the redacted data so as to be more efficient with their use (if necessary) of the remote data access.

Or, in the words of Dwork: "differential privacy requires a new way of interacting with data, in which the analyst accesses data only through a privacy mechanism, and in which accuracy and privacy are improved by minimizing the viewing of intermediate results." But this is a foreign concept to data analysts. A major challenge will be to convince researchers working on big data to agree to these kinds of scenarios, both because big data are much more complex than the data we have worked with before and because knowledge of the data-generating process matters. Research is necessary to ensure that access to big data preserves the advantages of big data – so that the act of data redaction does not negatively affect data volume, the time taken to redact does not significantly affect data velocity, and there are minimal consequences for data variety in terms of the ability to keep all different data sources in a redacted dataset.

Dwork's conclusion is clarity itself in establishing the future research agenda. By far the hardest problem is addressing "the *social* challenges of a changing world, in which highly detailed research datasets are expected to be shared and reused, linked and analyzed, for knowledge that may or may not benefit the subjects." Complexity of this type requires a rigorous theory of privacy and its loss. "Other fields – economics, ethics, policy – cannot be brought to bear without a 'currency', or measure of privacy, with which to work. In this connected world, we cannot discuss trade-offs between privacy and statistical utility without a measure that captures cumulative harm."

12 Extracting Information from Big Data: Issues of Measurement, Inference and Linkage

Frauke Kreuter and Roger D. Peng

Introduction

Big data pose several interesting and new challenges to statisticians and others who want to extract information from data. As Groves pointedly commented, the era is "appropriately called Big Data as opposed to Big Information,"[1] because there is a lot of work for analysts before information can be gained from "auxiliary traces of some process that is going on in the society." The analytic challenges most often discussed are those related to three of the Vs that are used to characterize big data.[2] The *volume* of truly massive data requires expansion of processing techniques that match modern hardware infrastructure, cloud computing with appropriate optimization mechanisms, and re-engineering of storage systems.[3] The *velocity* of the data calls for algorithms that allow learning and updating on a continuous basis, and of course the computing infrastructure to do so. Finally, the *variety* of the data structures requires statistical methods that more easily allow for the combination of different data types collected at different levels, sometimes with a temporal and geographic structure.

However, when it comes to *privacy* and *confidentiality,* the challenges of extracting (meaningful) information from big data are in our view similar to those associated with data of much smaller size, surveys being one example. For any statistician or quantitative working (social) scientist there are two main concerns when extracting information from data, which we summarize here as concerns about *measurement* and concerns about *inference.* Both of these aspects can be implicated by privacy and confidentiality concerns.

Measurement By questions of *measurement* we mean: Do the data contain the right key variables and all covariates to answer the research question? Are there any unobserved variables that confound the analysis when uncontrolled for? Is there systematic measurement error in the measures? What is the best unit of analysis on which the measures are taken?

Inference By questions of *inference* we mean: Do we understand the sampling process? Are all units we need in the analysis? Are certain units systematically missing? Do some units appear multiple times? Do we have all measures on all units that we need? Whom do the units represent?

In settings where data collection (big or small) is *designed* to answer specific scientific research questions, the answers to the questions above are (usually) relatively easy to determine or, better yet, are taken into account when the data are collected. But as soon as data are used for purposes other than those for which they were collected, or when the *data-generating process* breaks down – as is unfortunately often the case, both with data collected for experimental purposes and with data from sample surveys and censuses – this is no longer the case. Neither is it the case for most big data sources.

There are many reasons for breakdowns in the data-generating process of traditional data as well as big data; actual and perceived privacy and confidentiality threats can be one of them, though for traditional data usually more on the measurement side than on the inference side. This might be very different with big data. Irrespective of the size of the data at hand, researchers have to understand the data-generating process in order to determine whether statistical model assumptions are met and to know whether all relevant information is indeed included in the data for a given research question. Here is where we see a key difference between data commonly referred to as big data and more traditional data sources.

For traditional data, methodology has been developed to overcome problems in the data-generating process. Among survey methodologists, a guiding principle for detecting problems is the total survey error framework,[4] and statistical methods for weighting, calibration, and other forms of adjustment[5] are commonly used to mitigate errors in the survey process. Likewise for 'broken' experimental data, techniques like propensity score adjustment and principal stratification are widely used to fix flaws in the data-generating process.[6] Some of this methodology is likely transferable to big data, but some of it might need to be tweaked. Most likely solutions will lie in a combination of traditionally designed data and big data. However, such solutions of *data linkage and information integration* are themselves threatened by concerns about privacy and confidentiality.

Data-Generating Processes

When we look closely, it is the data-generating process that differentiates big data from more traditional types of data, such as censuses, sample surveys, and experimental data. Even with more traditional data, we would, at least once in a while, see data of large volume (e.g. the U.S. census), data of high velocity (e.g. Nielsen TV meter),[7] and data with a wide variety of different variables and a complex structure (e.g. National Health and Nutrition Examination Survey).[8] What is different is that the data collection for all three of these examples was specifically *designed* with a research question in mind.

In the case of designed data collection, the research questions are translated into concepts that need to be measured (e.g. employment), measurement instruments are developed to best capture these concepts (e.g. people that did any work for pay or profit during the survey week, or those with a job but not working during the survey week because of illness, vacation, leave, etc.),[9] populations are defined for which measures should be taken (e.g. U.S. civilian non-institutionalized population 16 years and older),[10] and (usually) samples are drawn to make data collection more efficient. If used for official statistics these design features are transparent and made public. For other data products the Transparency Initiative of the American Association for Public Opinion Research[11] now has guidelines in place to match such transparency. A slightly different approach is taken by ESOMAR but also with the goal of creating transparency for the data-generating process.[12]

It is this transparency that allows an evaluation of the quality and usefulness of the data for a given research question – for example, knowledge about people who were sampled to be in the survey but could not be reached, or decided not to participate. If the error mechanism is known, researchers can evaluate potential bias. Statisticians express such bias, for example for an estimated mean, as a function of the rate of response and the difference between respondents and nonrespondents or, if response is seen as a stochastic process, as $\mathrm{Bias}(\overline{\mathrm{y}}_{\mathrm{r}}) = \frac{\sigma_{\mathrm{yp}}}{\overline{\mathrm{p}}}$ where σ_{yp} is the covariance between y and the propensity to (in this case) respond to the survey, p, and $\overline{p}$ is the mean propensity over the sample. If, for example, sampled cases with privacy concerns are less likely to participate in a survey[13] but those privacy reasons are unrelated to the topic of the survey, then no bias will appear due to (this) breakdown in the process.[14]

In contrast, the data-generating process of many data streams is less transparent and much less under the control of the researcher. Many big data streams arise through systems that create data as a byproduct of human behavior, usually in interaction with a computer. Credit card transactions, loyalty cards, Twitter posts, Facebook entries, cellphone usage, all those behaviors leave electronic traces that arise *organically*.[15] It is this difference between *designed* and *organic* data that poses extra challenges to the extraction of information.

Note, some big data have features of designed data, such as environmental monitors which are built to measure specific types of pollution. Their locations are carefully selected and their measurement times are either selected by design or continuous, removing any 'organic' element from the data-generating process. However, even in this example, the data collected by environmental monitoring agencies are often used for purposes for which they were not designed. The most common example perhaps is estimation of the health impacts of ambient air pollution.

The use of data for other than the designed purpose is also reflected in a comment by Horrigan, who characterizes big data as "nonsampled data, characterized by the creation of databases from electronic sources whose primary purpose is something other than statistical inference."[16] In designed data (ideally) all measures taken have to fulfill a purpose with respect to the research question (and a related statistic of interest), whereas with big data, the data fulfill a purpose for a particular action (i.e. the credit card payment, the reception of a phone call, or the expression of 'liking'), but not for a particular research question unless the research question is a mere measure of the number of actions of that particular type.

The use of organic data for other than their immediate purpose within the system in which they arise is not peculiar to the big data area. For decades researchers have used administrative data in statistical analyses. Administrative data are the result of government or other administrative processes; prime examples of data often increasingly used for statistical analyses are tax data,[17] or data produced through social security notifications.[18] The variables contained in those data are usually those necessary for administering a program or administrative task, and they measure what is necessary to evaluate program-relevant features. When these data are used to extract information that go beyond the description of the program, researchers often quickly realize that the measures do not match the concepts they try to measure, and that often not all cases of interest are captured through these administrative data. For example, in the social security notifications that form the basis of the German administrative data used in a study of workplace heterogeneity and wage inequality,[19] education is known to have many missing values or conflicting data within a person across employment spells, and certain groups such as civil servants and the self-employed are not included in the database at all.

Another mechanism by which big data often arise is the merging of multiple databases, studies, surveys, or datasets. Computing power today has allowed 'mega-databases' to be formed to address new and interesting questions using data in novel ways. However, two key concerns when merging takes place is that (1) now multiple datasets are being used for purposes for which they were not designed and (2) data that once retained a subject's privacy may no longer do so when they are merged with another database.

Challenge: Measurement

There is much to say about possible measurement problems when using big data, for example challenges in text mining of Twitter or Facebook feeds to distinguish between sarcasm, irony, humor, actions vs. intentions, and so forth. Some recent examples have shown promise in solutions to automate such 'feature selection'.[20] Here we focus on those measurement problems that are likely induced by privacy, confidentiality, and related issues: (1) misreporting in anticipation of perceived data use and (2) lack of model-relevant variables.

We know from survey research that people misreport or underreport socially undesirable behavior and illegal activities to avoid unfavorable impressions and legal consequences should the information be disclosed and they overreport desirable behavior such as voting.[21] Preliminary research reports reasonably accurate reflections of user characteristics on social network sites,[22] though honesty is suspected to vary across personality types.[23] Not much is known about how reported attitudes and behaviors vary across networks, topics and (perceived) use of the data, and the associated *actual and perceived privacy* issues. In survey data collection self-administered modes have been shown to mitigate effects of social desirability compared to interviewer-administered modes.[24] But even if the mode is similar to those used in self-administered surveys (Web and text messages), the difference here is that the primary purpose of generating these data includes an intended readership. Also it is possible that what is shared is of good quality, but what matters most is not shared. Likewise a *privacy*-related observation issue can be the sharing/posting/reporting of information from people other than oneself. This trade-off between observation and non-observation error in big data deserves attention. It would be a sad state of affairs if increased research use of social media or other big data sources changes the very nature of these sites.

A second feature of many big data sources that can easily create measurement problems is the fact that while they are 'case rich' they tend to be 'variable poor'.[25] The Billion Prices Project,[26] for example, can give timely and detailed information if the purpose is only to measure prices of consumer goods, but as soon as the research interest shifts to understanding how price changes lead to shifts in household expenditures a single source of big data will no longer suffice.[27] To perform household-level analysis consumption and purchase data need to be linked up at a household level, which might be doable if confidential credit card billing information from websites such as Amazon and Paypal are shared, though this obviously would require a data infrastructure that is currently not in place (see Chapter 11 in this volume, by Wilbanks).

A recent project at the University of Michigan uses Twitter data to predict initial claims for unemployment insurance using the University of Michigan Social Media Job Loss Index.[28] The prediction is based on a factor analysis of social media messages mentioning job loss and related outcomes.[29] As part of this project, auxiliary information was linked to Twitter messages via geographic location information, a powerful tool but not without its *own threats to privacy*. O'Neil and Schutt also point to the usefulness of location data for the correct interpretation of timestamp data by determining the user's time zone.[30] Such information and other additional covariates can also be used to make sense of outliers that would otherwise be removed from analyses, a common strategy in dealing with these types of data.[31]

Challenge: Inference

One interesting feature of big data is that because there is so much data, it is easy to overlook what is missing. At first glance one might think that there is no problem of inference with big data. After all, some say, with a constant stream of data and the technological capacity to capture all the data, sampling is no longer needed because the population can be analyzed instead.[32] But what is the population? If Amazon is interested in the correlation between purchases of two different books on amazon.com, they can of course analyze the population and no inference problem appears. But if, for example, Twitter is used to measure health or political attitudes, the problem of inference is much harder and starts with a specification of the unit of analysis. Paul and Dredze reported good success with population-level metrics in their analysis of public health, but are hampered by the irregularity of Twitter postings when trying to answer questions on frequency of health problems, re-occurrence of health problems, and the like.[33]

As with any data source, researchers have to ask themselves whether the data-generating process is appropriate for the research question and captures the population to which they are trying to make inference. In statistical terms, the challenge is to assess undercoverage, overcoverage, multiplicity, as well as other forms of missing data and (here) their relationship to privacy and confidentiality.

Undercoverage

Whether or not one captures all available data (at a given time point) or just a sample of the data, the dataset from which one draws cases (the frame) is almost never perfect. It is not hard to imagine that big data taken from the Internet suffer from *undercoverage*, from the exclusion of units in the target population (those without Internet access). According to estimates from the Current Population Survey, the Internet penetration rate among U.S. households was still below 80% in 2011,[34] and almost 30% of individuals report no Internet use anywhere (irrespective of access to a computer at home or elsewhere). Likewise, an estimated 30% of American adults do not have a credit card,[35] which means again about a third of the population is not represented in data from commercial vendors.

The proportion of people with Internet access who could be covered (if they all used a particular service that is accessed for web scraping) is, however, only one consideration and translates into *coverage bias* only when there is a sizable difference between those covered (e.g. by having Internet access or a credit card) and those uncovered. Coverage bias can be expressed as follows:[36]

$$\text{bias}\left(\overline{Y}_{c}\right) = \frac{N_{uc}}{N_{pop}}\left[\overline{Y}_{c} - \overline{Y}_{uc}\right],$$

with N_{uc} reflecting the number of people who are undercovered, N_{pop} the total number of people in the population, $\overline{Y}_{c}$ the mean for the covered units, and $\overline{Y}_{uc}$ the mean for the undercovered units. Unfortunately, as can be seen from this equation, while the ratio of undercovered people to the population stays the same across all statistics derived from a given data source, the difference in means between covered and undercovered units can vary from variable to variable.

Several studies list potential indicators of bias when comparing those without Internet access to those with access.[37] In a German study, persons without Internet access are found to be less educated and slightly older than those with access.[38] In a U.S. study, Internet users are more likely to be in good health, to have health insurance, and to exercise regularly,[39] and for the elderly in the United States, significant differences between Internet users and non-users on financial and health variables were detected.[40] If we look at specific big data sources both the undercoverage rate and the potential for undercoverage bias is even bigger; in 2013 only 13% of the U.S. online population actively tweets[41] and nearly half of Twitter users are under age 35 and only 2% are 65 or older.[42]

Privacy and confidentiality concerns contribute to undercoverage, both at the person level but also at the level of individual postings or transactions.[43] Survey results from Google showed a strong correlation between privacy concerns and low engagement.[44] The concerns mentioned in the survey can be grouped into those about identity theft in the financial world, those about unwanted spam and solicitations in the digital world, as well as concerns about offline harm, stalkers, and employment risk in the physical world. A study of a selected group of college students found personality characteristics to be predictive of Facebook postings, with people who score high on the openness factor having less strict privacy settings, which may cause them to be susceptible to privacy attacks.[45] The individual perception of the risks of online participation can probably be characterized in terms of cost–utility equations, just like participation in surveys.[46] For many users the utility of connecting with friends, receiving and spreading information, and searching and purchasing online outweighs the costs and risks, but it remains interesting to see whether the recent privacy debates will change the equation for more people.[47]

Overcoverage and Multiplicity

Overcoverage refers to the inclusion of units in the data that are not members of the target population, while multiplicity refers to multiple appearances of a single person. On Facebook one can, for example, find pages for dead people or animals, and errors in geocodes[48] or disguised IP addresses might erroneously identify users as being part of a geographically defined population of interest when they are not. Similarly, Paul and Dredze observed in their analysis of Twitter data that "some tweets, particularly those about more serious illnesses like cancer, were in regards to family members rather than the user. This could throw off geographic statistics if the subject of a tweet lives in a different location than the author of the tweet."[49]

Multiplicity can occur for several reasons. Some people might open several Twitter accounts and Facebook pages, or use different credit cards to keep different aspects of their lives separate or to ease accounting. In simple scrapings of online data, those people would be included multiple times, which is no problem if a transaction is the unit of analysis but can be a problem if inference is made to a defined set of people. Depending on the research question, de-duplication is a problem most likely only solvable through data linkage or integration with other forms of additional identifying information, *creating again possible threats to privacy*.

Other Missing Data

Aside from undercoverage, data can be missing for a variety of other reasons. Infrequent postings by Facebook users would be an example of missing data. In principle those people are covered, but information on them is missing. Whether background information is missing because users choose to not reveal it, or because the data stream does not include these kinds of information, for analyses that try to explain behaviors or are interested in subgroup analyses, the lack of demographic and other background variables can be a big hindrance.[50]

In a presentation at the 2013 FedCasic Conference, Link reported that about one-third of demographic information is missing from Facebook users.[51] A study by Kosinski, Stillwell, and Graepel tried to infer demographic information from digital records of behavior – Facebook Likes – for 58,000 Facebook volunteers. They classified African Americans and Caucasian Americans correctly in 95% of cases, and males and females in 93% of cases, "suggesting that patterns of online behavior as expressed by Likes significantly differ between those groups, allowing for nearly perfect classification."[52] Good prediction accuracy was achieved for more private information such as relationship status (65%), substance use (73%), and sexual orientation (88% for males, 75% for females).

While it is unknown how well the prediction models would perform for those who did not volunteer to participate in the study, these results raise important concerns about the possibility of keeping information *purposefully private*. In the words of the authors:

> [T]he predictability of individual attributes from digital records of behavior may have considerable negative implications, because it can easily be applied to large numbers of people without obtaining their individual consent and without

them noticing. Commercial companies, governmental institutions, or even one's Facebook friends could use software to infer attributes such as intelligence, sexual orientation, or political views that an individual may not have intended to share. One can imagine situations in which such predictions, even if incorrect, could pose a threat to an individual's well-being, freedom, or even life.[53]

Other forms of missing data appear on an event level. Here too privacy concerns can be the explanation. Consider for example the use of cellphone data to analyze traffic patterns. Users of cellphones carry their devices with them and thus leave a trace in the data files of everywhere they go. However, some users might turn their devices off when they take trips that should not be known to anybody.

In his keynote lecture at the 2013 conference of the European Survey Research Association,[54] Couper points out that the behavior of people is likely to change based on *concerns about privacy*, and that this concern is likely to grow if the collection and use of big data become more broadly known. While some report behavioral changes in privacy and tracking settings[55] as result of concerns about data use, to our knowledge no data has been released by owners of major social network sites about the effects of increased awareness of data use. More research is needed on how to address issues of consent to data capturing and data use (see the discussion by Barocas and Nissenbaum in Chapter 2 of this volume). Given that the data generation process spans several countries, an exchange with professional organizations across the globe might be useful (see Chapter 8 in this volume, by Elias).

Solutions and Challenges: Data Linkage and Information Integration

Many of the measurement and inference problems – whether they occur because of privacy and confidentiality concerns or for other reasons – can be addressed through data linkage and other forms of information integration. Data linkage in particular has its own *privacy and confidentiality challenges*, but the other methods also have implications related to privacy and confidentiality.

In surveys, a variety of methods[56] that all fall under the heading of 'calibration' can correct for coverage and nonresponse errors by using auxiliary data. Auxiliary data can be any information that is available for the target population, either for each individual population unit or in aggregate form. In a household survey, counts of persons in groups defined by age, race/ethnicity, and gender may be published from a census or from population projections that are treated as highly accurate.[57] A necessary requirement for using those techniques is the *availability of comparable content* and *units of analysis* to link with information from more trusted sources. For example, if variables used for weighting are based on respondents' answers during an interview, then prior to data collection care should be taken that the questions asked in the survey match those of the benchmarking source. Even for demographic variables, although this seems like a straightforward task, it is often not that simple. The current U.S. census, for example, allows for multiple-race categories, a feature many surveys or social network sites don't share. Non-probability surveys have started using such methods to attempt to make population estimates.[58] For some big data sources this would require cooperation with companies to obtain sufficient access to common variables, which can raise issues of confidentiality pledges given by the respective data collectors.

Another solution increasingly popular among surveys severely challenged by falling response rates, hard-to-reach populations, and stagnant survey budgets[59] is use of or linkage to administrative data to gain information about coverage bias, nonresponse bias, or measurement error in the survey data. Likewise, policymakers ask for evidence-based program evaluation, ideally with controlled randomized trials, but those often fail to be fully randomized when the practical challenges of data collection kick in. And current movements toward comparative effectiveness mean increased interest in answering questions regarding effects outside the narrow study population.[60] Here too statisticians have started to combine multiple datasets or integrate information from different sources to mitigate the shortcomings of the original study, and high hopes are expressed that organic big data can help here even more.

The act of combining several data sources, for example survey and administrative data, is not without its own challenges, and in several areas research is needed to ensure that this approach is one that can lead to success. Direct record linkage and statistical matching are two of the most common procedures for linking different data sources. Both methods involve tradeoffs and rely on untested assumptions that threaten their practical use.

The quality of linked databases can be adversely affected by (a) inconsistencies between information collected from respondents in one dataset (e.g. the survey) and information contained in the other (e.g. an administrative database or social network site information) and (b) errors in the record linkage process itself.[61] Such errors can be due to erroneous linkage, imprecise matching variables, or the lack of respondent consent to link their survey responses to administrative data. The impact of these issues on the quality of the linked datasets is often unknown. Evaluation of the procedures and criteria used in combining data sources and assessment of their impact on data quality are clearly needed if combined data are to become a key source of statistical information in the future.

Several studies have examined inconsistencies between observations made with survey and administrative data. A common assumption in those studies is that the administrative records are error free and suitable as a 'gold standard' against which survey responses can be compared, though this assumption often holds only for variables necessary to administer a certain program. Similar studies are needed for the various big data sources as well.

Prior to linkage, informed consent is usually needed to ensure that respondents are aware of the risks and benefits involved in releasing information for research purposes, Research has shown that consent rates to linkage requests vary widely from study to study, by study population and country, and across the social, economic, and health fields, with percentages ranging anywhere from the mid-20s to the high-80s. But as in the discussion of coverage, here too it is not only the rate that matters but the difference between consenters and non-consenters. Studies have found gender, age, education, and wealth to be strongly related to the likelihood of consent; and health status, income item nonresponse, and prior-wave cooperation in longitudinal studies also correlate with linkage consent. Direct bias assessments with respect to a target variable, on the other hand, are rare.[62]

We know by now that consent rates do not only vary by personal characteristic and study topic but also as a function of the consent request itself. So far no studies have experimented with the stated risk of data linkage, but research on informed consent for study participation have shown that stronger assurances of data confidentiality lead to higher response rates, with the exception that stronger assurances can backfire and lead to lower participation rates and more expressions of suspicion when the survey topic is innocuous and the confidentiality risk is minimal.[63] Recent experiments on the wording of data linkage requests have shown that the framing of the request as well as the consent placement influences consent rates.[64] The linkage indicator itself also matters, and greater hesitation arises when Social Security numbers are used. For large-scale linking, other identifiers that do not require self-reporting might be better suited. Geocodes of addresses might be the most fruitful approach,[65] though here too one can find differential error depending on the type of address used.[66] Geography also allows augmentation with other geographic information, which is a variable available in many big data sources. If higher levels of aggregation are used, requirements for exact placement are less stringent.

In contrast to direct record linkage, statistical matching, or the act of matching 'similar' records (in a statistical sense) based on a set of variables common to both datasets, does not require consent from survey respondents and can be performed on the entire sample. Several matching procedures have been proposed. However, many of these procedures depend on the assumption of conditional independence, asserting that any systematic differences between two or more data sources are explained by the matching variables. Only when this assumption is met will the true underlying distribution of all variables in both datasets be reflected in the matched dataset; otherwise, statistical relationships estimated from the matched dataset will be misleading.

One problem that cannot be solved by either technique is if people are missing for privacy reasons from all data sources that are intended for linkage. For example, nonrespondents in surveys might also not be active on Facebook or decline the use of credit cards. To our knowledge there has so far been no systematic research on this issue.

Another real challenge to data integration is the linkage of data with different privacy restrictions. Several chapters in this volume, notably Chapter 9 by Greenwood et al. and Chapter 13 by Karr and Reiter, provide suggestions. Infrastructure needs to be in place that allows linkage and the addition of identifiable IDs to place people in locations, de-duplicate, and so forth.

As the former census director Robert Groves summarized on the *Director's Blog*:[67] "The challenge to the Census Bureau is to discover how to combine designed data with organic data, to produce resources with the most efficient information-to-data ratio. [. . .] Combining data sources to produce new information not contained in any single source is the future. I suspect that the biggest payoff will lie in new combinations of designed data and organic data, not in one type alone."

Discussion

Some of the challenges of big data may, in some sense, be addressed by adapting the approaches of 'small data' to a larger scale. Designed big data experiments are already commonplace, particularly so-called A/B testing in industry applications,[68] and likely will see much wider application. We will almost certainly encounter methodological challenges as we continue to move into big data and so the corresponding research will be essential.

Another development that may serve to simplify many big data problems is the continuous refinement and sharpening of scientific questions and hypotheses. While one of the key advantages of big data is that it allows one to explore and discover previously unobserved patterns, such free-form exploration can lead to numerous problems with respect to inference and measurement, some of which we have discussed in this chapter. Eventually, science will move beyond the fascination with terabytes and petabytes, and the ability to collect data, and will focus more on the ability to analyze data and address specific questions. As investigators become more comfortable and familiar with the scope of big data, they will develop a much better understanding of what kinds of hypotheses can be addressed by such data and, more important, what kinds cannot.

Some research questions can only be addressed through the combination of data products. Thus research on linkage and matching needs to be expanded, covering the issues raised above and preparing for the analysis of linkage (consent) bias. When done successfully, linkage can be a rich resource, not just for substantive analysis when surveys, experiments, administrative data, and sensor data or other organic big data are respectively enriched, but also to evaluate the quality of the linked sources, including research on the missing elements in both.

As the march of big data progresses, continuous effort will be needed to ensure standards of transparency in the collection of data, so that there may be appropriate accountability and so that valid statistical inferences can be made. Principles developed for traditional data collection efforts, such as the use of metadata and paradata to describe and monitor data-generating processes, will need to be translated to modern high-volume data collection methods such as sensor networks, real-time data streams, and imaging modalities. So far there is no context-independent framework for evaluating data-generating processes, and more thought in this area would be useful.

Transparency in data collection goes hand in hand with transparency on the back end. The issues related to the reproducibility of any scientific investigation are potentially magnified in a big data analysis. Naturally, any linkage, data preparation and processing, analysis, and reporting must be fully transparent so that the handling of these tasks may be open to critique. However, because of the increased complexity of documenting the actions applied to big data, we will tend to rely on computer code descriptions rather than more traditional written forms. Ultimately, ensuring reproducibility through the proper documentation of any statistical analysis will be critical because the uniqueness and size of big data studies will make them less likely to be replicated in other investigations. (See Chapter 5 in this volume, by Stodden.)

Because many datasets are too large to travel from one place to the next, or privacy concerns may prevent them from being examined at an insecure location, data processing at the originator's site or through trusted third parties[69] will be an essential element in working with these combined data products. Examples of such procedures are already in place (e.g. the National Center for Health Statistics's Research Data Centers), though more work is needed to create legal agreements that allow for data access and also solve issues of liability, international border crossing, and mismatches in confidentiality standards. However, one potential upside of 'taking computation to the data' is that such remote data processing will require people to document and share the code for their analyses. Hence, there may be a built-in reproducibility safeguard if datasets must be analyzed remotely. It is these changes in data processing that will, if implemented in the right way, ultimately lead to greater transparency and with that, hopefully, to greater trust by individuals in the use of big data (at least in the context of statistical analyses), and to better data quality.

Acknowledgements We thank Fred Conrad, Stephanie Eckman, Lars Lyberg, and Eleanor Singer for critical review, Martin Feldkircher and Richard Valliant for comments and suggestions, and Felicitas Mittereder for research assistance.

NOTES

1. Robert Groves's talk at the World Bank event on December 19, 2012, "What Happens when Big Data Meets Official Statistics," http://live.worldbank.org/what-happens-when-big-data-meets-official-statistics-live-webcast (accessed January 20, 2014).
2. C. O'Neil and R. Schutt, *Doing Data Science* (Sebastopol, CA: O'Reilly Media, 2014).
3. D. Agrawal, P. Bernstein, E. Bertino, S. Davidson, U. Dayal, M. Franklin, J. Gehrke, L. Haas, A. Halevy, J. Han, H. V. Jagadish, A. Labrinidis, S. Madden, Y. Papakonstantinou, J. Patel, R. Ramakrishnan, K. Ross, C. Shahabi, D. Suciu, S. Vaithyanathan, and J. Widom, "Challenges and Opportunities with Big Data 2011-1," Cyber Center Technical Reports, Paper 1, Purdue University, 2011, http://docs.lib.purdue.edu/cctech/1.
4. R. M. Groves and L. Lyberg, "Total Survey Error," *Public Opinion Quarterly* 74, no. 5 (2010): 849–879.
5. For an overview, see R. Valliant, J. A. Dever, and F. Kreuter, *Practical Tools for Sampling and Weighting* (New York: Springer, 2013).

6. R. Rosenbaum and D. B. Rubin, "The Central Role of the Propensity Score in Observational Studies for Causal Effects," *Biometrika* 70, no. 1 (April 1983): 41–55; C. Frangakis and D. Rubin, "Principal Stratification in Causal Inference," *Biometrics* 58 (2002): 21–29.
7. See http://www.nielsen.com/us/en/nielsen-solutions/nielsen-measurement/nielsen-tv-measurement.html.
8. See http://www.cdc.gov/nchs/nhanes.htm.
9. See http://www.bls.gov/cps/cps_htgm.pdf (accessed January 20, 2014).
10. See http://www.bls.gov/cps/cps_over.htm#coverage (accessed January 20, 2014).
11. See http://www.aapor.org/Transparency_Initiative.htm#.Ut2pHrQo7IU (accessed January 20, 2014).
12. See http://www.esomar.org/news-and-multimedia/news.php?idnews=104.
13. For studies that show this effect, see E. Singer, "Confidentiality, Risk Perception, and Survey Participation," *Chance* 17, no. 3 (2004): 30–34; E. Singer, N. Mathiowetz, and M. P. Couper, "The Role of Privacy and Confidentiality as Factors in Response to the 1990 Census," *Public Opinion Quarterly* 57 (1993): 465–482.
14. For a meta-analysis on nonresponse bias, see R. Groves and E. Peytcheva, "The Impact of Nonresponse Rates on Nonresponse Bias: A Meta-Analysis. *Public Opinion Quarterly* 72, no. 2 (2008): 167–189.
15. R. M. Groves, "Three Eras of Survey Research," *Public Opinion Quarterly* 75, no. 5 (2011): 861–871.
16. See http://magazine.amstat.org/blog/2013/01/01/sci-policy-jan2013/ (accessed January 20, 2014).
17. For a recent example, see the Equality of Opportunity project (Ray Chetty, Nathaniel Hendren, Patrick Kline, Emmanuel Saez, and colleagues), which uses the Earned Income Tax Credit to study the impact of tax expenditure on intergenerational mobility, http://www.equality-of-opportunity.org/.
18. J. Schmieder, T. von Wachter, and S. Bender, "The Effects of Extended Unemployment Insurance over the Business Cycle: Evidence from Regression Discontinuity Estimates over 20 Years," *Quarterly Journal of Economics* 127, no. 2 (2012): 701–752.
19. D. Card, J. Heining, and P. Kline, "Workplace Heterogeneity and the Rise of West German Wage Inequality," *Quarterly Journal of Economics* 128, no. 3 (2013): 967–1015.
20. For a discussion and examples of feature selection, see D. Antenucci, M. J. Cafarella, M. C. Levenstein, C. Ré, and M. D. Shapiro, "Ringtail: Feature Selection for Easier Nowcasting," in *16th International Workshop on the Web and Databases (WebDB 2013), New York,* http://web.eecs.umich.edu/~michjc/papers/ringtail.pdf.
21. R. Tourangeau and T. Yan, "Sensitive Questions in Surveys," *Psychological Bulletin* 133, no. 5 (2007): 859–883.
22. V. R. Brown and E. D. Vaughn, "The Writing on the (Facebook) Wall: The Use of Social Networking Sites in Hiring Decisions," *Journal of Business and Psychology* 26, no. 2 (2011): 219–225.
23. K. Karl, J. Peluchette, and C. Schlaegel, "Who's Posting Facebook Faux Pas? A Cross-Cultural Examination of Personality Differences," *International Journal of Selection and Assessment* 18, no. 2 (2010): 174–186.

24. F. Kreuter, S. Presser, and R. Tourangeau, "Social Desirability Bias in CATI, IVR, and Web Surveys: The Effects of Mode and Question Sensitivity," *Public Opinion Quarterly* 72, no. 5 (2008): 847–865.
25. A point made by M. P. Couper, "Is the Sky Falling? New Technology, Changing Media, and the Future of Surveys," *Survey Research Methods* 7, no. 3 (2013): 145–156; K. Prewitt, "The 2012 Morris Hansen Lecture: Thank you Morris, et al., for Westat, et al.," *Journal of Official Statistics* 29, no. 2 (2013): 223–231. This is different from the biomedical sciences where often a few cases have countless variables.
26. See Bpp.mit.edu.
27. M. Horrigan, "Big Data: A Perspective from the BLS," *AMSTATNews*, January 1, 2013.
28. See http://econprediction.eecs.umich.edu/.
29. D. Antenucci, M. Cafarella, M. Levenstein, and M. Shapiro, "Creating Measures of Labor Market Flows Using Social Media," presentation to the National Bureau of Economic Research, Cambridge, MA, July 16, 2012.
30. O'Neil and Schutt, *Doing Data Science*.
31. T. Yan and K. Olson, "Analyzing Paradata to Investigate Measurement Error," in *Improving Surveys with Paradata: Making Use of Process Information*, ed. F. Kreuter (Hoboken, NJ: Wiley, 2013).
32. V. Mayer-Schönberger and K. Cukier, *Big Data: A Revolution That Will Transform How We Live, Work and Think* (London: John Murray, 2013).
33. M. Paul and M. Dredze, "You Are What You Tweet: Analyzing Twitter for Public Health," in *5th International AAAI Conference on Weblogs and Social Media (July 2011)*, http://www.aaai.org/ocs/index.php/ICWSM/ICWSM11/paper/view/2880.
34. T. File, "Computer and Internet Use in the United States," Current Population Survey Report P20-568 (Washington, DC: U.S. Census Bureau, 2013).
35. See http://www.statisticbrain.com/credit-card-ownership-statistics/; http://www.creditcards.com/credit-card-news/credit-card-industry-facts-personal-debt-statistics-1276.php (accessed January 20, 2014).
36. J. T. Lessler and W. D. Kalsbeek, *Nonsampling Error in Surveys* (Hoboken, NJ: Wiley, 1992).
37. For a review, see S. Eckman, "Did the Inclusion of Non-Internet Households in the LISS Panel Reduce Coverage Bias?" Manuscript, Institute for Employment Research, Germany, 2014.
38. M. Bosnjak, I. Haas, M. Galesic, L. Kaczmirek, W. Bandilla, and M. P. Couper, "Sample Composition Discrepancies in Different Stages of a Probability-Based Online Panel," *Field Methods* 25, no. 4 (2013): 339–360.
39. J. A. Dever, A. Rafferty, and R. Valliant, "Internet Surveys: Can Statistical Adjustment Eliminate Coverage Bias?" *Survey Research Methods* 2, no. 2 (2008): 47–62.
40. M. P. Couper, A. Kapteyn, M. Schonlau, and J. Winter, "Noncoverage and Nonresponse in an Internet Survey," *Social Science Research* 36, no. 1 (2007): 131–148; M. Schonlau, A. Van Soest, A. Kapteyn, and M. Couper, "Selection Bias in Web Surveys and the Use of Propensity Scores," *Sociological Methods and Research* 37, no. 3 (2009): 291–318.

41. M. Link, "Emerging Technologies: New Opportunities, Old Challenges," presented at FedCASIC Workshop, Washington, DC, March 19, 2013.
42. Paul and Dredze, "You Are What You Tweet."
43. See R. Kirkpatrick, "Big Data and Real-Time Analytics for Agile Global Development," 2013, http://unstats.un.org/unsd/statcom/statcom_2013/seminars/Big_Data/BigData_UNGlobalPulse_Kirkpatrick.pdf (accessed January 20, 2014).
44. J. Staddon, D. Huffaker, L. Brown, and A. Sedley, "Are Privacy Concerns a Turn-off? Engagement and Privacy in Social Networks," in *Proc. 8th Symposium on Usable Privacy and Security (SOUPS '12),* article 12, https://static.googleusercontent.com/media/research.google.com/en/us/pubs/archive/38142.pdf.
45. T. Halevi, J. Lewis, and N. Memon, "Phishing, Personality Traits and Facebook," Preprint arXiv:1301.7643 [CS.HC] (2013).
46. E. Singer, "Toward a Benefit-Cost Theory of Survey Participation: Evidence, Further Tests, and Implications," *Journal of Official Statistics* 27, no. 2 (2011): 379–392.
47. See http://www.globalresearch.ca/nsa-spying-and-search-engine-tracking-technologies/5365435?utm_source=rss&utm_medium=rss&utm_campaign=nsa-spying-and-search-engine-tracking-technologies (accessed January 20, 2014).
48. P. A. Zandbergen, "Accuracy of iPhone Locations: A Comparison of Assisted GPS, WiFi and Cellular Positioning," *Transactions in GIS* 13, no. s1 (2009): 5–25.
49. Paul and Dredze, "You Are What You Tweet."
50. Couper, "Is the Sky Falling?"
51. Link, "Emerging Technologies."
52. M. Kosinski, D. Stillwell, and T. Graepel, "Private Traits and Attributes are Predictable from Digital Records of Human Behavior," *Proceedings of the National Academy of Sciences* 110, no. 15 (2013): 5802–5805.
53. Ibid.
54. Couper, "Is the Sky Falling?"
55. See http://www.fastcoexist.com/3015860/people-are-changing-their-internet-habits-now-that-they-know-the-nsa-is-watching (accessed January 20, 2014).
56. E.g. post-stratification, general regression estimation, and raking.
57. Valliant, Dever, and Kreuter, *Practical Tools for Sampling and Weighting.*
58. R. Valliant and J. Dever, "Estimating Propensity Adjustments for Volunteer Web Surveys," *Sociological Methods and Research* 40 (2011): 105–137; J. Dever, A. Rafferty, and R. Valliant, "Internet Surveys: Can Statistical Adjustments Eliminate Coverage Bias?" *Survey Research Methods* 2 (2008): 47–60. For a very accessible discussion of these approaches to non-probability samples, see Couper, "Is the Sky Falling," and AAPOR, "Report of the AAPOR Task Force on Non-Probability Sampling," *Journal of Survey Statistics and Methodology* 1 (2013): 90–143.
59. See the volume edited by Douglas S. Massey and Roger Tourangeau, *The Nonresponse Challenge to Surveys and Statistics*, ANNALS of the American Academy of Political and Social Science Series 645 (Thousand Oaks, CA: Sage, 2013).
60. E. A. Stuart, S. R. Cole, C. P. Bradshaw, and P. J. Leaf, "The Use of Propensity Scores to Assess the Generalizability of Results from Randomized Trials," *Journal of the Royal Statistical Society, Series A* 174, no. 2 (2011): 369–386; S. R. Cole and

E. A. Stuart, "Generalizing Evidence from Randomized Clinical Trials to Target Populations: The ACTG-320 Trial," *American Journal of Epidemiology* 172 (2010): 107–115.

61. T. Smith, "The Report of the International Workshop on Using Multi-Level Data from Sample Frames, Auxiliary Databases, Paradata and Related Sources to Detect and Adjust for Nonresponse Bias in Surveys," *International Journal of Public Opinion Research* 23 (2011): 389–402.
62. J. Sakshaug and F. Kreuter, "Assessing the Magnitude of Non-Consent Biases in Linked Survey and Administrative Data," *Survey Research Methods* 6, no. 2 (2012): 113–122.
63. E. Singer, H. J. Hippler, and N. Schwarz, "Confidentiality Assurances in Surveys: Reassurance or Threat?" *International Journal of Public Opinion Research* 4, no. 3 (1992): 256–268; N. Bates, J. Dalhammer, and E. Singer, "Privacy Concerns, Too Busy, or Just Not Interested: Using Doorstep Concerns to Predict Survey Nonresponse," *Journal of Official Statistics* 24, no. 4 (2008): 591–612; M. P. Couper, E. Singer, F. G. Conrad, and R. M. Groves, "Experimental Studies of Disclosure Risk, Disclosure Harm, Topic Sensitivity, and Survey Participation," *Journal of Offiical Statistics* 26, no. 2 (2010): 287–300.
64. J. Sakshaug, V. Tutz, and F. Kreuter, "Placement, Wording, and Interviewers: Identifying Correlates of Consent to Link Survey and Administrative Data," *Survey Research Methods* 7, no. 2 (2013): 133–144; F. Kreuter, J. Sakshaug, and R. Tourangeau, "Using Gain-Loss Framing to Obtain Respondent Consent to Link Survey and Administrative Data" (under review).
65. For an overview and new developments, see R. Schnell, "Combining Surveys with Non-Questionnaire Data: Overview and Introduction," in *Improving Surveys Methods: Lessons from Recent Research*, ed. U. Engel, B. Jann, P. Lynn, A. Scherpenzeel, and P. Sturgis (New York: Psychology Press, 2014); R. Schnell, "Getting Big Data but Preventing Big Brother: Entwicklung neuer technischer Lösungen für die datenschutzgerechte Zusammenführung personenbezogener Daten," UNIKATE 45: Fusionen – Universität Duisburg-Essen.
66. S. Eckman and N. English, "Creating Housing Unit Frames from Address Databases Geocoding Precision and Net Coverage Rates," *Field Methods* 24, no. 4 (2012): 399–408.
67. R. Groves, "Designed Data" and "Organic Data," *Director's Blog,* http://directorsblog.blogs.census.gov/2011/05/31/designed-data-and-organic-data/ (accessed January 20, 2014).
68. R. Kohavi, R. Longbotham, D. Sommerfield, and R. M. Henne, "Controlled Experiments on the Web: Survey and Practical Guide," *Data Mining and Knowledge Discovery* 18 (2009): 140–181.
69. See http://unstats.un.org/unsd/statcom/statcom_2013/seminars/Big_Data/Big Data_OECD_Wyckoff.pdf.

13 Using Statistics to Protect Privacy

Alan F. Karr and Jerome P. Reiter

Introduction

Those who generate data – for example, official statistics agencies, survey organizations, and principal investigators, henceforth all called *agencies* – have a long history of providing access to their data to researchers, policy analysts, decision makers, and the general public. At the same time, these agencies are obligated ethically and often legally to protect the confidentiality of data subjects' identities and sensitive attributes. Simply stripping names, exact addresses, and other direct identifiers typically does not suffice to protect confidentiality. When the released data include variables that are readily available in external files, such as demographic characteristics or employment histories, ill-intentioned users – henceforth called *intruders* – may be able to link records in the released data to records in external files, thereby compromising the agency's promise of confidentiality to those who provided the data.

In response to this threat, agencies have developed an impressive variety of strategies for reducing the risks of unintended disclosures, ranging from restricting data access to altering data before release. Strategies that fall into the latter category are known as statistical disclosure limitation (SDL) techniques. Most SDL techniques have been developed for data derived from probability surveys or censuses. Even in complete form, these data would not typically be thought of as big data, with respect to scale (numbers of cases and attributes), complexity of attribute types, or structure: most datasets are released, if not actually structured, as flat files.

In this chapter, we explore interactions between data dissemination and big data. We suggest lessons that stewards of big data can learn from statistical agencies' experiences. Conversely, we discuss how big data and growing computing power could impact agencies' future dissemination practices. We conclude with a discussion of research needed and possible visions of the future of big data dissemination.

Experiences from Agencies

When disseminating a dataset to the public, agencies generally take three steps. First, after removing direct identifiers like names and addresses, the agency evaluates the disclosure risks inherent in releasing the data 'as is'. Almost always the agency determines that these risks are too large, so that some form of restricted access or SDL is needed. We focus on SDL techniques here, because of the importance to researchers and others of direct access to the data. Second, the agency applies an SDL technique to the data. Third, the agency evaluates the disclosure risks and assesses the analytical quality of the candidate data release(s). In these evaluations, the agency seeks to determine whether the risks are sufficiently low, and the usefulness is adequately high, to justify releasing a particular set of altered data (Reiter 2012). Often, these steps are iterated multiple times; for example, a series of SDL techniques is applied to the data and subsequently evaluated for risk and utility. The agency stops when it determines that the risks are acceptable and the utility is adequate (Cox et al. 2011).

To set the stage for our discussion of SDL frameworks and big data releases, we begin with a short overview of common SDL techniques, risk assessment, and utility assessment. We are not comprehensive here; additional information can be found in, for example, Federal Committee on Statistical Methodology (1994), Willenborg and de Waal (2001), National Research Council (2005, 2007), Karr et al. (2010), Reiter (2012), and Hundepool et al. (2012).

Risk Assessment for Original Data

Most agencies are concerned with two types of disclosures, namely (i) identification disclosures, which occur when an intruder correctly identifies individual records in the released data, and (ii) attribute disclosures, which occur when an intruder learns the values of sensitive variables for individual records in the data (Reiter 2012). Often agencies fold assessment of attribute risk into assessment of identification risk. For concreteness, in this chapter, we focus on data regarding individuals. In the world of official statistics, many datasets contain information on establishments such as hospitals, manufacturers, and schools. Many of the problems we discuss here are significantly more challenging for establishment data (Kinney et al. 2011).

To assess identification disclosure risks, agencies make assumptions, either explicitly or implicitly, regarding what intruders know about the data subjects. Typical assumptions include whether the intruder knows that certain individuals participated in the survey, which quasi-identifying variables the intruder knows, and the amount of measurement error, or other error, in the intruder's data. For example, a common approach to risk assessment is to perform re-identification studies: the agency matches records in the original file with records from external databases that intruders could use to attempt identifications, matching on variables common to both files such as demographics, employment histories, or education. In such studies, the information on the external files operationally defines the agency's assumptions about intruder knowledge.

Agencies are particularly concerned about data subjects that are unique in the population with respect to characteristics deemed to be available to intruders, which often are called *keys* in the SDL literature. An intruder who accurately matches the keys of a record that is unique in the population (on those keys) to an external file is guaranteed to be correct. Typically agencies only know that a record is unique on the keys in the sample. They have to estimate the probability that a data subject is unique in the population given that the subject is unique in the sample. See Skinner and Shlomo (2008) and Manrique-Vallier and Reiter (2012) for reviews of such methods. We also note that intruders who know that a particular record was in the sample can identify that record easily if it is unique in the sample on the keys.

Almost surely, the agency does not know very precisely what information intruders possess about the data subjects. Hence, and prudently, they examine risks under several scenarios, for example, different sets of quasi-identifiers known by intruders, and whether or not intruders know who participated in the study.

Statistical Disclosure Limitation Techniques

Most public use datasets released by national statistical agencies have undergone SDL treatment by one or more of the methods below.

Aggregation Aggregation turns atypical records – which generally are most at risk – into typical records. For example, there may be only one person with a particular combination of keys in a county, but many people with those characteristics in a state. Releasing data for this person with county indicators would pose a high disclosure risk, whereas releasing the data at the state level might not. Unfortunately, such aggregation makes analysis at finer levels difficult and often impossible, and it creates problems of ecological inferences. Another example is to report exact values only below specified thresholds, for example, reporting all ages above 90 as '90 or older'. Such top coding (or bottom coding) eliminates detailed inferences about the distribution beyond the thresholds. Chopping off tails also negatively impacts estimation of whole-data quantities (Kennickell and Lane 2006).

Suppression Agencies can delete at-risk values, or even entire variables, from the released data (Cox 1980). Suppression of at-risk values creates data that are not missing at random, which are difficult to analyze properly.

Data Swapping Agencies can swap data values between selected pairs of records – for example, switch counties for two households with the same number of people – to discourage users from matching, because matches may be based on 'incorrect' data (Dalenius and Reiss 1982). Swapping at high levels destroys relationships involving both swapped and unswapped variables. Even at low levels of swapping, certain analyses can be compromised (Drechsler and Reiter 2010; Winkler 2007).

Adding Random Noise Agencies can add randomly sampled amounts to the observed numerical values, for example, adding a random deviate from a normal distribution with mean equal to zero (Fuller 1993). This reduces the potential to match accurately on the perturbed data and changes sensitive attributes. Generally, the amount of protection increases with the variance of the noise distribution; however, adding noise with large variance distorts marginal distributions and attenuates regression coefficients (Yancey et al. 2002).

Synthetic Data Agencies can replace original data values at high risk of disclosure with values simulated from probability distributions specified to reproduce as many of the relationships in the original data as possible (Reiter and Raghunathan 2007). Partially synthetic data comprise the original individuals with some subset of collected values replaced with simulated values. Fully synthetic data comprise entirely simulated records; the originally sampled individuals are not on the file. In both types, the agency generates and releases multiple versions of the data to enable users to account appropriately for uncertainty when making inferences. Synthetic data can provide valid inferences for analyses that are in accord with the synthesis models, but they may produce inaccurate inferences for other analyses. Despite being synthesized, synthetic data are not risk free, especially with respect to attribute disclosure.

Disclosure Risk and Data Utility Assessment after SDL

Disclosure Risk Assessment Many agencies perform re-identification experiments on SDL-protected data. In addition to matching records in the file being considered for release to external files, many agencies match the altered file against the confidential file. Agencies also specify conditional probability models that explicitly account for assumptions about what intruders might know about the data subjects and any information released about the disclosure control methods. For illustrative computations of model-based identification probabilities, see Duncan and Lambert (1986, 1989), Fienberg et al. (1997), Reiter (2005), and Shlomo and Skinner (2010).

It is worth noting that the concept of harm, such as a criminal act or loss of benefits, from a disclosure can be separated from the risk of disclosure (Skinner 2012). SDL techniques are designed to reduce risks, not harm. Agencies may decide to take on more risk if the potential for harm is low, or less risk if the potential for harm is high. We note that agencies could be concerned about the harm that arises from *perceived* identification or attribute disclosures – the intruder believes she has made an identification or learned an attribute, but is not correct – although in general agencies do not take this into account when designing SDL strategies. See Lambert (1993) and Skinner (2012) for discussion.

Data Utility Assessment Data utility is usually assessed by comparing differences in results of specific analyses between the original and the released data. For example, agencies look at the similarity between a set of quantities estimated with the original data and with the data proposed for release, such as first and second moments, marginal distributions, and regression coefficients representative of anticipated analyses. Similarity across a wide range of analyses suggests that the released data have high utility (Karr et al. 2006). Of course, such utility measures only reveal selected features of the quality of the candidate releases; other features could be badly distorted.

The SDL literature also describes utility measures based on global functions of the data, such as differences in distributions (Woo et al. 2009). Our sense is that these methods are not widely used by agencies.

Current SDL and Big Data

Can typical SDL techniques be employed to protect big data? To be blunt, we believe the answer is no, except in special cases. We reach this opinion via informal, but we think plausible, assessments of the potential risk–utility trade-offs associated with applying these methods.

Disclosure Risks in Original Files

Confidential big data carry greater disclosure risks than the typical survey sample. Often confidential big data come from administrative or privately collected sources so that, by definition, someone other than the agency charged with sharing the data knows the identities of data subjects. This is in contrast to small-scale probability samples, which agencies believe inherently have a degree of protection from the fact that they are random subsets of the population, and because membership in a survey is rarely known to intruders. Confidential big data typically include many variables that, because the data arguably are known by others, have to be considered as keys, so that essentially everyone on the file is unique in the population. Further, as the quality of administrative databases gets better with time (particularly as profit incentives strengthen their alignment with information collection), agencies cannot rely on measurement error in external files as a buffer for data protection.

Effectiveness of SDL

Because of the risks inherent in big data, SDL methods that make small changes to data are not likely to be sufficiently protective; there simply will be too much identifying information remaining on the file. This renders ineffective the usual implementations of data swapping, such as swapping entire records across geographies. On the other hand, massive swapping within individual variables, or even within many small sets of variables, would essentially destroy joint relationships in the data. Suppression is not a viable solution: so much would be needed to ensure adequate protection that the released data would be nearly worthless. Aggregation is likely to be problematic for similar reasons. When many variables are available to intruders, even after typical applications of aggregation, many data subjects will remain unique on the file. Very coarse aggregation/recoding is likely to be needed, which also leads to limited data utility. One potential solution is a fully synthetic, or barely partially synthetic, data release. With appropriate models, it is theoretically possible to preserve many distributional features of the original data. However, in practice it is challenging to find good-fitting models for joint distributions of large-scale data; to our knowledge there have been only a handful of efforts to synthesize large-scale databases with complex variable types (Abowd et al. 2006; Machanavajjhala et al. 2008; Kinney et al. 2011). Nonetheless, we believe that methods for generating massive synthetic databases merit further research.

Demands on Big Data

Through both necessity and desire, analyst demands on big data will be broader than what has been dealt with to date. We know some things about utility assessed in terms of 'standard' statistical analyses such as linear regressions (Karr et al. 2006), but almost nothing about utility associated with machine learning techniques such as neural networks or support vector machines, or data mining techniques such as association rules. Nor is it clear even what the right abstractions are. For instance, for surveys, SDL can to some extent be thought of as one additional source of error within a total survey error framework (Groves 2004), allowing use of utility measures that relate to uncertainties in inferences. For big data that are the universe (e.g. of purchases at Walmart), we do not yet know even how to think about utility, let alone measure it. To illustrate, consider partitioning analyses in which the data are split recursively into classes on the basis of a response and one or more predictor variables, producing a tree whose terminal nodes represent sets of similar data points. Measuring the nearness of two trees can be challenging, making it difficult to say how much SDL has altered a partitioning analysis.

Moreover, many demands on big data will be inherently privacy threatening. Most of today's statistical analyses require only that the post-SDL data sufficiently resemble the original data in some low-dimensional, aggregate sense. For instance, if means and covariances are close, so will be the results of linear regressions. On the other hand, data-mining analyses such as searching for extremely rare phenomena, like Higgs bosons or potential terrorists, require 'sufficiently resemble' at the individual record level. Current SDL techniques are, virtually universally, based on giving up record-level accuracy, which reduces disclosure risk, in return for preserving aggregate accuracy, which is the current, but not necessarily the future, basis of utility.

Vision for the Future

We now present a vision for the future, including (i) discussions of what disclosure risk might mean and how it might be assessed with big data and big computation; (ii) how methods based on remote access and secure computation might be useful; and (iii) a vision for a big data dissemination engine involving interplay between unrestricted access, verification of results, and trusted access.

Changing Views of Privacy

It is possible, if not likely, that concomitantly with the move to big data, there also will be changes in the legal, political, and social milieu within which data release lies. Of the authors of this chapter, one (AK) is a baby boomer, and the other (JR) is a genX-er. Perhaps as a result of our research on data confidentiality, our views on privacy do not differ dramatically. But, almost daily, we observe others whose views do seem quite different from ours. These include cellphone users who discuss intimate details of their lives within earshot (and 'lipshot', since many of them seem aware that there are skilled lip readers everywhere) of others, social media users who seem not to realize how much privacy-compromising information a photograph can contain, and others. Whether these behaviors represent true changes in thinking about privacy, or will change as the individuals mature or societal attitudes evolve, remains to be seen. If the former, less protection of data may be required, although "who doesn't care about privacy" may be a potent form of response bias in both surveys and administrative data. If the latter, less may change.

Also unclear is the denouement of the current trend of reluctance-to-refusal by individuals to provide data to government agencies. Some of the decline in response rates is the result of privacy concerns, some is the result of everyone's increasingly complicated lives, some represents a belief that the government already has the information, and some is political opposition to *any* government data collection. It is hard not to think that response rates will continue to decline, but might they stabilize at a level at which statistical fixes still work?

What seemingly must change no matter what is the nature of the compact between data collectors and data subjects. Currently there is a major disequilibrium: official statistics agencies collect and protect much information about individuals and organizations that is readily accessible elsewhere, albeit sometimes there is a cost. Data subjects are clueless as to whether their information is protected adequately. Incentives to data subjects are seen as a means of payment for burden; subjects could (but are not now) also be compensated for (actual or risked) loss of privacy (Reiter 2011). A supremely intriguing thought experiment is to ask "What would happen if data subjects were promised no privacy at all, and simply paid enough to get them to agree to participate?" Would data quality be destroyed? Would the cost be affordable? We do not know.

The connection between privacy and big data is likewise evolving. Answering a few questions on a survey does not generate big data, nor does it cause most people to think anew about privacy. Collecting entire electronic medical records, DNA sequences, or videotapes of two years of driving (as in the Naturalistic Driving Study of the Virginia Tech Transportation Institute) may generate big data, and may change attitudes about privacy. Big data may also change what information is considered sensitive. Forty years ago, most people would have considered salary to be the most sensitive information about them. Today, a significant fraction of salaries are directly available, or accurately inferable, from public information. Instead, medical records may be more sensitive for many people. Partly this is because (in the same way that salaries were once seen as protectable) they are perceived still to be protectable; in addition, the risk associated with knowledge of medical records may be greater (e.g. loss of insurance or a job), as well as more nebulous.

SDL of the Future: A Framework

A significant change that big data and big computing will produce is the capability to enumerate all possible versions of the original dataset that could plausibly have generated the released data. To understand what this means, we sketch here a framework for this 'SDL of the future'.

Let O be the original dataset and M be the released dataset after SDL is applied to O. Let $\mathcal{O}$ denote the set of all possible input datasets that could have been redacted to generate M. In general, the extent to which an analyst or intruder can specify $\mathcal{O}$ is a function of M, agency-released information about the SDL applied to O, and external knowledge. We denote this collective knowledge by (the σ-algebra) $\mathcal{K}$ and for the moment restrict it to consist only of M and agency-released information. External knowledge is addressed in the next section.

To illustrate, suppose that O is a categorical dataset structured as a multi-way contingency table containing integer cell counts. Suppose that M is generated from O by means of suppressing low-count cells deemed to be risky, but contains correct marginal totals. In this case, additional cells must almost always be suppressed in order to prevent reconstruction of the risky cells from the marginals. Figure 1 contains an illustration: in the table O, on the left, the four cells with counts less than 5 are suppressed because they are risky, and the cells with counts 5 and 6 are suppressed

1	18	6	25
13	5	2	20
4	1	10	15
18	24	18	60

*	18	*	25
13	*	*	20
*	*	10	15
18	24	18	60

Figure 1 *Left:* Original dataset *O*. *Right:* Masked data set *M*, after cell suppression.

to protect them. In M, on the right, there is no distinction between the 'primary' and 'secondary' suppressions. Minimally, $\mathcal{K}$ consists of M and the knowledge that cell suppression was performed; $\mathcal{K}$ might or might not contain the value of the suppression threshold or information distinguishing primary from secondary suppressions. In the minimal case, $\mathcal{O}$ consists of six tables: O and the tables obtained by putting 0, 2, 3, 4, and 5 as the upper left-hand entry and solving for the other entries. We denote these by $O_0, \ldots, O_5$, respectively. If the suppression threshold is known and zero is not considered risky, the first of these is ruled out because applying the rules to it does not yield M. Every one of the other four is ruled out if $\mathcal{K}$ distinguishes primary from secondary suppressions. Already one key implication for agencies is clear: the framework can distinguish what must be protected from what might be protected.

Equally important, the framework can distinguish analysts from intruders. The sardonic but apt comment that "One person's risk is another person's utility" demonstrates how subtle the issues are. Within our framework, both analysts and intruders wish to calculate the posterior distribution $P\{O = (\cdot)|\mathcal{K}\}$, but *use this conditional distribution in fundamentally different ways.*

Specifically, analysts wish to perform statistical analyses of the masked data M, as surrogates for analyses of O, and wish to understand how faithful the results of the former are to the results of the latter. (See also the section on microdata release, below.) Conditional on O, the results of an analysis are a deterministic (in general, vector-valued) function $\mathbf{f}(O)$. To illustrate, for categorical data, $\mathbf{f}(O)$ may consist of the entire set of fitted values of the associated contingency table under a well-chosen log-linear model. In symbols, given $P\{O = (\cdot)|\mathcal{K}\}$, *analysts integrate* to estimate $\mathbf{f}(O)$:

$$\widehat{\mathbf{f}(\mathrm{O})} = \int_{\mathcal{O}} \mathbf{f}(o) dP\{O = o|\mathcal{K}\}. \quad (1)$$

It is important to keep in mind that $\mathcal{O}$ depends on $\mathcal{K}$, even though the notation suppresses the dependence.

To illustrate with the example in Figure 1, if $\mathcal{K}$ is only the knowledge that cell suppression was performed, then $\mathcal{O} = \{O, O_0, O_2, O_3, O_4, O_5\}$ and $P\{O = (\cdot)|\mathcal{K}\}$ is the uniform distribution on this set. By contrast, if $\mathcal{K}$ contains in addition the suppression rules, then $\mathcal{O} = \{O, O_2, O_3, O_4, O_5\}$ and $P\{O = (\cdot)|\mathcal{K}\}$ is the uniform distribution on *this* set. Finally, if $\mathcal{K}$ distinguishes primary from secondary suppressions, then $\mathcal{O} = \{O\}$.

If the analysis of interest were a χ^2 test of independence, then, in the second case, the average of the five χ^2 statistics is 34.97, and independence would be rejected. Indeed, independence is rejected for all of O, O_2, O_3, O_4, and O_5 so the analyst can be certain, even without knowing O, that independence fails.

The point is that big computing makes this approach feasible in realistic settings.

By contrast, intruders are interested in global or local maxima in $P\{O = (\cdot)|\mathcal{K}\}$, which correspond to high posterior likelihood estimates of the original data O. In the extreme, *intruders maximize*, calculating

$$O^* = \arg\max_{o \in \mathcal{O}} P\{O = o\,|\mathcal{K}\}. \tag{2}$$

We do not prescribe what intruders would do using O^* but assume only that any malicious acts would be done using O itself, for instance, re-identifying records by means of linkage to an external database containing identifiers.

This distinction allows the agency to reason in principled manners about risk and utility, especially in terms of how they relate to $\mathcal{K}$. *High utility* means that the integration in (1) can be performed or approximated relatively easily. *Low risk* means that the maximization in (2) is difficult to perform or approximate.

A central question is then: How large is the set $\mathcal{O}$ of possible values of given $\mathcal{K}$? Of course, high utility and low risk remain competing objectives: when $\mathcal{O}$ is very large, then the maximization in (2) is hard, but so may be the integration in (1). Because of the integration in (1), it may be more natural to view $|\mathcal{O}|$ as a measure of disclosure risk than as an inverse measure of data utility.

Incorporating External Information

The framework in the preceding section meshes perfectly with a Bayesian approach to external knowledge possessed by analysts or intruders. Once $\mathcal{O}$ is known, such information exists independently of the knowledge embodied in $\mathcal{K}$, for instance, (1), so that it becomes completely natural to view as the product of a prior distribution on O and a likelihood function. See McClure and Reiter (2012a) for implementation of a related approach. In the example in the preceding section, the prior would simply weight the elements of $\mathcal{O}$ on the basis of external knowledge.

More important from a computational perspective is that the integration in (1) can be performed by sampling from the posterior distribution $P\{O = (\cdot)|\mathcal{K}\}$, which is exactly what (Markov chain) Monte Carlo methods do!

Operational Implications

Most of today's (2013) big data are physical measurements that seem to need no SDL. There are, of course, very large transaction databases held by e-commerce websites, as well as databases containing information about telephone or e-mail communications. The extent to which any of the latter will be shared in any form is not clear. What is clear is that, in the short run at least, local storage and computing power will be supplemented or even supplanted by 'cloud computing', in which, transparently to the user, data and cycles reside in multiple physical machines.

Some implications of cloud computing are troubling to official statistics agencies. They may lose control over who has physical possession of their data, over who can view the data, and over how access to the data is controlled. The number of vulnerabilities increases in the cloud model, as does the possibility of secondary disclosure. In today's model, someone seeking illicitly to access Census Bureau data must penetrate Census Bureau servers, all of which are physically and electronically controlled by the Bureau. What happens if Census Bureau data might 'accidentally' be seen by someone attempting to access credit card records? Can the Census Bureau legitimately promise confidentiality of records when 'transparency' means lack of knowledge rather than openness? Similar, and perhaps more challenging, issues arise for licensing of datasets.

These issues notwithstanding, we expect that the data access model of the future will be to take the analysis to the data rather than the data to the analyst or the analyst to the data. There are multiple reasons for this. Truly big data are too big to take to the users. Dataset size, coupled with the current impetus for availability of research datasets, seems to demand archives that can deal with complex issues of data format, metadata, paradata, provenance, and versioning. In our view, archives will also provide computational power. They will resemble today's remote access servers

(Karr et al. 2010), but with vastly increased computational power and flexibility.

Construction of such archives will require addressing issues we currently choose (mostly) to bypass by limiting server capability. If the data do require protection, perhaps the most pressing challenge is query interaction: both risk and utility increase in ways we do not currently understand when multiple queries are posed to the server. Answering one query may permanently preclude answering others (Dobra et al. 2002, 2003). Many current remote access servers in effect dodge this issue by severely limiting the space of allowable queries, for instance, by forbidding high-leverage variable transformations or limiting the degree of interactions. Others include manual review of both analyses and results, a strategy that is hopelessly non-scalable. Linkage to other datasets is rarely permitted, nor are exploratory tools such as visualizations. In virtually all of these instances, everything from sound abstractions to computational tools is lacking.

Because cloud data are distributed data, operational systems will require techniques for handling distributed data. A set of techniques from computer science known generically as secure multiparty computation (SMPC) has been shown to allow analyses based on sufficient statistics that are additive across component databases (Karr et al. 2005, 2007; Karr 2010; Karr and Lin 2010). These analyses include creation of contingency tables, linear and logistic regression (as well as extensions such as generalized linear models), and even iterative procedures such as numerical maximum likelihood estimation using Newton-Raphson methods. For almost all other analyses, the details remain to be worked out.

Is There a Future for Microdata Releases?

In view of the discussion in the preceding section, it is natural to ask whether there is a future for released microdata – individual records, as opposed to only summary statistics. We believe that there is, but that new tools will be required to attain it.

To begin, there *is* and will remain a case for releasing microdata. Microdata are essential to the education and training of early career researchers. Historically, there has been no substitute for working directly with data, and we do not believe that this will change. (Indeed, the risk that 'big data' means 'disconnected from the data' is both real and disconcerting.) Perhaps more important, even skilled, mature researchers rarely know in advance which are the right questions to ask, and exploratory analyses

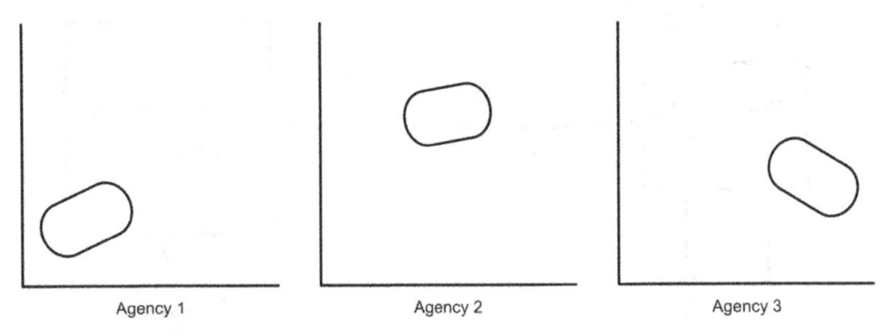

Figure 2 Three datasets, each with locally linear structure, but quadratic global structure.

dealing with the data themselves remain the best, if not only, path to the 'right' questions.

Currently available techniques for query-based analysis of distributed data using SMPC are notably poor in this respect. To illustrate, consider the example in Figure 2. There are three distributed datasets containing two variables, lying in the ranges shown. An analyst familiar with any one of the three databases would believe that the relationship between the two variables is linear, but, of course, it is quadratic instead. Existing query system models might thwart knowing the right question to ask. But even a small sample with intensive SDL from the integrated dataset would have made the right question apparent.

The question then: If highly redacted microdata are released publicly, for example, using novel methods of generating fully synthetic data, how can an analyst know whether he or she is on the right track to the right questions, which can then be posed to an archive/server? *Verification servers* are one technology that offers a solution (Reiter et al. 2009; McClure and Reiter 2012b). Briefly, a verification server is a web-accessible system based on a confidential database O with an associated public microdata release M – derived from O – that

- receives from the analyst a description of a statistical analysis Q performed on M;
- performs the analysis on both M and O;
- calculates one or more measures of the fidelity of $Q(M)$ to $Q(O)$;
- returns to the analyst the values of the fidelity measure(s).

The concept is illustrated pictorially in Figure 3. When the fidelity is high, the analyst may pose the query to a server, and receive a more detailed set of results.

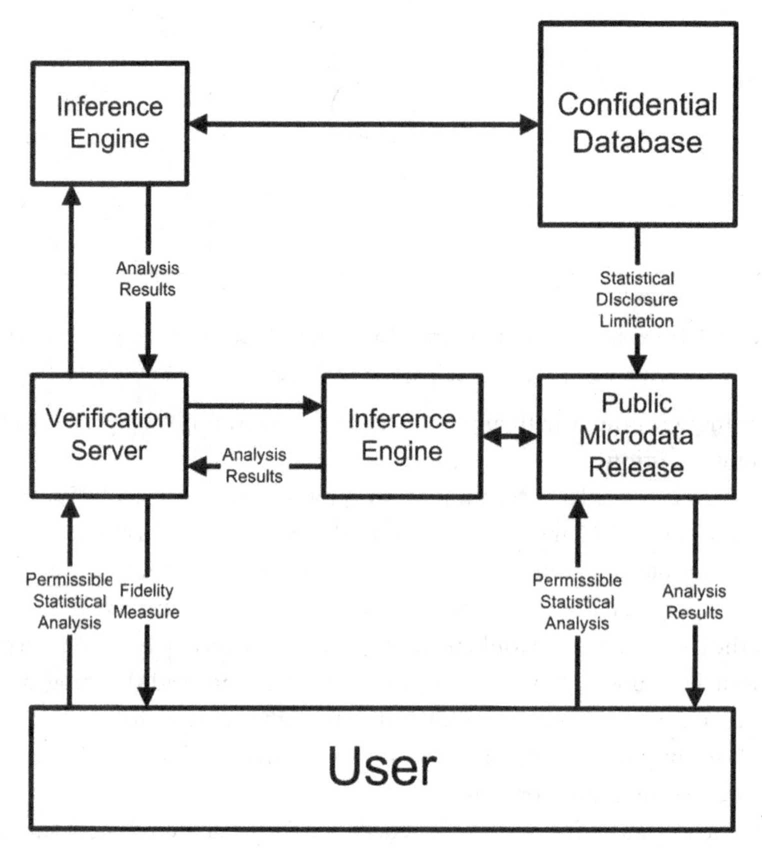

Figure 3 Pictorial schematic view of a verification server.

Verification servers also could help reduce costs of accessing servers that host confidential data. Currently, and we expect also in the future, users who want access to confidential data via virtual or physical data enclaves are vetted by the data stewards. This involves cost that often is passed to the user, for example in the form of fees to access data. With the output from a verification server, users can decide if analysis results based on the redacted data are of satisfactory quality for their particular purposes. If so, they may choose to forgo the dollar and time costs of gaining access. Even users who are not satisfied with the quality of the results can benefit from starting with the redacted data. Storage and processing of big data is costly to data stewards, who likely will pass some costs to users. Analysts who have an informed analysis plan can improve their efficiency when using the server, thereby saving dollars and time.

Although attractive conceptually, verification servers remain an untested technology with both known and to-be-discovered risks. The former include risks shared with remote access servers – unlimited and/or arbitrary queries, interaction among multiple queries, high-complexity variable transformations, subsetting of the data, and intruders with extreme computational resources. Too many, or too high precision, fidelity measures are among the latter. We do know that the latter *are* problems: if they are unaddressed, many SDL methods, including data swapping and top coding, can be reversed (Reiter et al. 2009). At the extreme, returning to the SDL framework we sketched above, with sufficiently many queries, sufficiently precise fidelity measures, and enough computational power, O can be recovered *exactly* from M.

Archive/server-based models also seem (at least currently) to be poor at handling record linkage, except in simple cases where the linkage amounts to a database join. Knowing which variables to link with, and understanding how uncertainties are affected by linkage, require – at least in exploratory stages – actual microdata.

One item of interest in this setting is that as a means of SDL, sampling is typically seen as ineffectual, at least by itself. If the goal is to produce an analysis-capable dataset M, most records must be retained. If no other SDL is performed and this information is known, then an intruder seeking to carry out the maximization in (2) need only worry about the possibility that the maximizer is not in M. Typically, this would be deemed insufficient protection. On other hand, if the goal is to produce an M that allows analysts to ask the right questions, small samples, especially if accompanied by weights, may be entirely adequate.

How Do Official Statistics Agencies Fit In?

Despite some of the challenges alluded to earlier (e.g. cloud computing), official statistics agencies are playing, and we expect will continue to play, significant roles in advancing methodology and practice for accessing big data. Many official statistics agencies that currently collect large-scale databases are experimenting with methods for providing access to these data. For example, as reported in presentations at the 2013 Joint Statistical Meetings, the Centers for Medicare and Medicaid Services (CMS) has contracted with the National Opinion Research Center (NORC) to develop an unrestricted-access, synthetic public use file for Medicare claims data. This file is intended to have limited analytic utility. It exists to help researchers develop methods and code to run on the actual data. After vetting, these researchers may be approved to access the restricted data in a data enclave setting. CMS and NORC had one team develop the synthetic datasets, while another team evaluated the disclosure risks, a separation we recommend as a general dissemination practice.

The Census Bureau has forged similar partnerships with researchers in academia to develop public use products for Longitudinal Employer-Household Dynamics (LEHD) data; see lehd.ces.census.gov for details.

At the same time, agencies are, properly and of necessity, conservative and slow to change. In particular, they must deal with extremely diverse sets of data users and other stakeholders. To illustrate, agencies have been slow to adopt multiple imputation as a means of dealing with item nonresponse, because not all users, even in the research community, are able to analyze such data. More 'exotic' technologies, such as synthetic data, other Bayesian methods, and differential privacy (Dwork 2006; also Chapter 14 in this volume), will replace existing methods, if at all, only at an evolutionary pace. One promising trend, however, is increasing agency attention to the fact that most people pay heed only to the decisions based on agency data, not to the data themselves (Karr 2012, 2013), which seems likely to yield important new insights about data utility.

Concluding Remarks

In spite of many steps toward wider data availability, legal, ethical, scale, and intellectual property restrictions are part of the foreseeable future (Karr 2014). "Make everything available to everyone" will not be ubiquitous, and SDL techniques are not likely to offer broadly the kind of one-off databases released by statistical agencies today. Statistical agencies already balance what to release to whom against other considerations, and this mode of thinking can, we believe, be crucial to big data.

For many big datasets, confidentiality risks of disseminating data may be so high that it is nearly impossible to share unrestricted-use microdata without massive data alterations, which call into question the usefulness of the released big data. We believe that methods for nonparametric estimation of distributions for large-scale data – a focus of significant research effort in the machine learning and statistical communities – offer potential to be converted to data synthesizers (Drechsler and Reiter 2011). Nonetheless, unrestricted-access, big datasets probably need to take on less ambitious roles than current agency practice permits; for example, they may serve as code testbeds or permit only a limited number of (valid) analyses. Verification servers, which promise to provide automated feedback on the quality of inferences from redacted data, could enhance the usefulness of such datasets, allowing users to determine when they can trust results and when they need to accept the costs of applying for access to the confidential data.

Highly redacted datasets also should help users of remote query systems to identify sensible queries.

To conclude, we believe a way forward for big data access is an integrated system including (i) unrestricted access to highly redacted data, most likely some version of synthetic data, followed with (ii) means for approved researchers to access the confidential data via remote access solutions, glued together by (iii) verification servers that allow users to assess the quality of their inferences with the redacted data so as to be more efficient in their use (if necessary) of the remote data access. We look forward to seeing how this vision develops.

Acknowledgement This work was partially supported by National Science Foundation grants CNS–1012141 and SES–1131897.

REFERENCES

Abowd, J., M. Stinson, and G. Benedetto. 2006. Final report to the Social Security Administration on the SIPP/SSA/IRS Public Use File Project. Technical report, U.S. Census Bureau Longitudinal Employer-Household Dynamics Program. Available at http://www.census.gov/sipp/synth_data.html.

Cox, L. H. 1980. Suppression methodology and statistical disclosure control. *Journal of the American Statistical Association* 75:377–385.

Cox, L. H., A. F. Karr, and S. K. Kinney. 2011. Risk-utility paradigms for statistical disclosure limitation: How to think, but not how to act (with discussion). *International Statistical Review* 79(2):160–199.

Dalenius, T., and S. P. Reiss. 1982. Data-swapping: A technique for disclosure control. *Journal of Statistical Planning and Inference* 6:73–85.

Dobra, A., S. E. Fienberg, A. F. Karr, and A. P. Sanil. 2002. Software systems for tabular data releases. *International Journal of Uncertainty, Fuzziness and Knowledge Based Systems* 10(5):529–544.

Dobra, A., A. F. Karr, and A. P. Sanil. 2003. Preserving confidentiality of high-dimensional tabular data: Statistical and computational issues. *Statistics and Computing* 13(4):363–370.

Drechsler, J., and J. P. Reiter. 2010. Sampling with synthesis: A new approach for releasing public use census microdata. *Journal of the American Statistical Association* 105:1347–1357.

Drechsler, J., and J. P. Reiter. 2011. An empirical evaluation of easily implemented, nonparametric methods for generating synthetic datasets. *Computational Statistics and Data Analysis* 55:3232–3243.

Duncan, G. T., and D. Lambert. 1986. Disclosure-limited data dissemination. *Journal of the American Statistical Association* 81:10–28.

Duncan, G. T., and D. Lambert. 1989. The risk of disclosure for microdata. *Journal of Business and Economic Statistics* 7:207–217.

Dwork, C. 2006. Differential privacy. In *Automata, Languages and Programming*, ed. M. Bugliesi, B. Preneel, V. Sassone, and I. Wegener, LNCS 4052, 1–12. Berlin: Springer.

Federal Committee on Statistical Methodology. 1994. *Report on Statistical Disclosure Limitation Methodology*. Statistical Policy Working Paper 22. Washington, DC: U.S. Office of Management and Budget.

Fienberg, S. E., U. E. Makov, and A. P. Sanil. 1997. A Bayesian approach to data disclosure: Optimal intruder behavior for continuous data. *Journal of Official Statistics* 13:75–89.

Fuller, W. A. 1993. Masking procedures for microdata disclosure limitation. *Journal of Official Statistics* 9:383–406.

Groves, R. M. 2004. *Survey Errors and Survey Costs*. New York: Wiley.

Hundepool, A., J. Domingo-Ferrer, L. Franconi, S. Giessing, E. Schulte-Nordholt, K. Spicer, and P.-P. de Wolf. 2012. *Statistical Disclosure Control*. New York: Wiley.

Karr, A. F. 2010. Secure statistical analysis of distributed databases, emphasizing what we don't know. *Journal of Privacy and Confidentiality* 1(2):197–211.

Karr, A. F. 2012. Discussion on statistical use of administrative data: Old and new challenges. *Statistica Neerlandica* 66(1):80–84.

Karr, A. F. 2013. Discussion of five papers on Systems and Architectures for High-Quality Statistics Production. *Journal of Official Statistics* 29(1):157–163.

Karr, A. F. 2014. Why data availability is such a hard problem. *Statistical Journal of the International Association for Official Statistics*, to appear.

Karr, A. F., W. J. Fulp, X. Lin, J. P. Reiter, F. Vera, and S. S. Young. 2007. Secure, privacy-preserving analysis of distributed databases. *Technometrics* 49(3):335–345.

Karr, A. F., S. K. Kinney, and J. F. Gonzalez, Jr. 2010. Data confidentiality – the next five years: Summary and guide to papers. *Journal of Privacy and Confidentiality* 1(2):125–134.

Karr, A. F., C. N. Kohnen, A. Oganian, J. P. Reiter, and A. P. Sanil. 2006. A framework for evaluating the utility of data altered to protect confidentiality. *The American Statistician* 60:224–232.

Karr, A. F., and X. Lin. 2010. Privacy-preserving maximum likelihood estimation for distributed data. *Journal of Privacy and Confidentiality* 1(2):213–222.

Karr, A. F., X. Lin, J. P. Reiter, and A. P. Sanil. 2005. Secure regression on distributed databases. *Journal of Computational and Graphical Statistics* 14(2):263–279.

Kennickell, A., and J. Lane. 2006. Measuring the impact of data protection techniques on data utility: Evidence from the Survey of Consumer Finances. In *Privacy in Statistical Databases 2006*, ed. J. Domingo-Ferrer and L. Franconi, LNCS 4302, 291–303. New York: Springer.

Kinney, S. K., J. P. Reiter, A. P. Reznek, J. Miranda, R. S. Jarmin, and J. M. Abowd. 2011. Towards unrestricted public use business microdata: The synthetic Longitudinal Business Database. *International Statistical Review* 79:363–384.

Lambert, D. 1993. Measures of disclosure risk and harm. *Journal of Official Statistics* 9:313–331.

Machanavajjhala, A., D. Kifer, J. Abowd, J. Gehrke, and L. Vilhuber. 2008. Privacy: Theory meets practice on the map. In *Proc. IEEE 24th International Conference on Data Engineering*, 277–286.

Manrique-Vallier, D., and J. P. Reiter. 2012. Estimating identification disclosure risk using mixed membership models. *Journal of the American Statistical Association* 107:1385–1394.

McClure, D., and J. P. Reiter. 2012a. Differential privacy and statistical disclosure risk measures: An illustration with binary synthetic data. *Transactions on Data Privacy* 5:535–552.

McClure, D., and J. P. Reiter. 2012b. Towards providing automated feedback on the quality of inferences from synthetic datasets. *Journal of Privacy and Confidentiality*, 4(1): article 8.

National Research Council. 2005. *Expanding Access to Research Data: Reconciling Risks and Opportunities*. Panel on Data Access for Research Purposes, Committee on National Statistics, Division of Behavioral and Social Sciences and Education. Washington, DC: The National Academies Press.

National Research Council. 2007. *Putting People on the Map: Protecting Confidentiality with Linked Social-Spatial Data*. Panel on Confidentiality Issues Arising from the Integration of Remotely Sensed and Self-Identifying Data, Committee on the Human Dimensions of Global Change, Division of Behavioral and Social Sciences and Education. Washington, DC: The National Academies Press.

Reiter, J. P. 2005. Estimating identification risks in microdata. *Journal of the American Statistical Association* 100:1103–1113.

Reiter, J. P. 2011. Commentary on article by Gates. *Journal of Privacy and Confidentiality* 3: article 8.

Reiter, J. P. 2012. Statistical approaches to protecting confidentiality for microdata and their effects on the quality of statistical inferences. *Public Opinion Quarterly* 76:163–181.

Reiter, J. P., A. Oganian, and A. F. Karr. 2009. Verification servers: Enabling analysts to assess the quality of inferences from public use data. *Computational Statistics and Data Analysis* 53:1475–1482.

Reiter, J. P., and T. E. Raghunathan. 2007. The multiple adaptations of multiple imputation. *Journal of the American Statistical Association* 102:1462–1471.

Shlomo, N., and C. J. Skinner. 2010. Assessing the protection provided by misclassification-based disclosure limitation methods for survey microdata. *Annals of Applied Statistics* 4:1291–1310.

Skinner, C. 2012. Statistical disclosure risk: Separating potential and harm. *International Statistical Review* 80:349–368.

Skinner, C. J., and N. Shlomo. 2008. Assessing identification risk in survey microdata using log-linear models. *Journal of the American Statistical Association* 103:989–1001.

Willenborg, L., and T. de Waal. 2001. *Elements of Statistical Disclosure Control*. New York: Springer.

Winkler, W. E. 2007. Examples of easy-to-implement, widely used methods of masking for which analytic properties are not justified. U.S. Census Bureau Research Report Series, No. 2007–21. Washington, DC: U.S. Census Bureau.

Woo, M. J., J. P. Reiter, A. Oganian, and A. F. Karr. 2009. Global measures of data utility for microdata masked for disclosure limitation. *Journal of Privacy and Confidentiality* 1:111–124.

Yancey, W. E., W. E. Winkler, and R. H. Creecy. 2002. Disclosure risk assessment in perturbative microdata protection. In *Inference Control in Statistical Databases*, ed. J. Domingo-Ferrer, 135–152. Berlin: Springer.

14 Differential Privacy: A Cryptographic Approach to Private Data Analysis

Cynthia Dwork

Introduction

Propose. Break. Propose again. So pre-modern cryptography cycled. An encryption scheme was proposed; a cryptanalyst broke it; a modification, or even a completely new scheme, was proposed. Nothing ensured that the new scheme would in any sense be better than the old. Among the astonishing breakthroughs of modern cryptography is the methodology of rigorously defining the goal of a cryptographic primitive – what it means to break the primitive – and providing a clear delineation of the power – information, computational ability – of the adversary to be resisted (Goldwasser and Micali 1984; Goldwasser et al. 1988). Then, for any proposed method, one proves that no adversary of the specified class can break the primitive. If the class of adversaries captures all feasible adversaries, the scheme can be considered to achieve the stated goal.

This does not mean the scheme is invulnerable, as the goal may have been too weak to capture the full demands placed on the primitive. For example, when the cryptosystem needs to be secure against a passive eavesdropper the requirements are weaker than when the cryptosystem needs to be secure against an active adversary that can determine whether or not arbitrary ciphertexts are well formed (such an attack was successfully launched against PKCS#1; Bleichenbacher 1998). In this case the goal may be reformulated to be strictly more stringent than the original goal, and a new system proposed (and proved). This strengthening of the goal converts the propose–break–propose again cycle into a path of progress.

'Big data' mandates a mathematically rigorous theory of privacy, a theory amenable to measuring – and minimizing – cumulative privacy loss, as data are analyzed and re-analyzed, shared, and linked. This chapter discusses *differential privacy*, a definition of privacy tailored to statistical analysis of data and equipped with a measure of privacy loss. We will motivate and present the definition, give some examples of its use, and discuss the scientific and social challenges to adoption. We will argue that, whatever the measure on which the community eventually settles, data usage should be accompanied by publication of the amount of privacy lost, that is, its privacy 'price'.

Toward Articulating a Privacy Goal

Following the paradigm of modern cryptography, we need to articulate what it is we are trying to prevent – what does it mean to 'break' privacy? What is a *privacy adversary* trying to achieve? Here, *adversary* is a term of art; we do not necessarily ascribe malicious intent to the government compliance monitor analyzing loan information data, to the citizen putting two and two together by looking at census results and the neighbors' blog posts, or to the research scientist poring over the published results of multiple studies of a small population of patients with a rare disease. We do, however, wish to understand and control what such parties can learn. Let us look at some examples of things that can go wrong when confidentiality is a stated goal.

Linkage Attacks In a linkage attack 'anonymized' records containing sensitive (say, medical encounter data) and 'insensitive' information (say, date of birth, sex, and ZIP code) are matched up, or 'linked', with records in a different dataset (say, voter registration records) on the basis of the insensitive fields. In fact exactly this happened, resulting in the identification of the medical records of the governor of Massachusetts (Sweeney 1997). In another famous example, the records of a Netflix user were identified among anonymized training data for a competition on movie recommendation systems, in this case by linkage with the Internet Movie Database (IMDb; Narayanan and Shmatikov 2008). Linkage attacks are powerful because a relatively small collection of seemingly innocuous facts often suffices to uniquely identify an individual. For example, among the observations in Narayanan and Shmatikov (2008): "with 8 movie ratings (of which we allow 2 to be completely wrong) and dates that may have a 3-day error, 96% of Netflix subscribers whose records have been released can be uniquely identified in the dataset." In other words, the removal of all 'personal information' ultimately did not 'anonymize' the records in the Netflix training dataset.

Succintly put, "'De-identified' data isn't," and the culprit is *auxiliary information* – that is, information from a source (voter registration records or IMDb) other than the database itself (HMO medical encounter data or Netflix Prize training dataset). This is not to say that many records, indeed most records, did not get identified *in these particular linkage attacks*. Rather, it demonstrates the fragility of the 'anonymization' protection. There are many ways of knowing that a family member, colleague, or public figure has watched a few movies on a few given days. Were the 'anonymized' viewing habits of such a person made accessible, it could be very easy to learn things about her viewing habits that she would prefer, and is arguably entitled, to conceal.

The Statistics Masquerade Privacy problems do not disappear if we give up on 'anonymizing' individual records and instead release only statistics. For example, the *differencing attack* exploits the relationships between certain pairs of statistics, such as:

1. the number of members of the U.S. House of Representatives with the sickle cell trait and
2. the number of members of the House of Representatives, other than the Speaker of the House, with the sickle cell trait.

When taken together, these two statistics, each of which covers a large set of individuals, reveal the sickle cell status of the Speaker. Such dangerous pairs of queries are not always so easy to spot; indeed, if the query language is sufficiently rich the question of whether two queries pose such a threat is *undecidable*, meaning that there provably cannot be any algorithmic procedure for determining whether the pair of queries is problematic.

A more general adversarial strategy may be called the 'Big Bang' attack.[1] Given a large dataset, the attacker focuses on a relatively small subset, say, members of his extended family, of some size k. For concreteness, let us set $k = 128$. The attacker's goal is to learn a single private bit – not necessarily the same bit – about each member of the extended family. For example, the attacker may wish to know if Aunt Wilma, who has two children, has had more than two pregnancies, and to know if Uncle William has a history of depression, and so on. Clearly, by asking k 'counting queries', each describing exactly one member of the extended family and the property in question, e.g. "How many people with the following identifying characteristics [description of Aunt Wilma and only Aunt Wilma] have had at least three pregnancies?" the attacker can learn the desired bits. But suppose the attacker does not receive perfectly accurate answers. Can introducing small inaccuracies into the query responses protect the family's privacy? Intuitively this approach seems perfect: it renders useless any query about an individual, while not significantly distorting 'statistical' queries whose answers are expected to be fairly large.

The degree to which small distortions can protect against arbitrary counting query sequences depends on the size of 'small' compared to the number of queries. For example, there is a sequence of 128 counting queries with the following property. If the errors introduced are always of magnitude at most 1, then the adversary can reconstruct at least 124 of the 128 private bits. If the errors have magnitude bounded by 3, the number accurately reconstructed is still at least 92. With these same bounds on the magnitudes of the errors, taking $k = 256$, the adversary can correctly reconstruct at least 252 and 220 bits, respectively.

The general form of the bound is: if the magnitudes of the errors are all bounded by E, then at least $k - 4E^2$ bits can be correctly reconstructed (Dwork and Yekhanin 2008).[2] This turns out to be essentially the 'right' answer for arbitrary counting queries.

The Big Bang attack is concrete. It gives a simple and computationally very efficient method by which information released by a disclosure control method that yields accurate answers to a relatively small number of apparently statistical queries can be abused to compromise privacy. The basic result is also very robust; with slight changes in bounds other attacks with similar outcomes can be launched using *random* linear queries, even if more than one-fifth of the responses are completely arbitrary (Dwork et al. 2007) and even under more general types of queries (Kasiviswanathan et al. 2012). The attacks are generally efficient and so, at least for non-Internet-scale datasets, can be launched against the entire database (i.e. $k = n$).

The Kindness of Strangers Now that the era of 'big data' is upon us, personal information – our searching, traveling, purchasing, and entertainment histories – flows from one individual to another via statistical learning systems. The set of search hits that receive clicks from one user affects which hits are returned to the next user; our presence on the road affects congestion, which in turn affects route suggestions; recommendation systems propose products based on observed paired purchases; Last.fm recommends music based on preferences of 'similar' users. Can these flows be used to compromise privacy?

Astonishingly, despite any potential adversary's tremendous uncertainty regarding the dataset, such attacks are possible. The currently known examples require a small amount of auxiliary information. For instance, a blog post about a recent purchase, taken together with the vendor's (e.g. Amazon's) continually changing public lists of 'similar items', can reveal purchases by the blogger that are not disclosed in the blog (Calandrino et al. 2011).

Smoking Causes Cancer Defining a *query* to be a function mapping datasets to some output range, we can view everything discussed so far – the production of microdata, statistics, predictors, classifiers, and so on – as queries. A user's interaction with a dataset can be viewed as receiving responses to queries, and a natural attempt to articulate a privacy goal tries to relate what is known about a member of the dataset before, versus after, obtaining the response to a query or sequence of queries. Ideally, nothing would be learned about an individual from such an interaction.

This turns out to be unachievable if the responses are *useful*, in that they teach us things we did not know (Dwork 2006; Dwork and Naor 2010): We would like to learn facts such as "smoking causes cancer," but in doing so our views and beliefs about individuals whom we know to smoke will change; for example, we will revise our predictions about their health. On the other hand, statistical analysis is meaningless without this type of *generalizability* – the whole point of a statistical database is to learn useful facts like "smoking causes cancer," not just for the participants in the study but for human beings in general. Our definition of privacy must take into account this desired utility.

Framing Our Goal: In/Out vs. Before/After If the database teaches that smoking causes cancer, the bad (pays higher insurance premiums) and good (joins smoking cessation program) consequences for an individual smoker will be incurred *independent of whether or not the particular smoker is in the database*. This suggests a new privacy goal: to ensure that, by participating in a dataset, one will be no worse off than one would be had one declined to participate. This is the heart of differential privacy.

Informed by our examples of attacks, we want this 'In/Out' privacy guarantee to hold regardless of the sources of auxiliary information – detailed information about family members and co-workers, blogs, other datasets, product recommendations, etc. – to which an attacker may have access.

An Ideal Scenario

Most of the literature on differential privacy assumes an ideal scenario in which the data are all held by a trusted and trustworthy *curator*, who carries out computations on the entire dataset and releases the results to the data analyst. Not to put too fine a point on it, the *data* (and the processing time and the power consumption, etc.) remain secret, the *responses* are published.

In reality, the data may not all reside in the same place – for example, the analyst may wish to study the combined medical records of multiple hospitals which do not choose to share their data with one another – or the data may reside, encrypted, in a semi-trusted cloud, where the cloud is trusted to keep data intact and to run programs, but it is desired that the cloud operator not have access to unencrypted data. For these situations, cryptography comes to the rescue. For example, the first may be addressed through *secure multiparty computation* (Prabhakaran and Sahai 2013) and the second through *fully homomorphic encryption* (Gentry 2009; Brakerski and Vaikuntanathan 2011). The role of cryptography in these cases is to abstract away the details and ensure that the system *looks just like,* or emulates, the ideal scenario.

Privacy-preserving data analysis is difficult even in the ideal scenario, but of course in any real implementation of differential (or other) privacy, whether in differentially private generation of synthetic data that are released to the public or in differentially private query/response systems, questions of physical security of the data, protection against timing and power consumption attacks (Kocher 1996; Kocher et al. 1999), errors in floating point implementations (Mironov 2012) do not go away, and must be addressed with additional technology.

Adversaries

Who are the 'adversaries' and what motivates them? To what kinds of information do they have access? Do they collude, intentionally or accidentally? Here it seems we are limited only by our imagination. We list a few examples.

- An abusive and controlling partner has copious auxiliary information about the victim, including dates and details of abuse, rendering useless anonymization of medical records. Privacy of medical and police records may be a question of life or death in such a situation.
- Snake oil salesmen who prey on the desperate are financially motivated to find very sick individuals. Purchasing, through an online advertising system, the ability to track individuals based on the issuing of certain search queries could be very lucrative, and very easy.
- Blackmailers are motivated to find the unfaithful, for example, by analysis of telephony and mobility records.
- Learning the reading preferences of an employee or a prospective employee, via recommendation systems, can enable discrimination, or can inhibit intellectual exploration.

- A thief, observing patterns of power consumption through improperly aggregated smart grid data, learns good times to break into homes.
- Medical insurance companies wish to charge higher rates for customers with less healthy, or more risky, eating, exercise, and sexual habits, which may be revealed by purchase, search, and advertising click histories.
- A member of a middle-class community might find her relationships with her neighbors significantly altered were they able to deduce from an interactive census database that, despite her modest living style, she has a seven-figure income.
- 'Anonymized' social networks, published to enable social science research, may be vulnerable to 'structural steganography', revealing private social connections (Backstrom et al. 2007).
- Allele frequency statistics from a medical study, when combined with a DNA sample obtained on a date, can reveal membership in a case group (Homer et al. 2008).

1 Differential Privacy

On an intuitive level, the goal of differential privacy is to obscure the presence or absence of any individual, or small group of individuals, while at the same time preserving statistical utility. A little thought (and perhaps a lot of experience) shows that, absent constraints on auxiliary information, any such method must employ *randomization*. The introduction of randomness for preserving privacy is not new. For example, *randomized response* (Warner 1965) is a well-known technique from the social sciences used to survey respondents about embarrassing or illegal behavior. One version of randomized response goes as follows. Fix a specific yes/no question. The subject is told to flip a coin. If the outcome is heads, the subject answers the yes/no question honestly. If the outcome is tails, the subject flips a second coin and answers yes or no depending on the outcome of the second coin. Thus, in randomized response the subject randomizes his or her answers before handing them over to the researcher.[3] Nonetheless, the researcher can recover statistics such as the (approximate) fraction of subjects who engage in the behavior in question.

In our setting, the raw data have been collected by a trusted curator, who can therefore compute exact answers to these sorts of statistical queries. The twin concerns are privacy loss to potentially untrustworthy data users and privacy loss due to bad interactions among statistics (or other data products) published by virtuous, well-intentioned, users. We saw already in the case of the differencing and Big Bang attacks that exact answers can lead to loss of privacy, so, as in randomized response, our algorithms will inject carefully designed random noise. Algorithms that flip coins are said to be *randomized*.

Given a randomized data protection algorithm, a database, and a query, we get a probability distribution on query responses, where the probabilities come from the coin flips of the randomized algorithm. Similarly, given a randomized data protection algorithm, a database, and an adversary issuing queries and receiving responses, once we fix the randomness of the adversary, we get a probability distribution on transcripts, where the probabilities come from the coin flips of the data protection algorithm. Differential privacy says that the distribution on the outcome of any analysis is essentially unchanged independent of whether any individual opts into or opts out of the dataset. 'Essentially' is formalized by a parameter, usually called epsilon (ε), measuring privacy loss.

The property of being differentially private depends *only* on the data protection algorithm – something that the data curator, or 'good guy', controls. Thus, if an algorithm is differentially private then it remains differentially private no matter what an adversarial data analyst knows – including to which other datasets he or she has access. So differentially private algorithms automatically protect against linkage attacks. Differential privacy even guarantees that, if the analyst knows that the dataset is either D or $D' = D \cup \{p\}$, the outcome of the analysis will give at most an ε advantage in determining which of D, D' is the true dataset.

1.1 Formal Definition of Differential Privacy

A database is modeled as a collection of *rows*, with each row containing the data of a different individual. Differential privacy will ensure that the ability of an adversary to inflict harm (or good, for that matter) – of any sort, on any set of people – will be essentially the same, independent of whether any individual opts into, or opts out of, the dataset. This is achieved indirectly, simultaneously addressing all possible forms of harm and good, by focusing on the probability of any given output of a privacy mechanism and how this probability can change with the addition or deletion of any one person. Thus, we concentrate on pairs of databases (D, D') differing only in one row, meaning one is a subset of the other and the larger database contains just one additional row. (Sometimes it is easier to think about pairs of databases D, D' of the same size, say, n, in which case they agree on $n - 1$ rows but one person in D has been replaced, in

D', by someone else.) Databases differing in at most one row are said to be *adjacent*.

Definition 1 (Dwork 2006; Dwork et al. 2006a, 2006b) A randomized mechanism M gives (ε, δ)-*differential privacy* if for all pairs of adjacent datasets D and D' and all $S \subseteq \text{Range}(M)$,

$$\Pr[M(D) \in S] \leq e^{\varepsilon} \times \Pr\left[M(D') \in S\right] + \delta,$$

where the probability space in each case is over the coin flips of M.[4]

δ should always be very small, preferably less than the inverse of any polynomial in the size of the dataset.

For most of this chapter, we will take $\delta = 0$. This is sometimes referred to in the literature as 'pure differential privacy'. Consider any possible set S of outputs that the mechanism might produce. Then the probability that the mechanism produces an output in S is essentially the same – specifically, to within an e^{ε} factor – on any pair of adjacent databases. This means that, from the output produced by M, it is hard to tell whether the database is D which, say, contains the data of the Speaker of the House, or the adjacent database D', which does not contain the Speaker's data. The intuition for privacy is: if you cannot even tell whether or not the database contains the Speaker's data, then you cannot learn anything about the Speaker's data (other than what you can learn from the data of the rest of the House). As this example shows, differential privacy defeats the differencing attack, even if the adversary knows everyone's data except the Speaker's!

1.2 Properties of Differential Privacy

Why is this a strong definition? We describe some properties implied by the definition itself; that is, *any* differentially private algorithm will enjoy the properties listed here.

Addresses Arbitrary Risks Any data access mechanism satisfying differential privacy should satisfy all concerns a participant might have about the leakage of her personal information, regardless of any auxiliary information – other databases, newspapers, websites, and so on – known to an adversary: even if the participant removed her data from the dataset, no outputs (and thus consequences of outputs) would become significantly more or less likely. For example, if the database were to be consulted by an insurance provider before deciding whether to insure a given individual, then the presence or absence of *any* individual's data in the database will not significantly affect her chance of receiving coverage. Protection against arbitrary risks is, of course, a much stronger promise than the often-stated goal in anonymization of protection against re-identification. And so it should be! Without re-identifying anything, an adversary studying anonymized medical encounter data could still learn that a neighbor, observed to have been taken to the emergency room (the ambulance was seen), has one of only, say, three possible complaints.[5]

Quantification of Privacy Loss Differential privacy is not binary; rather, privacy loss is *quantified*. For adjacent databases D, D' and any $y \in \text{Range}(M)$, the privacy loss incurred by observing y when the dataset is D is

$$L^{(y)}_{(M(D)||M(D'))} = \ln\left[\frac{\Pr[M(D) = y]}{\Pr[M(D') = y]}\right].$$

In particular, $(\varepsilon, 0)$-differential privacy ensures that this *privacy loss* is bounded by ε, and in general (ϵ, δ)-differential privacy ensures that this holds with probability at least $1 - \delta$. This quantification permits comparison of algorithms: given two algorithms with the same degree of accuracy (quality of responses), which one incurs smaller privacy loss? Or, given two algorithms with the same bound on privacy loss, which one permits the more accurate responses?

Automatic and Oblivious Composition Returning to the example of the sickle cell status of the Speaker of the House, we see that the method of ensuring privacy by only presenting counts of large groups *fails to compose:* each of the two counts in itself may not compromise privacy, but as a general method the approach cannot even tolerate an unfortunate set of two queries. In contrast, differential privacy immediately offers some composition guarantees. For example, given two differentially private computations, on the same or on different, possibly overlapping, databases, where one is $(\varepsilon_1, 0)$-differentially private and the other is $(\varepsilon_2, 0)$-differentially private, the cumulative privacy loss incurred by participating in (or opting out of) both databases is at worst $\varepsilon_1 + \varepsilon_2$. This is true even if the responses are generated obliviously of one another. This also teaches us one way to cope with high demand; for example, to ensure a cumulative loss bounded by ε^* over k computations, it is enough to ensure that each computation is $(\varepsilon^*/k, 0)$-differentially private. Composition bounds are what allow us to reason about cumulative privacy loss of complex differentially private algorithms built from simple differentially private primitives (see e.g. Blum et al. 2005; Dwork et al. 2006b). This 'programmability' also enables the construction of differentially private programming platforms (McSherry 2009; Roy et al. 2010).

Group Privacy Every $(\varepsilon, 0)$-differentially private algorithm is *automatically* $(k\varepsilon, 0)$-differentially private for groups of k individuals, for all k. This protects small groups, such as families. It will not necessarily offer protection for large groups, and indeed it should not! If two databases differ significantly, their statistics are expected to be different, and this should be observable if the databases are to be useful.

Closure under Post-Processing Differential privacy is immune to post-processing. A data analyst, without additional knowledge about the private database, cannot compute a function of the output of a differentially private algorithm M and make it less differentially private. That is, a data analyst cannot increase privacy loss, either under the formal definition or even in any intuitive sense, simply by sitting in a corner and thinking about the output of the algorithm, *no matter what auxiliary information is available.*

1.3 Achieving Differential Privacy

The differential privacy literature contains many astonishingly beautiful and powerful algorithmic techniques, some of which have given impressive results even on small datasets. For the most part, we will confine ourselves in this chapter to some simple techniques that, nonetheless, have nontrivial applications; the power of these techniques is illustrated in Section 2.

Differentially private algorithms hide the presence or absence of a single row. Consider a real-valued function f. The (worst-case, or global) *sensitivity* of f is the maximum absolute value by which the addition or deletion of a single database row can change the value of f:

$$\Delta f = \max_{D,D'} |f(D) - f(D')|,$$

where the maximum is taken over all pairs of adjacent databases. For vector-valued functions mapping databases to points in $\mathbb{R}^k$ we extend this to the L_1-norm:

$$\Delta f = \max_{D,D'} ||f(D) - f(D')||_1 = \sum_{i=1}^{k} |f(D)_i - f(D')_i|.$$

Speaking intuitively, Δf is the worst-case difference that a differentially private algorithm for the function f will have to 'hide' in order to protect the presence or absence of an individual.

Definition 2 (The Laplace Distribution) The *Laplace distribution* (centered at 0) with scale b is the distribution with probability density function:

$$\mathrm{Lap}(x|b) = \frac{1}{2b}\exp\left(-\frac{|x|}{b}\right).$$

The variance of this distribution is $\sigma^2 = 2b^2$. We will sometimes write $\mathrm{Lap}(b)$ to denote the Laplace distribution with scale b, and will sometimes abuse notation and write $\mathrm{Lap}(b)$ simply to denote a random variable $X \sim \mathrm{Lap}(b)$.

We will now define the *Laplace mechanism*. As its name suggests, the Laplace mechanism will simply compute f, and perturb each coordinate with noise drawn from the Laplace distribution. The scale of the noise will be calibrated to the sensitivity of f (divided by ε).

Definition 3 (The Laplace Mechanism) Let f be a function mapping databases to $\mathbb{R}^k$. The *Laplace mechanism* is defined as

$$M(D, f(\cdot), \varepsilon) = f(D) + (Y_1, \ldots, Y_k),$$

where Y_i are i.i.d. random variables drawn from $\mathrm{Lap}(\Delta f/\varepsilon)$.

Theorem 4 (Dwork et al. 2006b) *The Laplace mechanism preserves* $(\varepsilon, 0)$-*differential privacy.*

Example 5: Counting Queries Queries of the form "How many people in the database are over six feet tall?" have sensitivity $\Delta f = 1$, since the presence or absence of any individual in D can affect the true answer by at most 1. Thus, the Laplace mechanism will return the true count perturbed by a random draw from $\mathrm{Lap}(1/\varepsilon)$.

One way to handle $k > 1$ counting queries is via composition: by running each individual query with parameter ε/k we ensure that the cumulative privacy loss due to k queries is bounded by $k \cdot \varepsilon/k = \varepsilon$.

A second approach permits us to take advantage of the special properties of the particular set of counts we wish to compute, which may have lower sensitivity than the worst-case $\Delta f = k$, leading to better accuracy for the same privacy loss. An extreme case is illustrated in the next example.

Example 6: Histograms In a histogram query, the universe of possible database rows is partitioned into a fixed set of bins, say k, so that every database row belongs in exactly one bin. The true answer to the histogram query H when the database is D is, for each of the k bins in H, the number of rows in D that are in the given bin. For example, the bins may be income ranges [0, 25K), [25K, 50K), ..., [$\geq$ 1,000,000] for the year 2011, so the query is asking about the distribution on incomes for the sample of the population that database D comprises. The sensitivity of a histogram query is 1, since the addition or deletion of one individual can change the count of at most one bin, and that change will have magnitude at most 1. Thus $||H(D) - H(D')||_1 \leq 1$ for all adjacent D, D'. Theorem 4 says that $(\varepsilon, 0)$-differential privacy can be achieved by adding independently generated draws from $\text{Lap}(1/\varepsilon)$ to each output of $H(D)$. Compare this to the accuracy we would have obtained naïvely, by viewing the histogram query as k independent counting queries (one per bin) and applying the composition result mentioned in Section 1.2, which would have suggested adding noise drawn from $\text{Lap}(k/\varepsilon)$ to the count for each bin – a factor of k worse than what we get by thinking carefully about sensitivity.

The Laplace mechanism provides *one* method for ensuring $(\varepsilon, 0)$-differential privacy for any value of $\varepsilon > 0$. It does not necessarily give the best method for every setting. For example, the method gives poor responses if we seek answers to a superlinear (in the size of the database) number of queries, but other algorithms[6] give meaningful responses even for a number of queries that grows *exponentially* in the size of the database! In Section 2 we briefly mention empirical results for computation of marginals using one of these algorithms.

Differentially private algorithms will typically be composed of several steps, and the Laplace mechanism is frequently employed at one or more of these individual steps. It is therefore an important *primitive*, or building block. We will see an example of extensive use of this primitive in constructing differentially private probability distributions in Section 2.

The Exponential Mechanism Differential privacy can be ensured for discrete output ranges by the *exponential mechanism* (McSherry and Talwar 2007). This mechanism uses a computation-specific *quality function* mapping (dataset, output) pairs to a real number. It assigns to each output a probability that grows exponentially in the quality, and then selects an output according to the resulting probability distribution. The exponential mechanism is another very important primitive. We will see an example of its use in Section 2 in selecting a 'best' (or perhaps just sufficiently good) distribution from a family of distributions.

The Gaussian Mechanism The Gaussian distribution is more tightly concentrated than the Laplace distribution, and may also be closer in form to other sources of variation in the data. The addition of Gaussian noise yields (ε, δ)-differential privacy with $\delta > 0$. Roughly speaking, (ε, δ)-differential privacy ensures that, given a pair D, D' of adjacent databases, with probability $1 - \delta$ the privacy loss on D with respect to D' will be bounded by ε. Recall that typically we have in mind cryptographically small values of δ.

Redefining sensitivity to be the maximum L_2 (i.e. Euclidean) norm $||f(D) - f(D')||_2$ on pairs of adjacent databases D, D' (rather than the L_1 difference $||f(D) - f(D')||_1$ we have discussed until this point), we obtain the following theorem for the *Gaussian mechanism*.

Theorem 7 *Let f be a function mapping databases to $\mathbb{R}^k$, and let Δ denote the L_2-sensitivity of f. The 'Gaussian mechanism' that adds i.i.d. noise drawn from $\mathcal{N}(0, 2\Delta^2 \ln(2/\delta)/\varepsilon^2)$ to each of the k coefficients of f is (ε, δ)-differentially private.*

Like the Laplace mechanism, the Gaussian mechanism is an important primitive, especially in geometric algorithms for ensuring differential privacy (Hardt and Talwar 2010; Nikolov et al. 2013).

Beyond supporting the addition of Gaussian noise, this relaxation to $\delta > 0$ is also useful in differentially private programming. For example, suppose we have two methods for differentially private release of a given statistic, say, the median income. The first method, A, always maintains $(\varepsilon, 0)$-differential privacy, but has poor accuracy on some inputs; the second method, B, has excellent accuracy, but its privacy loss exceeds ε on pathological inputs of a certain type, and only on these pathological inputs. We can use a differentially private test to determine whether it is safe to use algorithm B on the given dataset; but even if designed correctly there will be some very small probability, say, γ, that the test will erroneously indicate it is safe to use method B, potentially yielding probability γ of a large privacy loss. The best we can do in this case is to achieve (ε, γ)-differential privacy. We can make γ suitably small by designing the test to have an extremely small probability of error.

Remark 8 (Technical Remark) A more sophisticated analysis than that mentioned in Section 1.2 shows that the composition of k mechanisms, each of which is (ε, δ')-differentially private, $\varepsilon \leq 1$, satisfies $(\sqrt{2k \ln(1/\delta)}\varepsilon + k\varepsilon(e^{\varepsilon} - 1), k\delta' + \delta)$-differential privacy for all $\delta > 0$ (Dwork et al. 2010). This translates into much better *accuracy* when $\varepsilon \leq 1/k$.

1.4 An Aside

The version of randomized response described at the very start of this section is $(\varepsilon, 0)$-differentially private for $\varepsilon = \ln 3$. It is instructive to compare randomized response to the Laplace mechanism. For a single query such as "How many people in the dataset ingested a controlled substance in the past week?" randomized response will yield an error on the order of $\sqrt{n}$, while the Laplace mechanism will yield an error on the order of $1/\varepsilon$, which is a constant independent of n.

What about multiple queries? Suppose we have a database with a single 'sensitive' binary attribute, and that attribute is recorded using randomized response. In this case the population can be sliced and diced at will, and privacy of this single attribute will be maintained. In contrast, the Laplace and Gaussian mechanisms appear to cease to give meaningful responses after just under n^2 queries.[7] In this special case of a single sensitive attribute, randomized response is preferable once we require answers to a linear number of queries.[8]

2 Empirical Results

Since its inception, differential privacy has been the subject of intensive algorithmic research. There is also work on formal methods (e.g. Barthe et al. 2012) and a few programming platforms (e.g. McSherry 2009; Roy et al. 2010) that permit online interaction with the data.

OnTheMap (Machanavajjhala et al. 2008; Abowd et al. 2009), a privacy-preserving U.S. Census Bureau web-based mapping and reporting application that shows where people work and where workers live, and provides companion reports on age, earnings, industry distributions, and local workforce indicators, satisfies *probabilistic differential privacy*. While interactive, responding to queries issued by users of the site, the system gives exact answers computed from a privately generated *synthetic dataset* that was constructed offline (that is, before the website went live) from U.S. census data. To our knowledge, this is the only online system permitting anonymous members of the general public to issue queries while ensuring some form of differential privacy.[9]

We can think of a synthetic dataset as a collection of records with the same structure as real records, so that, for example, off-the-shelf software running on the original dataset could also run on the synthetic dataset. Given a (public) set $\mathcal{Q}$ of queries and a (private) database D, the goal is to produce a synthetic dataset y with the property that for all $q \in \mathcal{Q}$, $q(y)$ yields a good approximation to $q(D)$.

A synthetic dataset does not preserve privacy simply by virtue of being synthetic. The *process* for generating the synthetic dataset matters. Moreover, it follows from the Big Bang attack that it is impossible to simultaneously preserve any reasonable notion of privacy and to release a synthetic dataset that answers 'too many' queries with 'too much' accuracy. Finally, there are also considerations of *computational complexity*, that is, the computational difficulty of creating a synthetic dataset with the desired properties. Two factors come into play here: the size of the set $\mathcal{Q}$ of queries for which the curator promises correct answers, and the size of $\mathcal{U}$, the *universe* of possible data items. For example, if we wish to describe humans by their DNA sequences, the size of the universe is exponential in the length of the DNA sequence; if instead we describe the humans in our datasets by 6 binary attributes, the size of the universe is only $2^6 = 64$. Although theoretical results suggest formidable computational barriers to building synthetic datasets for certain large $\mathcal{Q}$ or large $\mathcal{U}$ cases (Dwork et al. 2009; Ullman and Vadhan 2011), the literature also contains some counterpoints with very encouraging experimental validation. We give two examples.

2.1 The MWEM Algorithm

The Multiplicative Weights with Exponential Mechanism (MWEM) algorithm (Hardt et al. 2012) optimizes an offline variant (Gupta et al. 2011) of the Private Multiplicative Weights update technique (Hardt and Rothblum 2010). A description of the techniques involved in these works is, unfortunately, beyond the scope of this book. The MWEM algorithm was evaluated on *tables of marginals*. These are tables that answer counting queries of a special form. The universe $\mathcal{U}$ of possible database elements is the set of d-bit strings, representing, for each individual, the values of d binary attributes. A k-way marginal is specified by a set S of k of these d attributes together with an assignment to these attributes. Assuming binary attributes, there are $\binom{d}{k}2^k$ k-way marginals. MWEM was evaluated on the sets of all 3-way marginals for three datasets discussed by Fienberg et al. (2011), for several values of $\varepsilon \in [0, 1]$. That is, $\mathcal{Q}$ is the set of all $\binom{d}{3}2^3$ 3-way marginals. The smallest dataset consisted of only 70(!) 6-attribute records. Of the $2^6 = 64$ possible settings of these bits, 22 appeared in the dataset (so the contingency table had 22 non-zero entries).

A byproduct of the MWEM algorithm is a synthetic database created solely from the privacy-preserving responses to the queries in $\mathcal{Q}$. In each case, the synthetic dataset was evaluated by computing the relative entropy, or Kullback-Leibler (KL) divergence, with respect to the real dataset and comparing this measurement with a report in the literature (Fienberg et al. 2011) that, roughly speaking, captures the best that can be done non-privately.[10]

Remarkably, even on the smallest dataset the relative entropy closely approaches the ideal when ε reaches about 0.7. For the other two datasets (665 records, 8 attributes, 91 non-zero cells; 1841 records, 6 attributes, 63 non-zero cells), the differentially private algorithms beat the non-private bounds once $\varepsilon \approx 0.7$ and $\varepsilon \approx 0.5$, respectively.

2.2 DP-WHERE

In this section we describe DP-WHERE (Mir et al. 2013), a differentially private version of the WHERE (Work and Home Extracted REgions) approach to modeling human mobility based on cellphone call detail records (Isaacman et al. 2012). For each individual, simultaneously, DP-WHERE protects *all* call detail records in the dataset; this is known in the literature as *user*-level privacy (here 'user' refers to a telephone user, not the data analyst).[11] The output of the system is a collection of *synthetic* call detail records. Example uses of synthetic call detail records include estimating daily ranges (the maximum distance a person travels in one day), modeling epidemic routing, and the modeling of hypothetical cities, in which the analyst creates a parameterized model of a city and user behavior patterns that cannot be observed in the real world, yielding the power to experiment with the effects of modifications to reality such as telecommuting (Isaacman et al. 2012).

2.3 A Sketch of DP-WHERE

Each call detail record corresponds to a single voice call or text message. Users making more than 120 calls in any hour are filtered out,[12] and it is assumed that the number of remaining users, denoted n, is known. Each of the n users is identified by an integer in $\{1, \ldots, n\}$. The calls were made in a metropolitan area divided by a grid into smaller geographic areas. Each call detail record is augmented with inferred home and work locations obtained by a combination of clustering and regression (Isaacman et al.

2011).[13] Thus, in DP-WHERE each element in the dataset contains an id (number between 1 and n), date, time, latitude, longitude, and the inferred home and work locations.

The approach is to create several probability distributions, all in a differentially private manner. The synthetic call detail records are generated by appropriate sampling from these distributions.

Description of the Distributions

First, we list the distributions, briefly commenting on some of the differential privacy techniques used in their construction.

Home and Work For each of Home and Work, DP-WHERE computes a probability distribution on a square grid covering the metropolitan area in question, with a simple histogram query (Example 6). For example, for the Home distribution, the histogram reports, for each grid cell, the approximate (that is, noisy) number of users in the dataset whose home location is in this grid cell. In order to be able to transform this to a probability distribution – for example, to remove negative counts – post-processing techniques are applied that require no additional access to the true data (Hay et al. 2010).

Commute Distance DP-WHERE uses a coarser 'commute grid' for this computation. For each cell in the commute grid, the system computes an empirical distribution on commute distances for people whose home location falls in this cell.

We briefly describe the construction of one of these cumulative distribution functions, say, for the ith grid cell. The algorithm creates a *data-dependent* histogram of commute distances for the residents in this cell. Each histogram bin is a range of distances, and the (true, non-noisy) count in bin j is the number of users living in the ith grid cell whose true commute distance is in the range associated with the jth bin. There is some subtlety in determining the 'right' set of bins for this histogram. This is done by assuming that the commute distances for the residents of grid cell i are modeled by an exponential distribution of the form $\eta(x) = \lambda e^{-\lambda x}$, which has median $\lambda^{-1} \ln 2$ (each grid cell has its own distribution on commute distances, i.e. its own λ). The approach is to approximate the median using the exponential mechanism, set λ to be $\ln 2$ divided by this approximate median, and then define the histogram bins according to the deciles of the exponential distribution with parameter λ.

Calls per Day per User In this step, for a fixed, discrete, set of potential means $M = \{\mu_1, \mu_2, \ldots, \mu_m\}$ and standard deviations $S = \{\sigma_1, \sigma_2, \ldots, \sigma_s\}$, the algorithm computes a probability distribution on $M \times S$ from a histogram query (the cells of the histogram correspond to (mean, deviation) pairs (μ_i, σ_j)).

2-Means Clustering DP-WHERE runs a privacy-preserving 2-means clustering algorithm (McSherry 2009) to classify users based on a 24-dimensional probability vector describing their daily calling patterns. For each user i, using only the call records for this user, DP-WHERE first constructs a *non-private* probability vector P_i, describing for each hour $j \in \{1, 2, \ldots, 24\}$ the fraction of i's calls made in the jth hour of the day. These 24-dimensional P_i are clustered into two clusters, using a differentially private algorithm.

Hourly Calls per Location The last set of probability distributions generated by DP-WHERE yield, for each hour of the day, a probability distribution over grid cells, describing where the population as a whole is likely to be during the given hour. That is, the Hourly Location distribution for hour $j \in \{1, 2, \ldots, 24\}$ yields a probability distribution on locations (grid cells) for the population as a whole during the jth hour of every day covered by the dataset. Ideally, this would be done by counting, for each grid cell and hour, the number of calls made from that grid cell during that hour of the day, summed over the different days covered by the dataset.

These counts are highly sensitive: for each hour, the total sensitivity of this computation is 120 times the number of days![14] Thus, even though, for a fixed $j \in \{1, 2, \ldots, 24\}$, DP-WHERE builds something like a histogram, with one cell for each cell of the geographic grid, the L_1-sensitivity of this data structure is 120 times the number of days. Applying the Laplace mechanism would add noise of this magnitude to *each* of the grid cells, which makes for too much distortion overall. This difficulty is addressed using a *grouping and smoothing* technique (Papadopoulos and Kellaris 2013), in which geographically close grid cells are 'merged', essentially coarsening the geographic grid, to give a data structure with fewer cells, yielding lower overall distortion.

This completes the overview of the (differentially privately generated) distributions used in generating synthetic call detail records. The differential privacy techniques used are the Laplace mechanism for histogram queries, post-processing to transform counts to distributions, the exponential mechanism, differentially private k-means clustering, and grouping and smoothing.

Generation of Synthetic Call Detail Records

Once the distributions have been generated in a differentially private fashion, privacy under post-processing tells us that sampling these distributions presents no additional risk to privacy. Thus, although DP-WHERE generates synthetic users and synthesizes movements for these users, the system can also publish the distributions and the data analysts can sample from them at will.

A synthetic user is generated by sampling a home location h from the Home distribution, sampling a commute distance d from the Commute Distance distribution for the commute grid cell corresponding to h, and weighting the cells at distance d from h according to their distribution under the Work distribution and sampling from the resulting distribution on cells at distance d to obtain a work location w. Having determined the home and work locations of the synthetic user, the final steps are to sample (μ, σ) according to the Calls per Day per User distribution and finally to sample one of the two calling pattern clusters obtained in the 2-means clustering.

Having generated the synthetic users, DP-WHERE 'moves' them between their home and work locations. Fix $i \in \{1, 2, \ldots, n\}$. The procedure described next generates a day in the life of synthetic user i.

1. Generate a number N of calls to be made during the day by sampling from a normal distribution with mean μ and variance σ^2.
2. Allocate the total number N of calls to be made in this day to the 24 different hours of the day, according to the calling pattern (cluster) to which synthetic user i was assigned. Assign the exact time within the hour by interpolating between the beginning and end of the hour.
3. For each call made by user i during hour j, choose the location – select between user i 's home (h) and work (w) locations – by sampling according to the (differentially privately generated) Hourly Calls per Location densities for these two locations during hour j.

Experimental Validation of DP-WHERE

Experiments were carried out using call detail records from actual cellphone use over 91 consecutive days. The dataset contains over one billion records involving over 250,000 unique phones chosen at random from phones billed to ZIP codes within 50 miles of the center of New York City.

As in WHERE, accuracy of DP-WHERE is measured by a 'normalized' Earth Mover's Distance. The results vary according to the coarseness of the commute grid (in both WHERE and DP-WHERE) and the choice of the *total, cumulative* privacy loss (in DP-WHERE). For a commute grid cell size of $0.01°$, (non-private) WHERE yields an average hourly error of 3.2150; when $\varepsilon = 0.33$ (respectively, 0.23 and 0.13) this quantity is 3.5136 (respectively, 3.4066 and 5.3391). A coarser grid cell size of $0.05°$ yields 3.0871 for WHERE and, respectively for these same values of ε, 4.5687, 5.1691, and 5.2754 for DP-WHERE.

Experiments showed that, in all cases, DP-WHERE as described above[15] performed better than (non-private) WHERE based on public data, such as U.S. census data (and not call detail records; Isaacman et al. 2012). Thus, if the choice is between unfettered access to public data and differentially private access to the call detail records, these experiments show that differential privacy, even with $\varepsilon = 0.13$, has the better utility.

Experiments were also carried out to measure the daily range, or the maximum distance between any two points visited by an individual in a day. The boxplots for daily range in DP-WHERE ($\varepsilon = 0.23$), WHERE, and the real call detail records are qualitatively similar, with differences of 0.5–1.3 miles across the middle two quartiles (the smallest interquartile range of the three sets is 5.2 miles).

3 Challenges for Differential Privacy

The greatest *scientific* challenge for differential privacy is that, for a given computational task and a given value of ε, finding a low-error, differentially private algorithm can be hard. An analogy may be made to numerical analysis. Suppose, in the non-private world, we wish to compute a matrix decomposition. A naïve algorithm for the decomposition may be numerically unstable, so we first consult a textbook on numerical algorithms and write our program based on the stable algorithm in the text. It is easy now – but quite possibly the algorithm in the text was a PhD thesis when it was developed in the 1970s.

A different sort of challenge is posed by 'non-algorithmic' thinking in data analysis. From data cleaning through detailed investigation, many researchers who work with data do not, indeed cannot, provide an algorithmic description of their interactions with the data. With no algorithm for the non-private case, there is essentially no hope of finding a differentially private alternative. This is less of an issue in machine learning and the

VLDB (Very Large Data Bases) communities, where the sheer volume of data rules out non-algorithmic approaches.

Differential privacy requires a new way of interacting with data, in which the analyst accesses data only through a privacy mechanism, and in which accuracy and privacy are improved by minimizing the viewing of intermediate results. But query minimization is a completely foreign concept to data analysts. A good analogy might be running an industrial scale database without the benefit of query planning, leading to (literally) prohibitive computational costs.

By far the hardest to grapple with are the *social* challenges of a changing world, in which highly detailed research datasets are expected to be shared and reused, linked and analyzed, for knowledge that may or may not benefit the subjects, and all manner of information exploited, seemingly without limit. That this is fundamentally incompatible with privacy is proved by a host of lower bounds and attacks.[16] What are we to make of this state of affairs? To paraphrase Latanya Sweeney (2012), computer science got us into this mess, can computer science get us out of it?

One thing seems certain: complexity of this type requires a mathematically rigorous theory of privacy and its loss. Other fields – economics, ethics, policy – cannot be brought to bear without a 'currency', or measure of privacy, with which to work. In this connected world, we cannot discuss trade-offs between privacy and statistical utility without a measure that captures cumulative harm over multiple releases.

Publish the Loss, and Pay a Fine for Infinity What should be the value of ε? How should ε be meted out? How should ε depend on the nature of the data, the nature of the queries, the identity or affiliation of the data analyst? The anticipated social value of the investigation? The commercial utility? These questions will take time to sort out.

> On a more philosophical level, consider an analogy to time: there are only so many hours in your lifetime, and once they are consumed you die. (This is sometimes worse than someone learning private information about you.) Yet, somehow, we as a society have found ways to arrive at values for an individual's time, and a fundamental part of that is the ability to quantitatively measure it. (Dwork et al. 2011)

Differential privacy provides a measure that captures cumulative privacy loss over multiple releases. Whatever the measure of privacy loss on which the community ultimately settles, we should take a page from data breach and environmental law and require data usage to be accompanied by publication of privacy loss.[17] In differential privacy, simply ensuring that the loss is finite helps to protect against many common avenues of attack. So let this be our starting point: publication of privacy losses – and a fine for infinite loss. If the data analyst cannot function without seeing raw data then, once the analyst has determined (in a non-private fashion) which statistics (or other data products) are to be released, the chosen statistics should be published using differential privacy (or whatever the community settles on), together with the privacy losses incurred in those calculations. Note that this is still infinite loss – and should still incur a fine – because the *choice* of what to publish was made in a non-private fashion. The attention to privacy loss will raise awareness and lead to innovation and competition, deploying the talents and resources of a larger set of researchers and other marketers and consumers of data in the search for private algorithms.

NOTES

1. These are also known variously as *blatant non-privacy*, *reconstruction*, or *Dinur-Nissim* attacks, the last in homage to the computer scientists who first demonstrated attacks of this kind (Dinur and Nissim 2003).
2. The attack involves computation of a Fourier transform and does not require knowledge of E.
3. Randomized response is also known as *the local model*, and it does not require a trusted curator; see Evfimievski et al. (2003), Blum et al. (2005), Dwork et al. (2006a), Kasiviswanathan et al. (2011), Hsu et al. (2012), and Duchi et al. (2013).
4. If $\varepsilon \ll 1$ then $e^{\varepsilon} \approx (1+\varepsilon)$; for example, $e^{1/10} \approx 1.1052$.
5. Two of which, say, a broken limb and heart attack, might be ruled out when the neighbor is seen the next day, leaving only 'panic attack'.
6. See Blum et al. (2008), Roth and Roughgarden (2010), Hardt and Rothblum (2010), Dwork et al. (2010), Hardt et al. (2012), and Nikolov et al. (2013).
7. Appearances can be deceptive: correlations between responses generated with independent noise can be exploited to extract surprisingly accurate approximate answers *on average*, even for a superpolynomial number of queries (Nikolov et al. 2013)!
8. See the literature on local learning mentioned in n.3, and in particular the careful treatment in Kasiviswanathan et al. (2011), for further understanding of the power and subtleties of randomized response.
9. The entry point is http://lehdmap3.did.census.gov/.
10. Any set of marginals determines the maximum likelihood estimator $\hat{p}$, which is the unique probability distribution in the log-linear model encoded by the given set of marginals that makes the observed dataset D the 'most likely' sample to have been observed. The bounds on KL divergence for the non-private case come from the KL divergence between $\hat{p}$ and D. While considered important,

small KL divergence is not necessarily viewed as a sufficient criterion for model selection.

11. This is in contrast to *event*-level privacy, which would only hide the presence or absence of a single (or small number of) call record.
12. These are assumed to be auto-dialers.
13. Since the clustering uses only information specific to the given user, together with global information about the locations of cell towers, the determination of these fields will not affect the privacy guarantee.
14. Recall that only users making more than 120 calls in a single hour have been filtered out.
15. That is, starting from real call detail records, generating the distributions in a differentially private fashion, creating synthetic users, and 'moving' these users.
16. It is also fundamentally incompatible with statistical power, where issues of false discovery arise.
17. See Hirsch (2006) for an investigation of basing privacy regulation on environmental law.

REFERENCES

Abowd, J., J. Gehrke, and L. Vilhuber. 2009. Parameter exploration for synthetic data with privacy guarantees for OnTheMap. In *Proc. Joint UNECE/Eurostat Work Session on Statistical Data Confidentiality (Bilbao, Spain, 2–4 December)*. Available at http://www.unece.org/fileadmin/DAM/stats/documents/ece/ces/ge.46/2009/wp.12.e.pdf (accessed January 9, 2014).

Backstrom, Lars, Cynthia Dwork, and Jon Kleinberg. 2007. Wherefore art thou r3579x? Anonymized social networks, hidden patterns, and structural steganography. In *Proc. 16th International Conference on World Wide Web*, 181–190.

Barthe, Gilles, Boris Köpf, Federico Olmedo, and Santiago Zanella Beguelin. 2012. Probabilistic relational reasoning for differential privacy. In *Proc. POPL 2012*.

Bleichenbacher, D. 1998. Chosen ciphertext attacks against protocols based on the RSA encryption standard PKCS# 1. In *CRYPTO '98*, LNCS 1462, 1–12.

Blum, A., C. Dwork, F. McSherry, and K. Nissim. 2005. Practical privacy: The SuLQ framework. In *Proc. 24th ACM Symposium on Principles of Database Systems (PODS)*, 128–138.

Blum, A., K. Ligett, and A. Roth. 2008. A learning theory approach to non-interactive database privacy. In *Proc. 40th ACM SIGACT Symposium on Theory of Computing (STOC)*, 609–618.

Brakerski, Z., and V. Vaikuntanathan. 2011. Efficient fully homomorphic encryption from (standard) LWE. In *Proc. 52nd Annual IEEE Symposium on Foundations of Computing (FOCS)*, 97–106.

Calandrino, J., A. Kilzer, A. Narayanan, E. Felten, and V. Shmatikov. 2011. You might also like: Privacy risks of collaborative filtering. In *Proc. IEEE Symposium on Security and Privacy (SP)*, 231–246.

Dinur, I., and K. Nissim. 2003. Revealing information while preserving privacy. In *Proc. 22nd ACM Symposium on Principles of Database Systems (PODS)*, 202–210.

Duchi, John, Michael Jordan, and Martin Wainwright. 2013. Local privacy and statistical minimax rates. In *Proc. 54th Annual IEEE Symposium on Foundations of Computer Science (FOCS)*.

Dwork, C. 2006. Differential privacy. In *Proc. 33rd International Colloquium on Automata, Languages and Programming (ICALP)*, 2:1–12.

Dwork, C., K. Kenthapadi, F. McSherry, I. Mironov, and M. Naor. 2006a. Our data, ourselves: Privacy via distributed noise generation. In *Advances in Cryptology: Proc. EUROCRYPT*, 486–503.

Dwork, C., F. McSherry, K. Nissim, and A. Smith. 2006b. Calibrating noise to sensitivity in private data analysis. In *Proc. 3rd Theory of Cryptography Conference (TCC)*, 265–284.

Dwork, C., F. McSherry, and K. Talwar. 2007. The price of privacy and the limits of LP decoding. In *Proc. 39th ACM Symposium on Theory of Computing (STOC)*, 85–94.

Dwork, C., and M. Naor. 2010. On the difficulties of disclosure prevention in statistical databases or the case for differential privacy. *Journal of Privacy and Confidentiality* 2. Available at http://repository.cmu.edu/jpc/vol2/iss1/8.

Dwork, C., M. Naor, O. Reingold, G. Rothblum, and S. Vadhan. 2009. When and how can privacy-preserving data release be done efficiently? In *Proc. 41st ACM Symposium on Theory of Computing (STOC)*, 381–390.

Dwork, C., and S. Yekhanin. 2008. New efficient attacks on statistical disclosure control mechanisms. In *Proc. CRYPTO*, 468–480.

Dwork, Cynthia, Frank McSherry, Kobbi Nissim, and Adam Smith. 2011. Differential privacy – a primer for the perplexed. In *Joint UNECE/Eurostat Work Session on Statistical Data Confidentiality*. Available at http://www.unece.org/fileadmin/DAM/stats/documents/ece/ces/ge.46/2011/26_Dwork-Smith.pdf.

Dwork, Cynthia, Guy N. Rothblum, and Salil P. Vadhan. 2010. Boosting and differential privacy. In *Proc. 51st Annual IEEE Symposium on Foundations of Computer Science (FOCS)*, 51–60.

Evfimievski, Alexandre, Johannes Gehrke, and Ramakrishnan Srikant. 2003. Limiting privacy breaches in privacy preserving data mining. In *Proc. 22nd ACM SIGMOD-SIGACT-SIGART Symposium on Principles of Database Systems (PODS)*.

Fienberg, Stephen, Alessandro Rinaldo, and Xiaolin Yang. 2011. Differential privacy and the risk-utility tradeoff for multi-dimensional contingency tables. In *Privacy in Statistical Databases*, LNCS 6344, 187–199.

Gentry, C. 2009. A fully homomorphic encryption scheme. PhD thesis, Stanford University.

Goldwasser, S., and S. Micali. 1984. Probabilistic encryption. *Journal of Computer and Systems Sciences* 28:270–299.

Goldwasser, Shafi, Silvio Micali, and Ron Rivest. 1988. A digital signature scheme secure against adaptive chosen-message attacks. *SIAM Journal on Computing* 17:281–308.

Gupta, Anupam, Moritz Hardt, Aaron Roth, and Jonathan Ullman. 2011. Privately releasing conjunctions and the statistical query barrier. In *Proc. 43rd Annual ACM Symposium on Theory of Computing (STOC)*, 803–812.

Hardt, M., K. Ligett, and F. McSherry. 2012. A simple and practical algorithm for differentially private data release. *Advances in Neural Information Processing Systems* 25:2348– 2356.

Hardt, M., and K. Talwar. 2010. On the geometry of differential privacy. In *Proc. 42nd ACM Symposium on Theory of Computing (STOC)*, 705–714.

Hardt, Moritz, and Guy Rothblum. 2010. A multiplicative weights mechanism for privacy-preserving data analysis. In *Proc. 51st Annual IEEE Symposium on Foundations of Computer Science (FOCS)*, 61–70.

Hay, Michael, Vibhor Rastogi, Gerome Miklau, and Dan Suciu. 2010. Boosting the accuracy of differentially private histograms through consistency. *Proc. VLDB Endowment* 3(1–2):1021–1032.

Hirsch, D. 2006. Protecting the inner environment: What privacy regulation can learn from environmental law. *Georgia Law Review* 41.

Homer, N., S. Szelinger, M. Redman, D. Duggan, W. Tembe, J. Muehling, J.V. Pearson, D.A. Stephan, S.F. Nelson, and D.W. Craig. 2008. Resolving individuals contributing trace amounts of dna to highly complex mixtures using high-density snp genotyping microarrays. *PLoS Genetics* 4(8):e1000167.

Hsu, Justin, Sanjeev Khanna, and Aaron Roth. 2012. Distributed private heavy hitters. In *Proc. 39th International Colloquium Conference on Automata, Languages, and Programming (ICALP)(Track 1)*, 461–472.

Isaacman, Sibren, Richard Becker, Ramón Cáceres, Stephen Kobourov, Margaret Martonosi, James Rowland, and Alexancer Varshavsky. 2011. Identifying important places in people's lives from cellular network data. In *Pervasive Computing*, LNCS 6696, 133–151.

Isaacman, Sibren, Richard Becker, Ramón Cáceres, Margaret Martonosi, James Rowland, Alexander Varshavsky, and Walter Willinger. Human mobility modeling at metropolitan scales. 2012. In *Proc. 10th International Conference on Mobile Systems, Applications, and Services*, 239–252.

Kasiviswanathan, Shiva, Homin K. Lee, Kobbi Nissim, Sofya Raskhodnikova, and Adam Smith. What can we learn privately? *SIAM Journal on Computing* 40:793–826.

Kasiviswanathan, Shiva, Mark Rudelson, and Adam Smith. 2012. The power of linear reconstruction attacks. arXiv:1210.2381.

Kocher, P., J. Jaffe, and B. Jun. 1999. Differential power analysis. In *Advances in Cryptology: Proc. CRYPTO'99*, 388–397.

Kocher, Paul. 1996. Timing attacks on implementations of Diffie-Hellman, RSA, DSS, and other systems. In *Advances in Cryptology: Proc. CRYPTO'96*, 104–113.

Machanavajjhala, Ashwin, Daniel Kifer, John Abowd, Johannes Gehrke, and Lars Vilhuber. 2008. Privacy: Theory meets practice on the map. In *Proc. International Conference on Data Engineering (ICDE)*, 277–286.

McSherry, F. 2009. Privacy integrated queries (codebase). Available on Microsoft Research downloads website. See also *Proc. SIGMOD 2009*, 19–30.

McSherry, F., and K. Talwar. 2007. Mechanism design via differential privacy. In *Proc. 48th Annual Symposium on Foundations of Computer Science (FOCS)*, 94–103.

Mir, Darakhshan, Sibren Isaacman, Ramón Cáceres, Margaret Martonosi, and Rebecca N. Wright. 2013. DP-WHERE: Differentially private modeling of human mobility. In *Proc. IEEE Conference on Big Data,* 580–588.

Mironov, Ilya. 2012. On significance of the least significant bits for differential privacy. In *Proc. ACM Conference on Computer and Communications Security (CCS),* 650–661.

Narayanan, Arvind, and Vitaly Shmatikov. 2008. Robust de-anonymization of large sparse datasets. In *Proc. IEEE Symposium on Security and Privacy (SP),* 111–125.

Nikolov, Aleksandar, Kunal Talwar, and Li Zhang. 2013. The geometry of differential privacy: The sparse and approximate cases. In *Proc. 45th ACM Symposium on Theory of Computing (STOC),* 351–360.

Papadopoulos, S., and G. Kellaris. 2013. Practical differential privacy via grouping and smoothing. In *Proc. 39th International Conference on Very Large Data Bases,* 301–312.

Prabhakaran, Manoj, and Amit Sahai. 2013. *Secure Multi-party Computation.* Washington, DC: IOS Press.

Roth, Aaron, and Tim Roughgarden. 2010. Interactive privacy via the median mechanism. In *Proc. 42nd ACM Symposium on Theory of Computing (STOC),* 765–774.

Roy, Indrajit, Srinath Setty, Ann Kilzer, Vitaly Shmatikov, and Emmett Witchel. 2010. Airavat: Security and privacy for MapReduce. In *Proc. 7th USENIX Symposium on Networked Systems Design and Implementation (NSDI),* 10:297–312.

Sweeney, L. 1997. Weaving technology and policy together to maintain confidentiality. *Journal of Law, Medicine and Ethics* 25:98–110.

Sweeney, Latanya. 2012. Keynote Lecture, Second Annual iDASH All-Hands Symposium, UCSD, La Jolla, CA.

Ullman, Jonathan, and Salil P. Vadhan. 2011. PCPs and the hardness of generating private synthetic data. In *Proc. 8th Theory of Cryptography Conference (TCC),* 400–416.

Warner, S. 1965. Randomized response: A survey technique for eliminating evasive answer bias. *Journal of the American Statistical Association* 60:63–69.